Engineering Drawing

This valuable textbook offers detailed discussion of the fundamental concepts of engineering drawing in an easy to understand manner. Important topics including projections of solids, auxiliary projections, sections of solids, isometric projections, orthographic projections and projection of planes are discussed comprehensively. Multi-aspect pedagogical features—more than 400 solved examples, 275 practice problems and 250 short answer questions—will help students in learning fundamental concepts. The text is written to cater to the needs of undergraduate students of all branches of engineering for an introductory course in engineering drawing/engineering graphics/ computer aided engineering drawing.

Lakhwinder Pal Singh is an Associate Professor in the Department of Industrial and Production Engineering at Dr B. R. Ambedkar National Institute of Technology, Jalandhar. Besides authoring the book *Work Study and Ergonomics*, he has published more than 100 papers in international conferences and journals. His areas of interest include human factors engineering, production and operations management, lean manufacturing, occupational health and safety, logistics and supply chain management.

Harwinder Singh is a Professor in the Department of Mechanical Engineering at Guru Nanak Dev Engineering College, Ludhiana. He has taught courses including engineering drawing and computer graphics, quantitative methods and operations research at undergraduate and graduate levels. His areas of interest include optimisation of production systems and human factors engineering.

Engineering Drawing

Principles and Applications

Lakhwinder Pal Singh

Harwinder Singh

CAMBRIDGE
UNIVERSITY PRESS

CAMBRIDGE
UNIVERSITY PRESS

University Printing House, Cambridge CB2 8BS, United Kingdom

One Liberty Plaza, 20th Floor, New York, NY 10006, USA

477 Williamstown Road, Port Melbourne, vic 3207, Australia

314 to 321, 3rd Floor, Plot No.3, Splendor Forum, Jasola District Centre, New Delhi 110025, India

79 Anson Road, #06–04/06, Singapore 079906

Cambridge University Press is part of the University of Cambridge.

It furthers the University's mission by disseminating knowledge in the pursuit of
education, learning and research at the highest international levels of excellence.

www.cambridge.org
Information on this title: www.cambridge.org/9781108707725

© Cambridge University Press 2021

First published 2021

Printed in India

A catalogue record for this publication is available from the British Library

Library of Congress Cataloging-in-Publication Data

Names: Singh, L. P., author. | Singh, Harwinder, author.
Title: Engineering drawing: principles and applications / L. P. Singh, and Harwinder Singh.
Description: Cambridge, United Kingdom New York, NY, USA: Cambridge University Press, 2020.
Identifiers: LCCN 2020002157 (print) | LCCN 2020002158 (ebook) |
    ISBN 9781108707725 (paperback) | ISBN 9781108659437 (ebook)
Subjects: LCSH: Mechanical drawing—Textbooks.
Classification: LCC T351.5 .S56 2020 (print) | LCC T351.5 (ebook) | DDC 604.2—dc23
LC record available at https://lccn.loc.gov/2020002157
LC ebook record available at https://lccn.loc.gov/2020002158

ISBN 978-1-108-70772-5 Paperback

Additional resources for this publication at www.cambridge.org/9781108707725

To Our Parents and Teachers

Contents

Preface		*xv*
Chapter 1	**Drawing Instruments**	**1**
1.1	Introduction	1
1.2	List of Draughting Tools	1
1.3	Drawing Board	2
1.4	Mini-Draughter	3
1.5	Small Instrument Box	4
1.6	Set-Squares	7
1.7	Set of Scales	7
1.8	Protractor	8
1.9	French Curves	8
1.10	Drawing Sheets	9
1.11	Drawing Pencils	10
1.12	Paper Fasteners	12
1.13	Sand Paper Pad	12
1.14	Eraser	13
1.15	Duster	13
	Exercises	13
	Objective Questions	13
	Answers	14
Chapter 2	**Lines, Lettering and Layout of Sheet**	**15**
2.1	Introduction	15
2.2	Lines	15
2.3	Lettering	17
2.4	Single Stroke Letters	17
2.5	Gothic Letters	20
2.6	General Proportions of Letters	21

2.7	Drawing Sheet Layout	21
2.8	Title Block	22
	Exercises	23
	Objective Questions	24
	Answers	24
Chapter 3	**Principles of Dimensioning**	**25**
3.1	Introduction	25
3.2	Types of Dimensions	25
3.3	Elements of Dimensioning	26
3.4	Execution of Dimensions	26
3.5	Placing Dimensions	28
3.6	Methods of Dimensioning	29
3.7	Principles of Dimensioning	30
	Exercises	35
	Objective Questions	36
	Answers	37
Chapter 4	**Sections and Conventions**	**38**
4.1	Introduction	38
4.2	Cutting Plane or Sectional Plane	39
4.3	Section Lines or Hatching Lines	40
4.4	Types of Sections	42
4.5	Conventions for Various Materials	45
4.6	Conventional Breaks	46
4.7	Conventional Representation of Common Features	46
	Exercises	49
	Objective Questions	50
	Answers	50
Chapter 5	**Geometrical Constructions**	**51**
5.1	Introduction	51
5.2	Bisection of a Straight Line	51
5.3	Dividing a Line into Equal Parts	52
5.4	Drawing a Line Parallel to a Given Straight Line	53
5.5	Bisecting an Angle	54
5.6	Finding the Centre of an Arc	55
5.7	Constructing an Equilateral Triangle	56
5.8	Constructing Squares	57
5.9	Constructing Regular Polygons	58
5.10	Drawing Tangents	63
5.11	Inscribed Circles	64
	Exercises	65
	Objective Questions	66
	Answers	66

Chapter 6	**Scales**	**67**
6.1	Introduction	67
6.2	Representative Fraction or Scale Factor	68
6.3	Scales on Drawings	69
6.4	Types of Scales	69
6.5	Plain Scales	69
6.6	Diagonal Scales	79
	Exercises	93
	Objective Questions	94
	Answers	94
Chapter 7	**Orthographic Projections**	**95**
7.1	Introduction	95
7.2	Methods of Projections	95
7.3	Planes of Projection	97
7.4	Four Quadrants	98
7.5	First-Angle Projection	98
7.6	Third-Angle Projection	99
7.7	Symbols Used for First-Angle Projection and Third-Angle Projection Methods	100
	Exercises	101
	Objective Questions	102
	Answers	102
Chapter 8	**Projections of Points**	**103**
8.1	Introduction	103
8.2	Projection of a Point Lying in the First Quadrant	103
8.3	Projection of a Point Lying in the Second Quadrant	104
8.4	Projection of a Point Lying in the Third Quadrant	105
8.5	Projection of a Point Lying in the Fourth Quadrant	106
8.6	Special Cases	106
8.7	A Point is Situated in the Three Planes of Projection	107
	Exercises	113
	Objective Questions	114
	Answers	114
Chapter 9	**Projections of Lines**	**115**
9.1	Introduction	115
9.2	Position of a Straight Line	115
9.3	Line Parallel to Both HP and VP	116
9.4	Line Inclined to One Plane and Parallel to the Other	117
9.5	Line Perpendicular to One of the Planes	123
9.6	Line Contained by One or Both of the Principal Planes	125
9.7	Line Inclined to Both HP and VP	130

9.8   Line Contained by a Profile Plane (PP) or Line Contained by a Plane,
         Perpendicular to Both HP and VP   152
9.9   Traces of a Line   156
         Exercises   176
         Objective Questions   178
         Answers   179

Chapter 10   Projections of Planes   **180**
10.1   Introduction   180
10.2   Types of Planes   181
10.3   Traces of Planes   184
10.4   A Secondary Plane in Different Positions with Respect to the
         Principal Planes   185
10.5   Projections of Plane Parallel to One of the Principal Planes   185
10.6   Projections of Plane Perpendicular to Both HP and VP   192
10.7   Projections of Plane Inclined to One of the Principal Planes and
         Perpendicular to the Other Plane   194
10.8   Projections of Plane Inclined to Both the Principal Planes   206
         Exercises   219
         Objective Questions   221
         Answers   221

Chapter 11   Auxiliary Projections   **222**
11.1   Introduction   222
11.2   Types of Auxiliary Planes and Views   222
11.3   Projections of Points   222
11.4   Projections of Straight Lines   226
11.5   Projections of Planes   230
11.6   Shortest Distance between Two Skew Lines   234
         Additional Problems   236
         Exercises   239
         Objective Questions   239
         Answers   240

Chapter 12   Projections of Solids   **241**
12.1   Introduction   241
12.2   Types of Solids   241
12.3   Projections of Solids in Different Positions   247
12.4   Axis Perpendicular to One of the Principal Planes and Parallel to the Other   247
12.5   Axis Parallel to Both HP and VP   266
12.6   Axis Inclined to One of the Principal Planes and Parallel to the Other   270
12.7   Axes Inclined to Both HP and VP   307
         Exercises   328
         Objective Questions   330
         Answers   331

Chapter 13	**Sections of Solids**	**332**
13.1	Introduction	332
13.2	Section Planes	332
13.3	Sections	332
13.4	Frustum of a Solid and a Truncated Solid	333
13.5	Classification of Sections of Solids	334
13.6	Section Plane Parallel to the HP	334
13.7	Section Plane Parallel to the VP	346
13.8	Section Plane Perpendicular to the VP and Inclined to the HP	355
13.9	Section Plane Perpendicular to the HP and Inclined to the VP	372
13.10	Section Plane Perpendicular to Both HP and VP	387
	Exercises	389
	Objective Questions	391
	Answers	392
Chapter 14	**Development of Surfaces**	**393**
14.1	Introduction	393
14.2	Methods of Development	394
14.3	Parallel Line Method	394
14.4	Radial Line Method	413
14.5	Triangulation Method	446
14.6	Approximate Method	448
	Exercises	451
	Objective Questions	456
	Answers	456
Chapter 15	**Intersection of Surfaces**	**457**
15.1	Introduction	457
15.2	Methods of Determining Line of Intersection	457
15.3	Intersection of Two Prisms	458
15.4	Intersection of Cylinder and Cylinder	462
15.5	Intersection of Cylinder and Prism	470
15.6	Intersection of Cylinder and Cone	473
15.7	Intersection of Cone and Prism	479
	Exercises	483
	Objective Questions	485
	Answers	485
Chapter 16	**Isometric Projection**	**486**
16.1	Introduction	486
16.2	Classification of Pictorial Drawings	486
16.3	Axonometric Projection	487
16.4	Isometric Projection	487
16.5	Terms Connected with Isometric Projection	489
16.6	Isometric Scale	489

16.7	Isometric Drawing	491
16.8	Isometric Dimensioning	491
16.9	Hidden and Centre Lines on an Isometric Projection	492
16.10	Isometric Drawing or Projection of Plane Figures	492
16.11	Isometric Drawings or Projections of Prisms, Pyramids, Cylinders and Cones	498
16.12	Isometric Projection of a Sphere	501
	Exercises	523
	Objective Questions	527
	Answers	527
Chapter 17	**Conversion of Pictorial Views into Orthographic Views**	**528**
17.1	Introduction	528
17.2	Direction of Sight	528
17.3	Orthographic Views	528
17.4	Spacing of Views	532
17.5	Procedure for Preparing Orthographic Views	532
17.6	Identification of Surfaces	539
17.7	Missing Lines and Missing Views	541
	Exercises	552
	Objective Questions	561
	Answers	561
Chapter 18	**Freehand Sketching**	**562**
18.1	Introduction	562
18.2	Sketching Materials	562
18.3	Uses of Sketches	563
18.4	Sketching Straight Lines	563
18.5	Sketching Circles	564
18.6	Sketching an Ellipse	566
18.7	Sketching Arcs and Curves	566
18.8	Sketching Angles	567
18.9	Types of Freehand Sketches	568
18.10	Sketching Orthographic Views	568
18.11	Sketching Isometric Views	569
	Exercises	570
	Objective Questions	571
	Answers	571
Chapter 19	**Computer Graphics**	**572**
19.1	Introduction	572
19.2	Computer Graphics	572
19.3	Requirements for Computer Graphics	573
19.4	Getting Started with AutoCAD	573

19.5	Saving a Drawing	575
19.6	Command Entry	576
19.7	Drawing Limits	576
19.8	Units	576
19.9	Draw Commands	577
19.10	Modify Commands	587
19.11	More Advanced Commands	597
	Exercises	613
	Objective Questions	617

Preface

This text book is an endeavour to deal with the subject of Engineering Graphics and Drawing in such a way that students can understand the subject thoroughly. The subject of Engineering Drawing and (Computer) Graphics is a core course taught to the first year students of all disciplines of engineering in all engineering colleges and universities. Therefore, this text book is an attempt to help students grasp the basic concepts of engineering drawing clearly and easily.

The book has many distinguishing features. It covers the fundamental concepts of manual and computer aided drafting. It includes more than 400 solved examples and 275 practice exercises along with 250 short answer questions, i.e., objective type questions with answers.

The organisation of this text is done very clearly and logically. Each chapter is organised as basic theory, solved problems, exercises and objective type questions. Each chapter contains a large number of worked examples, the problems for which have been selected from examinations of different universities. This book comprises 19 chapters, starting with an introduction to drawing instruments and their uses, followed by Chapter 2 on various types of lines and their uses and layout of a drawing sheet. Chapter 3 attempts to give knowledge about the various methods and principles of dimensioning of a drawing. The description about sections and conventions is given in Chapter 4, followed by Chapter 5 on geometrical constructions. From here onward, various scales are described in Chapter 6 and then the concept of orthographic projection is described in Chapter 7. Chapter 8 of the text is dedicated to projections of points, followed by projections of lines in Chapter 9. Once students are clear about the basic concepts of projections of points and lines, then this book introduces them to the projections of planes and auxiliary projections in Chapters 10 and 11, respectively.

Chapter 12 discusses the projections of solids followed by a lucid explanation of the sections of solids in Chapter 13: this will imbibe an imagination of the sections view of a solid object or engineering component. Students of first year in almost all branches of engineering are also offered a course on manufacturing processes and practices, where they need to do some jobs on sheet metal. Chapter 14 covers development of surfaces that will help students develop understanding of a pattern and calculate the requirement of sheet metal for mass production. Allied to this, students will get exposure to intersection of solids in Chapter 15.

It is very important to introduce students to isometric projections, so that they can imagine any shape from 2D surface to 3D; the same is explained in Chapter 16. Similarly conversion of pictorial view (3D) into orthographic view (2D) is described in Chapter 17. The same is followed by free hand sketches in Chapter 18 and basic principles and commands of computer graphics in Chapter 19.

Figures constitute the main feature of any engineering drawing book. Therefore, care has been taken that the figures are easily understood and students find it 'difficult to forget' them. Most of the problems are solved in the first-angle projection method; however, a few problems are also solved using the third-angle projection method. This book attempts to acquaint students with different types of questions. The contents of the book are in line with the syllabi of many universities, colleges and polytechnics in India.

A chapter on 'Computer Graphics' is given to explain the preparation of figures using 'AutoCAD'. The AutoCAD section of the book describes all the menu and commands items of the graphics package.

Overall, sincere efforts have been made to make this book student-friendly and self-explanatory. We are grateful to the Almighty for blessing us with good health and high spirits to take up this book as a project. We must concede that this book would never have been written without the constant support and encouragement of our family members, especially our children. We are extremely thankful to our respective heads of the institutions: Professor L. K. Awasthi, Director, NIT Jalandhar, and Professor Sehijpal Singh, Principal, GNDEC Ludhiana, for their continuous motivation and support to pursue such endeavours. We also owe an enormous debt to our colleagues and students for much valued assistance in the form of discussions and feedback. We express our gratitude to the editorial team at Cambridge University Press, especially to Ms Taranpreet Kaur (Commissioning Editor) for her excellent ground work and syllabus research that helped us in deciding the table of contents. We always turned to her for suggestions wherever we were stuck and she was always available to answer our queries. Last but not least we are thankful to all our students and teachers who have taught us and made us what we are today.

We devote our work to the Almighty, whose blessings are always with us.

Any suggestion or criticism for further improvement of the book will be gratefully acknowledge and highly appreciated.

Lakhwinder Pal Singh
Harwinder Singh

Chapter 1

Drawing Instruments

1.1 Introduction

Drawing is a graphic language by which communication is accomplished through sketches (drawings). Drawings can be of various types. If compared with verbal or written description, drawings provide a far better idea about the shape, size and appearance of any object or situation or location, that too fairly quickly. From a manufacturing point of view, 2D and 3D drawings are very important and commonly used in engineering industry. Drawings are prepared manually or by using a computer. In 2D view (orthographic projection), one view is not enough to get all the details of the object. So it is necessary to draw the front view, top view, bottom view, right side view and left side view.

To be more useful, in particular from the manufacturing point of view, the sketch or drawing should include dimensions, manufacturing details, materials used, etc. Drawing equipment and instruments are needed to record information on drawing paper or any other suitable surface. Drawing, mainly consisting of straight lines, curves, circles and arcs, is prepared with certain instruments. The quality of the drawing mainly depends on the quality of the instruments, their adjustment, handling and care. Therefore, the equipment must be reliable and accurate, as the same will result in good quality drawing, which will further enhance a student's interest. The correct selection and use of pencils and drawing instruments should be taught throughout the course. Beginners certainly need guidance on the selection and purchase of the drawing instruments and equipment essential for drafting. The various instruments and other drafting equipment are described below.

1.2 List of Draughting Tools

The following is the list of draughting tools which every student must possess:

1. Drawing Board
2. Mini-Draughter

3.   Small Instrument Box, containing the following:
- Large Size Compass
- Small Bow Compass
- Large Size Divider
- Small Bow Divider

4.   Set-Squares
5.   Set of Scales
6.   Protractor
7.   French Curves
8.   Drawing Sheets
9.   Drawing Pencils
10. Paper Fasteners
11. Pencil Sharpener
12. Sand Paper Pad
13. Eraser
14. Duster

Some of these have been explained in greater detail in the subsequent sections.

1.3 Drawing Board

A drawing board is usually made of well-seasoned soft wood. To prevent warping, narrow strips of wood are glued together. Prevention of warping will also be aided by two battens cleated at the bottom side of the board. These battens, also, help give rigidity to the board. One of the edges of the board is used as the working edge (generally made of hard and durable wood) on which the T-square is made to slide. See Fig. 1.1.

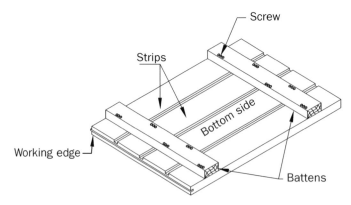

Fig. 1.1 Drawing board

The size of the drawing board will depend upon the size of the drawing sheet to be used. The Bureau of Indian Standards (BIS) recommends the sizes of the drawing board, as given in the Table 1.1.

Table 1.1 Sizes of the drawing board

S.No.	Designation	Size (Length × Breadth × Thickness), mm
1.	B_0	1500 × 1000 × 25
2.	B_1	1000 × 700 × 25
3.	B_2	700 × 500 × 15
4.	B_3	500 × 350 × 15

The following points need to be taken care of while handling a drawing board:
- Handle the drawing board carefully so that no dents or holes are made on its surface.
- Check the working edge at regular intervals and correct it, whenever it is found defective.
- Fasten a sheet of paper on the drawing board to keep its surface clean.

1.4 Mini-Draughter

It consists of an angle formed by two arms with scales marked and set exactly at right angle to each other. It is designed to combine the functions of T-square, set-squares, protractor and scales. The arms may also be set and clamped at any desired angle by means of an adjusting screw, which has a protractor. See Fig. 1.2.

Mini-Draughter has a mechanism which keeps the two blades always parallel to their original position, whenever they may be moved on the board. Thus by means of it, horizontal, vertical or inclined parallel lines of desired lengths can be drawn anywhere on the sheet, with considerable ease and saving of time.

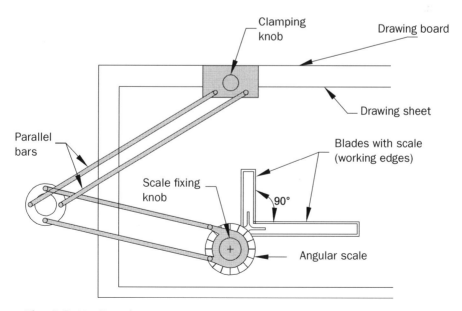

Fig. 1.2 Mini-Draughter

1.5 Small Instrument Box

It contains the following instruments:

• **Large Size Compass**: It is used for drawing circles and arcs up to 200 mm. It consists of two legs hinged at one end; one leg is so arranged that it can receive either a lead* or a pen, as shown in Fig. 1.3. The needle point should be at least 1 mm larger than the lead or pen point. This is because, when the compass is used the needle point penetrates into the paper and thus the pen or lead touches it at same level (Fig. 1.4). The hardness of the lead used should be the same as that of the pencil work of the drawing. The lead end may be sharpened as shown in Fig. 1.5.

(a) Lead attachment
(b) Needle attachment

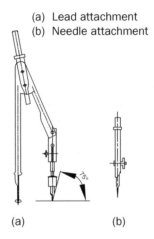

(a)   (b)

Fig. 1.3 Large size compass with different attachments

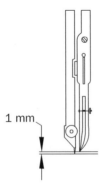

1 mm

Fig. 1.4 Relative position of compass point

---

* Although the actual drawing material is graphite (a non-poisonous substance), the popular term used for the material is 'lead'. We are retaining the popular term in this book.

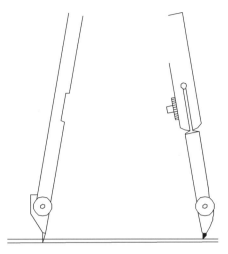

Fig. 1.5 Sharpening of compass lead

- **Small Bow Compass**: It is used for drawing small circles of radius up to 25 mm. This operates on the jackscrew principle by turning a knurled nut at its centre, as shown in Fig. 1.6. In some designs the adjusting nut may be on one side of the compass, as shown in Fig. 1.7.

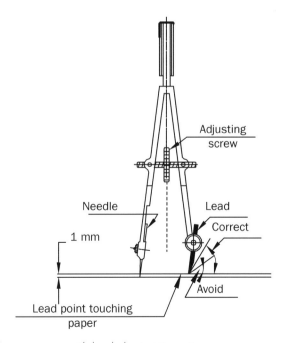

Fig. 1.6 Small bow compass with knurled nut at its centre

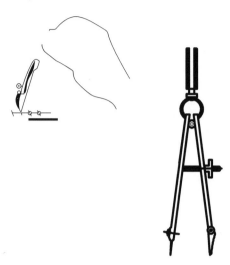

Fig. 1.7 Small bow compass with adjusting nut on one side

- **Large Size Divider:** The divider is used for transferring measurements from one part of the drawing to another part and also for dividing curved or straight lines into any number of equal parts. It is similar to the large size compass except that both the legs contain needle points, as shown in Fig. 1.8.

Fig. 1.8 Large size divider

Fig. 1.9 Small bow divider

- **Small Bow Divider:** It is similar to the large size divider, except that the distance between the legs is adjusted by a knurled nut as shown in Fig. 1.9. It is more convenient for transferring smaller distances.

1.6 Set-Squares

Set-squares are made of transparent celluloid or plastic material and retain their shape and accuracy for a long time, as shown in Fig. 1.10. Set-squares are available in two forms:

(i)   30°–60° set-square
(ii) 45° set-square

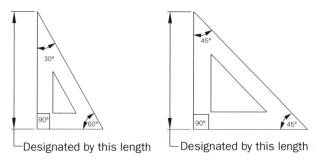

Fig. 1.10 Set-squares

1.7 Set of Scales

Scales are made of wood, steel celluloid, card board, etc. Scales may be flat or of triangular cross section. Both the longer edges of the scales are marked with divisions of centimetres, which are further sub-divided into millimetres, as shown in Fig. 1.11.

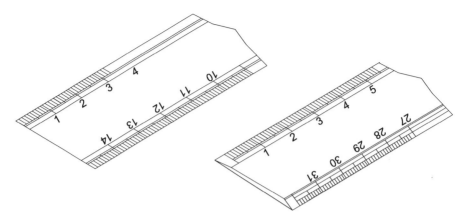

Fig. 1.11 Set of scales

Scales are used to transfer the true or relative dimensions of an object on to the drawing. Scales are required to make drawings accurately to any desired scale. Various other types of scales are discussed in Chapter 6.

1.8 Protractor

Protractor is made of either wood or tin or transparent celluloid material. It is used to draw or measure such angles which cannot be drawn with the set squares, as shown in Fig. 1.12.

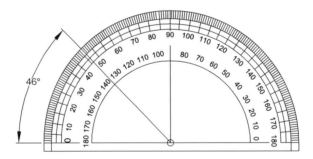

Fig.1.12 Protractor

1.9 French Curves

French curves are generally made of wood, plastic or celluloid material. These are used for drawing curves which cannot be drawn with a compass. These curves are available in various shapes and sizes; a few of them are shown in Fig. 1.13. While using, first of all a series of points are plotted along the desired path and then the most suitable curve is made to fit along it. A smooth curve is then drawn along the edge of the curve. Care should be taken to see that no corner is formed anywhere within the drawn curve.

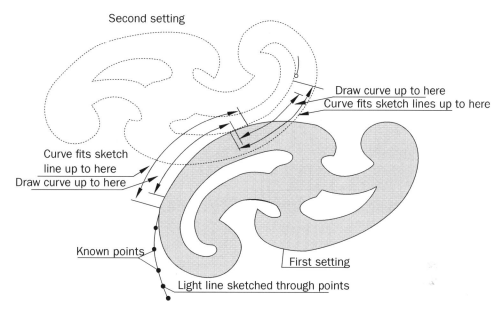

Fig. 1.13 French curves

1.10 Drawing Sheets

Drawing sheets are available in six standard sizes, specified by the Bureau of Indian Standards (BIS), as shown in Table 1.2. The sheet should be tough and strong, and when an eraser is used on it, its fibres should not disintegrate.

Table 1.2 Preferred drawing sheet sizes

S. No.	Designation	Trimmed size (mm)	Untrimmed size (mm)
1.	A_0	841 × 1189	880 × 1230
2.	A_1	594 × 841	625 × 880
3.	A_2	420 × 594	450 × 625
4.	A_3	297 × 420	330 × 450
5.	A_4	210 × 297	240 × 330
6.	A_5	148 × 210	165 × 240

Figure 1.14 shows the untrimmed size and trimmed size of the drawing sheet. The trimmed sizes of drawing sheets available are 841 × 1189 (A_0), 594 × 841 (A_1), 420 × 594 (A_2), etc., as shown in Fig. 1.15 The sizes of the successive sheets are obtained by folding the preceding size. The original ratio of sides needs to be constantly maintained when the sheet is halved (y/x). Therefore $y/x = x/(y/2)$ or $y/x = \sqrt{2} = 1.4142$. To have an area of 1 sqm (A_0) and at the same time to have $y/x = 1.4142$, $y = 1189$ mm and $x = 841$ mm.

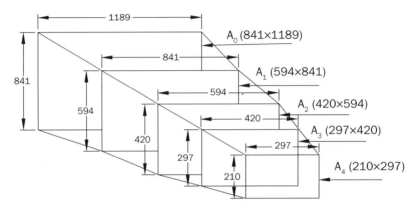

Fig. 1.14 Trimmed and untrimmed size drawing sheet

Fig. 1.15 Trimmed drawing sheet sizes

1.11 Drawing Pencils

The accuracy and appearance of a drawing depends upon the quality of the pencils used. The grade of a pencil is indicated by letters and numerals. The grade HB denotes medium hardness of the lead. The hardness of a pencil increases as the value of the numeral put before the letter H increases. Similarly, the lead becomes softer as the value of the numeral put before the letter B increases; so there are eighteen grades of pencil, which are as follows:

9H, 8H, 7H, 6H, 5H, 4H, 3H, 2H, H, F, HB, B, 2B, 3B, 4B, 5B, 6B, 7B. Out of these 9H is the hardest and 7B the softest pencil, as shown in Fig. 1.16.

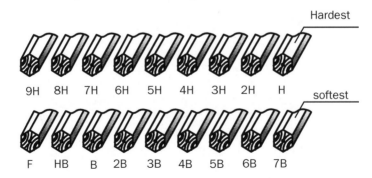

Fig. 1.16 Grades of pencil

Pencils of grades H and HB are more suitable for lettering and dimensioning. For sketching and artistic work, soft grade pencils such as HB should be used.

Both wood and mechanical pencils are available in the market. Wood pencils require frequent sharpening and shaping of the lead, so great care should be taken. The lead may be sharpened into two different forms:

- Conical point
- Chisel edge

For general purposes, a conical point is used, whereas the chisel edge is used for drawing straight lines, as it requires less sharpening and makes a dense line.

To prepare the pencil lead for drawing work, the wood around the lead from the end other than that on which the grade is marked is removed with a pen-knife, leaving about 8–10 mm of lead projecting out, as shown in Fig. 1.17. As shown in Fig. 1.18, the chisel edge is prepared by rubbing the lead on a sand paper pad, making it flat, first on one side and then on the other, by turning the pencil through 180°.

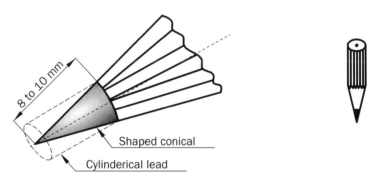

Fig. 1.17 Conical point of a pencil

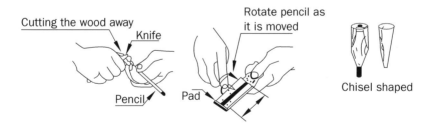

Fig. 1.18 How to prepare chisel edge

Mechanical pencils can hold leads of very small diameter (0.3 mm and above). These pencils are cleaner to use, much more convenient and are likely to produce a line of consistent thickness. A typical mechanical pencil is shown in Fig. 1.19.

Fig. 1.19 Mechanical pencil

1.12 Paper Fasteners

Clips, adhesive tapes and drawing pins are used for fixing drawing sheet on the drawing board. See Fig. 1.20.

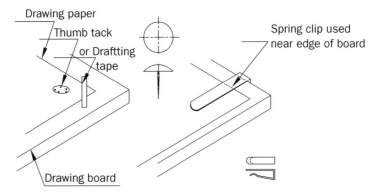

Fig. 1.20 Paper fasteners

1.13 Sand Paper Pad

A piece of sand paper is pasted on a wooden block as shown in Fig. 1.21. When it becomes dirty or worn out, it should be replaced by another one.

Fig. 1.21 Sand paper pad

1.14 Eraser

A soft pencil eraser should be used for erasing pencil lines. To avoid frequent erasing, careful planning in drawing is needed. See Fig. 1.22.

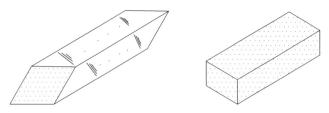

Fig. 1.22 Eraser

1.15 Duster

It should, preferably, be of towel cloth of convenient size. All instruments should be thoroughly cleaned with duster before starting the work.

Exercises

1.1   Using mini-draughter, set 15°, 30°, 45°, 60° and 75° angles with a horizontal line.
1.2   Using protractor, set 10°, 20°, 40°, 80° and 160° angles with a horizontal line.
1.3   Draw a line 150 mm long and divide it into ten equal parts by using a divider.
1.4   Draw a circle of 100 mm diameter and divide it into:
(i)   Eight,
(ii)   Twelve,
(iii) Sixteen and
(iv)   Twenty equal parts without using a protractor.
1.5   Draw two lines parallel to each other, separated by 30 mm and making an angle of 30° with the horizontal.

Objective Questions

1.1   The working edge of the drawing board should be on the .... side of the draughtsman.
1.2   To prevent warping of the drawing board, ... are cleated at its back.
1.3   To draw circles and arcs of circles, ... is used.
1.4   What are the main parts of a mini-draughter?
1.5   Why cellotape is preferred to drawing pins for fixing the drawing sheet to the board?
1.6   Why pencil should be rotated in fingers while drawing a line?

1.7   Measurements from the scale to the drawing are transferred with the aid of a ....

1.8   For drawing thin lines of uniform thickness, the pencil should be sharpened in the form of ....

1.9   Mention the uses of a divider.

1.10 To remove a particular spot on the drawing, ... is used.

1.11 What is a French curve and where is it used?

1.12 Why is the needle point of a compass slightly more in length than the lead point?

1.13 Uses of the T-square, set-squares, scale and protractor are combined in the ....

1.14 To draw or measure angles, ... is used.

1.15 What are the standard sizes of drawing sheets?

1.16 .... are used for drawing irregular curves.

1.17 What should be the grade of pencil used for lettering?

1.18 Define engineering drawing. Why is drawing called the universal language of engineers?

1.19 Name different types of drawing instruments.

1.20 The softness of the pencil lead .... as numeral letter B increases.

1.21 What are the standard sizes of drawing sheets according to BIS and what is suitable for drawing work?

1.22 The hardness of a pencil increases as the value of the numeral put before the letter H increases.

(True/False)

1.23 The lead becomes softer as the value of the numeral put before the letter B decreases.

(True/False)

1.24 The length to width ratio of all the drawing sheet sizes is ....

Answers

1.1	Left	1.14	Protractor
1.2	Battens	1.16	French curves
1.3	Compass	1.20	Increases
1.7	Divider	1.22	True
1.8	Chisel edge	1.23	False
1.10	Eraser	1.24	$\sqrt{2}:1$
1.13	Mini-Draughter		

Chapter 2

Lines, Lettering and Layout of Sheet

2.1 Introduction

The topics covered in this chapter are meant to introduce the readers to various types of lines, lettering and layout of sheets, which are used in engineering drawing.

2.2 Lines

A line is the basis of engineering drawing; a set of conventional symbols representing all the lines needed for different purposes may well be called an alphabet of lines. In engineering drawing each line has specific measurements and functions. The Bureau of Indian Standards (BIS) has recommended various types of lines and their applications in technical drawing. These are exhibited in Table 2.1.

Table 2.1 Various types of lines and their applications

S. No.	Type of line	Conventional Representation	Applications or Uses
1.	Continuous thick		Visible outline and edges
2.	Continuous thin		Dimension lines, extension lines, construction lines, leader lines, section lines
3.	Continuous thin wavy		Short break lines or irregular boundaries
4.	Continuous thin with zig-zag		Long break lines
5.	Short dashed thin	2 to 3 mm 1 to 2 mm	Hidden outlines and edges
6.	Long chain thin	15 to 30 mm 1 to 2 mm 2 to 3 mm	Centre lines, pitch circles, locus lines

S. No.	Type of line	Conventional Representation	Applications or Uses
7.	Long chain thin but thick at the ends		Cutting plane lines
8.	Long chain thick	—— —— — —— —— —	For indicating lines or surfaces with special treatment
9.	Long chain thin with double dashed	—— — — —— —— —	Outline of neighbouring parts, alternative and extreme positions of movable parts

Distinctive presentations of these different types of lines are shown in Fig. 2.1.

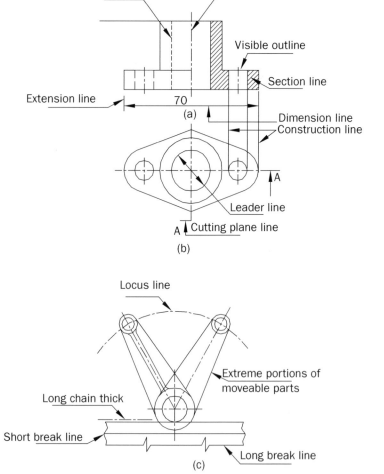

Fig. 2.1 Applications of different types of lines

2.3 Lettering

Lettering is an important feature of all engineering drawings, which is meant for indicating notes, dimensions, etc., on a drawing. Despite the fact that a drawing may be made accurately and neatly, poor lettering may lead to deteriorated appearance and sometimes make the drawing useless. Therefore, lettering on a drawing needs to be legible, neat in appearance and in appropriate style. The subsequent sections describe various types of letters.

2.4 Single Stroke Letters

BIS has recommended single stroke lettering for use in engineering drawing. In single stroke letters, the thickness of a letter is such that it is obtained in one stroke of the pencil. Horizontal line (stroke) is from left to right, and vertical or inclined lines (strokes) are from top to bottom.

Single stroke letters are of two types:

- Upper case or capital letters
- Lower case or small letters

Again both of them are further sub-divided into two categories:

- Vertical
- Inclined

In vertical lettering, stems are perpendicular to the line of lettering, whereas in inclined lettering, stems make a slope of 75° to the horizontal. Normally, letters are nominated by their height; therefore height of the letters enables us to classify the letters as

- Lettering 'A'
- Lettering 'B'

The standard ratios for line thickness (d) to height (h), d/h, as 1/14 and 1/10 are most prevalent as they result in a minimum number of thicknesses, as given in Table 2.2.

Table 2.2 Dimensions of lettering

Characteristic		Ratio	Dimensions (mm)						
Lettering A (d = h/14)									
Height of capital letters	h	$\left(\dfrac{14}{14}\right)$ h	2.5	3.5	5	7	10	14	20
Height of lower-case letters	c	$\left(\dfrac{10}{14}\right)$ h	–	2.5	3.5	5	7	10	14
Spacing between characters	a	$\left(\dfrac{2}{14}\right)$ h	0.35	0.5	0.7	1	1.4	2	2.8

Characteristic		Ratio		Dimensions (mm)					
Minimum spacing of base lines	b	$\left(\dfrac{20}{14}\right)$ h	3.5	5	7	10	14	20	28
Minimum spacing between words	e	$\left(\dfrac{6}{14}\right)$ h	1.05	1.5	2.1	3	4.2	6	8.4
Thickness of lines	d	$\left(\dfrac{1}{14}\right)$ h	0.18	0.25	0.35	0.5	0.7	1	1.4
Lettering height B (d = h/10)									
Height of capital letters	h	$\left(\dfrac{10}{10}\right)$ h	2.5	3.5	5	7	10	14	20
Height of lower-case letters	c	$\left(\dfrac{7}{10}\right)$ h	–	2.5	3.5	5	7	10	14
Spacing between characters	a	$\left(\dfrac{2}{10}\right)$ h	0.5	0.7	1	1.4	2	2.8	4
Minimum spacing of base lines	b	$\left(\dfrac{14}{10}\right)$ h	3.5	5	7	10	14	20	28
Minimum spacing between words	e	$\left(\dfrac{6}{10}\right)$ h	1.5	2.1	3	4.2	6	8.4	12
Thickness of lines	d	$\left(\dfrac{1}{10}\right)$ h	0.25	0.35	0.5	0.7	1	1.4	2

The height 'h' of capital letters is considered as the base of dimensioning as shown in Fig. 2.2.

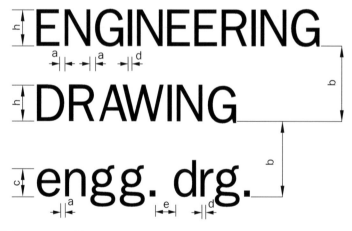

Fig. 2.2 Dimensions of lettering

Table 2.3 Recommended heights of letters

S. No.	Purpose	Size in mm
1.	Main titles, drawing numbers	6, 8, 10 and 12
2.	Subtitles	3, 4, 5 and 6
3.	Dimensions and notes	3, 4 and 5
4.	Alteration entries and tolerances	2 and 3

Lettering is normally used in capital letters; however, different sizes of letters are used for different purposes, as exhibited in Table 2.3.

Figures 2.3 and 2.4 show single stroke vertical and inclined letters, respectively.

Fig. 2.3 Single stroke vertical letters

Fig. 2.4 Single stroke inclined letters

While drawing letters, in order to maintain uniformity in size, guiding lines (light and thin) may be drawn first, followed by drawing the letter in between them.

2.5 Gothic Letters

The gothic letters are similar to the single stroke letters. The only change gothic letters have over single stroke letters is that if the stems of single stroke letters are given more thickness, they will produce gothic letters. These are regularly used for main titles of ink drawings. For these, the thickness of the stem may vary from 1/5 to 1/0 of the height of the letters. Figs. 2.5 and 2.6 illustrate vertical and inclined gothic letters, respectively.

Fig. 2.5 Gothic vertical letters

Fig. 2.6 Gothic inclined letters

2.6 General Proportions of Letters

According to the ratio of width and height, letters may be divided into three categories:

1. Normal letters
2. Compressed or condensed letters
3. Extended letters

Normal letters, which are commonly used or used for general purposes, have normal height and width, and are written in usual space. Compressed or condensed letters are those which are written narrow in their proportion of width to height, the height being proportionately more than the width. They are more useful for cases of limited space. Extended letters are written wide in their proportion of width to height, that is, the height of these letters is less or equal to the width. See Fig. 2.7.

NORMAL LETTERS
EXTENDED LETTERS
CONDENSED LETTERS

Fig. 2.7 Normal, extended and condensed letters

2.7 Drawing Sheet Layout

The preferred standard sizes for drawing sheets, as specified by the BIS, have been discussed in Chapter 1, Section 1.10. The layout of drawing sheet will facilitate easy reading and understanding of the drawings presented therein. A standard arrangement should make sure that all the required information is included and facilitate easy filing and binding. Fig. 2.8 illustrates a typical layout of a drawing sheet, following the grid reference system. The grid reference system is recommended to easily locate the drawing details, amendments, etc.

It is endorsed that the borders have a minimum width of 20 mm for A_0 and A_1 sheet sizes and a minimum width of 10 mm for A_2, A_3, A_4 and A_5 sheet sizes. A filing margin of 20 mm on the left hand side has to be provided (including the border), irrespective of the size of the sheet.

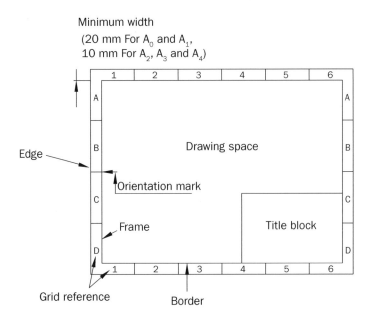

Fig. 2.8 Typical layout of a drawing sheet

2.8 Title Block

The drawing sheet layout should also provide the title block which is an important feature in a drawing. Appropriate space for the title block should be provided at the bottom right hand corner of the drawing sheet. Figure 2.9 illustrates typical layouts of the title blocks. All title blocks contain at least the particulars shown in Table 2.4.

Table 2.4 Particulars of title block

S. No	Particular
1.	Name of firm.
2.	Title of the drawing.
3.	Scale.
4.	Symbol for the method of projection.
5.	Drawing number.
6.	Initials, with dates, of persons who have designed, drawn and checked.
7.	No. of sheet and total number of sheets of the drawing of the object.

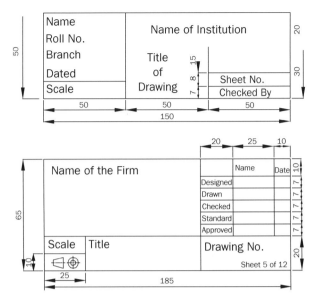

Fig. 2.9 Typical layouts of title blocks

Exercises

2.1 Print the following words freehand in a height of 12 mm and vertical
   (i)   WORK IS WORKSHIP
   (ii) ENGINEERING DRAWING
   (iii) INDIAN SOCIETY FOR TECHNICAL EDUCATION
   (iv) GOLDEN TEMPLE
   (v)   TIT FOR TAT

2.2 Print the following words freehand in a height of 12 mm and inclined at 75°
   (i)   HONESTY IS THE BEST POLICY
   (ii) MY BEST TEACHER
   (iii) HEALTH IS WEALTH

2.3 Print freehand, in vertical and inclined lower case letters, taking height 9 mm and using ratio 7:4.

2.4 Print freehand, in vertical and inclined numerals, taking height 10 mm, 6 mm and 4 mm.

2.5 Draw the conventional representation of various types of lines used in engineering practice. Also mention its uses or applications.

2.6 List out the contents of the title block. Also prepare a title block for use in classroom.

2.7 Write freehand the following sentences using inclined capital letters of 8 mm size in single stroke using 7:5 ratio.
   (i)   NATURE IS BEAUTIFUL
   (ii) PRACTICE MAKES A MAN PERFECT

2.8 Write 'ENGINEER" using single stroke capital letters.

Objective Questions

2.1   Outlines are drawn as ....

2.2   Dimension lines, hatching and construction lines are drawn as ....

2.3   The cutting plane line is shown by ....

2.4   Long break lines are shown by ....

2.5   Efficiency in the art of lettering is achieved by continuous ....

2.6   For maintaining uniformity in size, ... may be drawn.

2.7   The size of the letter is described by its ....

2.8   Lettering is generally done in ....

2.9   Two types of single stroke letters are .... and ....

2.10   What do you understand by normal and compressed letters?

2.11   What are the requirements of lettering?

2.12   Why is layout of sheet necessary?

2.13   What is the use of long chain thick line in engineering drawing?

2.14   What do you mean by single stroke letters?

2.15   Name different types of lines.

2.16   Sketch the hidden line, short break line, centre line, cutting plane line and long break line.

2.17   Provide the suitable location for the title block in a drawing sheet.

Answers

2.1	Continuous thick lines	2.6	Thin and light guidelines
2.2	Continuous thin lines	2.7	Height
2.3	Long chain thin but thick at the ends	2.8	Capital letters
2.4	Continuous thin with zig-zag	2.9	Vertical, inclined
2.5	Practice	2.17	Right hand bottom corner

Chapter 3

Principles of Dimensioning

3.1 Introduction

Every drawing consists of the views necessary to describe the shape of an object, and must provide its exact length, width, height, size, position of holes and any other details required for the manufacturing of the object. Providing this information on a drawing is called dimensioning.

3.2 Types of Dimensions

A drawing usually requires two types of dimensions:

(i)   Size or functional dimension
(ii) Location or datum dimension

(i)   **Size or functional dimension:** It gives the size of a piece or component as shown in Fig. 3.1. It is usually represented by the letter *S*.

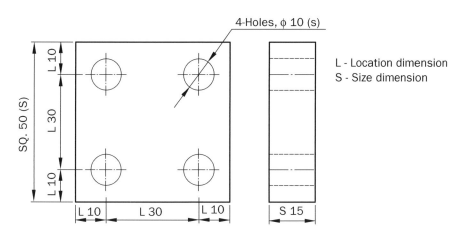

Fig. 3.1 Types of dimensions

(ii) **Location or datum dimension:** It fixes the relationship of the component parts (holes, slots, etc.) of a piece of structure. It may be from centre to centre, surface to surface or surface to centre, as shown in Fig. 3.1. Generally it is denoted by the letter *L*.

3.3 Elements of Dimensioning

The elements of dimensioning include the extension or projection line, dimension line, leader line, arrowhead and the dimension itself. The various elements of dimensioning are shown in Fig. 3.2.

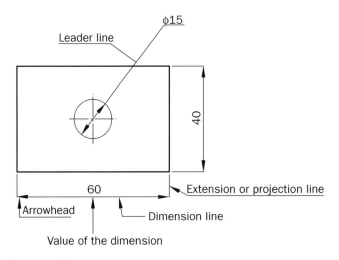

Fig. 3.2 Elements of dimensioning

3.4 Execution of Dimensions

The points given below have to be always followed while executing dimensions:

(i) Projection lines should be drawn from the visible features of the object and extend slighly beyond the dimension line, as shown in Fig. 3.3.

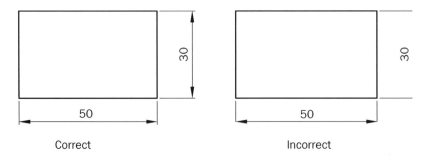

Fig. 3.3 Placing of projection lines

(ii)   Mutual crossing of dimension and projection lines should be avoided. See Fig. 3.4.

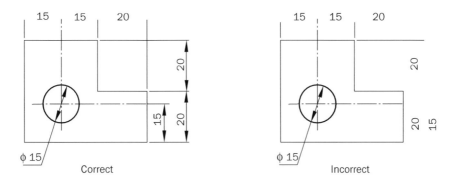

Fig. 3.4 Placing of projection and dimension lines

(iii)   Projection or extension line must be drawn perpendicular to the part to which it is to be dimensioned. See Fig. 3.5.

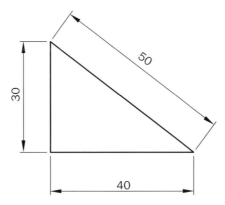

Fig. 3.5 Projection lines on a tapered surface

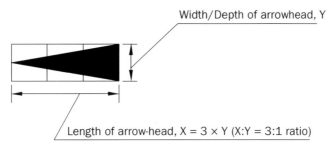

Fig. 3.6 Proportions of an arrowhead

(iv)   For arrowheads used at the ends of dimension and leader lines, the length may be taken as three times the width/depth as shown in Fig. 3.6. The size of the arrowhead should be proportionate to the size of the drawing.

3.5 Placing Dimensions

Dimensions may be placed according to either of the following two systems:

(i)   **Aligned system:** In an aligned system, all the dimensions are placed above the dimension line such that they may be read either from the bottom or from the right hand side of the drawing. See Fig. 3.7.

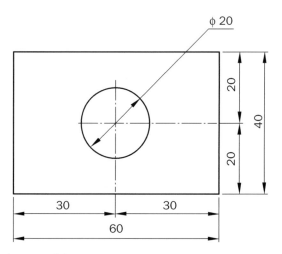

Fig. 3.7 Aligned system of dimensioning

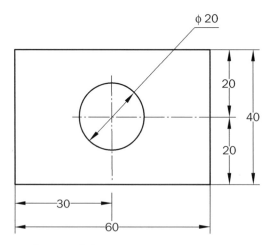

Fig. 3.8 Unidirectional system of dimensioning

(ii) **Unidirectional system:** In unidirectional system, all the dimensions are placed in one direction such that they may be read from the bottom of the drawing. Also in this system, the dimension lines are broken to insert the dimensions, as shown in Fig. 3.8. This system is mainly used on large drawings as of aircraft, automobiles, etc., where it is inconvenient to read dimensions from the right-hand side.

3.6 Methods of Dimensioning

(i) **Chain dimensioning:** It should be used only where the, possible, accumulation of tolerances does not endanger the functional requirement of the object. See Fig. 3.9.

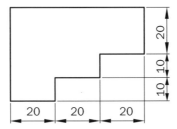

Fig. 3.9 Chain dimensioning

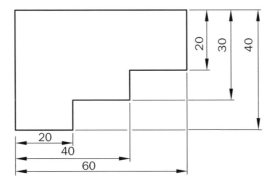

Fig. 3.10 Parallel dimensioning

(ii) **Parallel dimensioning:** When a number of dimensions of a part have a common datum feature. See Fig. 3.10.

(iii) **Combined dimensioning:** In combined dimensioning, both chain and parallel dimensioning are followed. See Fig. 3.11.

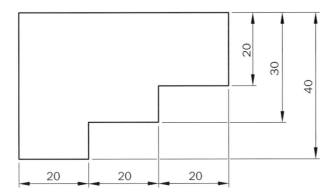

Fig. 3.11 Combined dimensioning

3.7 Principles of Dimensioning

The following are some of the principles to be applied in dimensioning:

(i)   Any given dimension must be clear and permit only one interpretation. See Fig. 3.12.

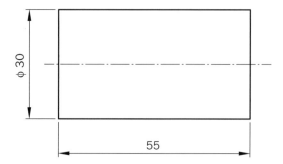

Fig. 3.12 Interpretation of the part to be dimensioned

(ii) Dimensions indicated in one view need not be repeated in another view, as shown in Fig. 3.13.

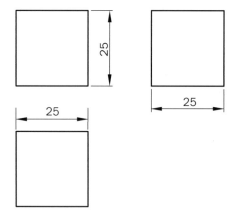

Fig. 3.13 Repetition of dimensioning in various views should be avoided

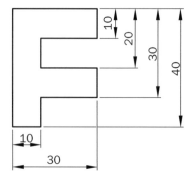

Fig. 3.14 Spacing of the dimension lines

(iii) Dimension lines should be drawn at least 5 to 8 mm away from the outlines and from each other as shown in Fig. 3.14.

(iv) As far as possible, dimensions should be placed outside the views, as shown in Fig. 3.15.

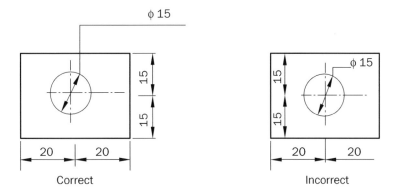

Fig. 3.15 Dimensions should be placed outside the view

(v) Smaller dimensions should be placed nearer the view and the larger ones farther away so that the extension lines do not intersect dimension lines. See Fig. 3.16.

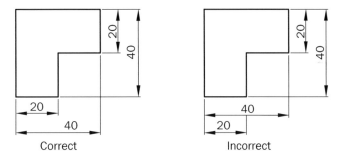

Fig. 3.16 Larger dimensions should be placed after the smaller dimensions

(vi) Dimensions should be marked from visible outlines rather than from hidden lines, as shown in Fig. 3.17.

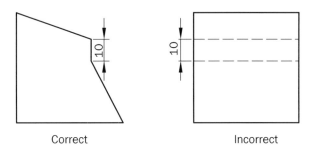

Fig. 3.17 Dimensions should be marked from the visible outlines

(vii) Arrowheads should normally be drawn within the limits of the dimensioned feature. But when the space is too narrow, they may be placed outside. A dot may also be used to replace an arrowhead. Sometimes due to lack of space, the dimension figure may be written above the extended portion of the dimension line. See Fig. 3.18.

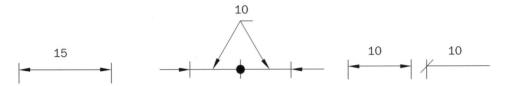

Fig. 3.18 Different ways of dimensioning of length when the space is insufficient

(viii) Dimensioning should be marked from a base line or centre line of a hole or cylindrical part or finished surface, etc., which may readily be located. See Fig. 3.19.

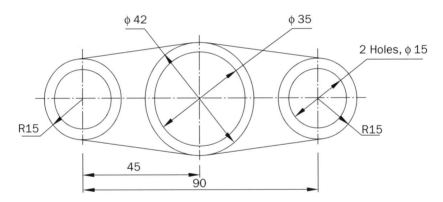

Fig. 3.19 Dimensioning should be marked from the centre of a hole

(ix) Dimensioning to a centre line should be avoided except when the centre line passes through the centre of a hole or a cylindrical part. See Fig. 3.20.
(x) Dimensioning of cylindrical parts should as far as possible be placed in the views in which they are seen as rectangles. See Fig. 3.21.
(xi) Dimensions indicating a diameter should always be preceded by the symbol φ. Fig. 3.22 shows various methods of dimensioning different sizes of circles.

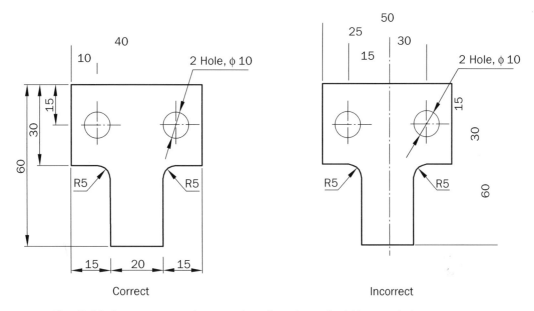

Correct   Incorrect

Fig. 3.20 Dimensioning to the centre line of an object should be avoided

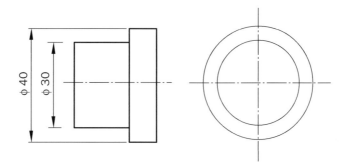

Fig. 3.21 Dimensioning of a cylindrical part

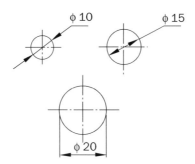

Fig. 3.22 Various methods of dimensioning the circles

(xii) Arcs of circles should be dimensioned by their respective radii. Dimension line for the radius should pass through the centre of the arc. Fig. 3.23 illustrates different methods of showing the radii of arcs.

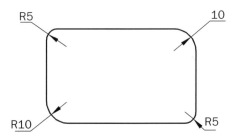

Fig. 3.23 Different methods of showing the radii of arcs

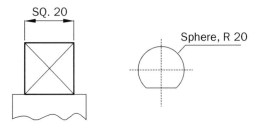

Fig. 3.24 Dimensioning of a square rod and a sphere

(xiii) The letters *SQ* should precede the dimension for a square rod, whereas the letter *R* should precede the dimension for a sphere. See Fig. 3.24.

(xiv) Angular dimensions may be given by any one of the methods as shown in Fig. 3.25.

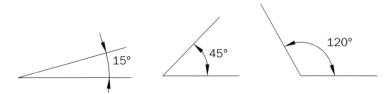

Fig. 3.25 Dimensioning angles

(xv) Holes should be dimensioned in the view in which they appear as circles. See Fig. 3.26.

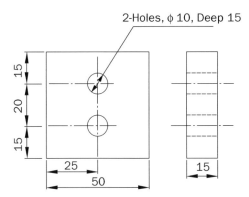

Fig. 3.26 Typical case of dimensioning of a hole size

(xvi) Preferably dimensions should be expressed in millimetres. The unit mm can then be dropped at the end of each dimension by adding a note separately that all dimensions are in millimetres.

Exercises

3.1 For the view, as shown in Fig. 3.27, indicate the chain, parallel and combination dimensioning by using any sizes.

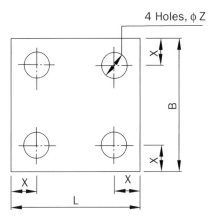

Fig. 3.27

3.2 Show a rectangle of 60 mm × 40 mm by using two systems of dimensioning.

3.3 Figure 3.28 shows the view of an object. Redraw the view and give all the necessary dimensions.

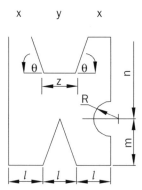

Fig 3.28

3.4   Explain the size and location dimensions with suitable sketches.

3.5   Write dimensions to a square view of an object in two different ways. Each side of the square is 40 mm.

3.6   Write dimensions to a triangular view of an object in two different ways. Each side of the triangle is 55 mm.

3.7   Write dimensions to a semi-circular view of an object in two different ways. Radius of the semi-circle is 25 mm.

3.8   Explain the methods of dimensioning of circles and radii with simple sketches.

3.9   List out the various principles to be followed while dimensioning a drawing.

3.10   Show by means of suitable examples aligned and unidirectional systems of dimensioning.

Objective Questions

3.1   Two systems of placing dimensions on a drawing are .... and ....

3.2   As far as possible, dimensions should be given in .... unit only, preferably in ....

3.3   Two types of dimensions required on a drawing are .... and .... dimensions.

3.4   The extension line should extend about 3 mm beyond the ....

3.5   The dimension indicating a diameter should be .... by the symbol ....

3.6   In .... system, the dimension is placed .... to the dimension line.

3.7   In .... system, all dimensions are so placed that they should be readable from the .... edge of the drawing sheet.

3.8   Dimension lines should not .... each other.

3.9   Dimension lines should be drawn about 8 mm away from the ....

3.10   An .... or .... should never be used as a dimension line.

3.11   Dimensions of cylindrical parts should be shown in the views in which they appear as ....

3.12   The line connecting a view to a note is called a ....

3.13 Dimensions should be taken from visible .... rather than from hidden lines.

3.14 What is the importance of dimensioning?

3.15 What is a leader line?

3.16 How are diameters and radii designated?

3.17 What are the general rules of dimensioning?

3.18 What is the necessity of dimensioning the drawing of an object?

3.19 Dimensions may be marked from hidden lines.   (True/False)

3.20 What is the proportion of an arrowhead?

Answers

3.1	Aligned, unidirectional	3.9	Outlines
3.2	One, millimetres	3.10	Outline, centre line
3.3	Size, location	3.11	Rectangles
3.4	Dimension line	3.12	Leader
3.5	Preceded, φ	3.13	Outlines
3.6	Aligned, perpendicular	3.19	False
3.7	Unidirectional, bottom	3.20	Length: Depth = 3: 1
3.8	Cross		

Chapter 4

Sections and Conventions

4.1 Introduction

In engineering drawing, it is often required to make drawings that show the interior details of an object, which are not visible to the observer from outside. The method of showing the interior details by hidden lines makes the view confusing and difficult to understand, as shown in Fig. 4.1. To overcome this difficulty, complicated objects are assumed to be cut by an imaginary plane. The part of the object between the imaginary cutting plane and the observer is assumed to be removed, as shown in Fig. 4.2. The exposed or cut surfaces are identified by section lines or hatching lines. The view obtained is called the sectional view, as shown in Fig. 4.3. Hidden lines and details behind the section are usually omitted from the sectional view, unless they are required for clarity. The sectional view is very important in daily life, as it helps in manufacturing and explaining the construction of complicated machines and their parts.

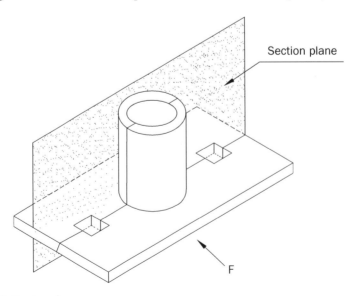

Fig. 4.1 Section plane

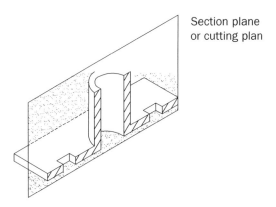

Fig. 4.2 Part of object left

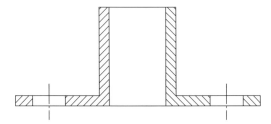

Fig. 4.3 Sectional view

4.2 Cutting Plane or Sectional Plane

The imaginary plane by which an object is assumed to be cut is called the cutting plane or the sectional plane. The cutting plane used in sectioning is indicated by a line in a view adjacent to the sectional view. This line is called the cutting plane line. It is a long chain that is thin, but is thick at the ends. Arrowheads, put at the ends, indicate the direction in which the cut away object is viewed, i.e. the direction of sight as shown in Fig. 4.4. At the end of the cutting plane line, capital letters such as $X – X$, $Y – Y$ or $A – A$ are often marked to identify the cutting plane line with corresponding section. When the cutting plane line coincides with a centre line, the cutting plane line takes precedence over a centre line. In case of symmetrical objects, the cutting plane line is omitted, unless it is needed for clarity.

Fig. 4.4 Cutting plane line

4.3 Section Lines or Hatching Lines

The lines used to represent the material which has been cut by a cutting plane are called section lines. These are also called hatching lines. Section lines are thin, equally spaced, parallel and inclined lines drawn in the area of the view where a cutting plane cuts the material of the object. The following points need to be noted while drawing section lines:

(i)   Generally section lines are drawn at an angle of 45° to the horizontal line or axis of the object. The correct and incorrect methods of sectioning an object are shown in Fig. 4.5.

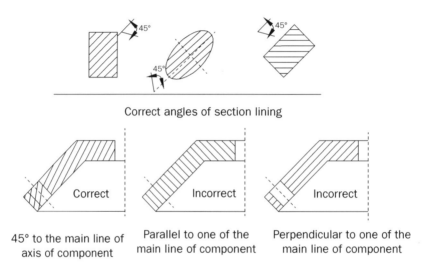

Fig. 4.5 Correct and incorrect section lining

(ii)   The spacing between section lines depends upon the area to be sectioned. For an ordinary working drawing, it varies from 1 mm to 3 mm. Some common errors in section lining are shown in Fig. 4.6.

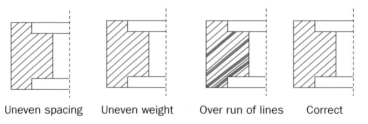

Fig. 4.6 Common errors in section lining

(iii) When two adjacent parts are to be shown in section, the section lines should be drawn in opposite directions as shown in Fig. 4.7.

Fig. 4.7 Hatching of adjacent part

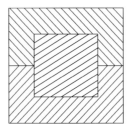

Fig. 4.8 Hatching of more than two adjacent parts

A third part to the first two parts should be sectioned at 30° or 60° to the horizontal as shown in Fig. 4.8.

    (iv)   Sometimes different materials are identified by varying the type of section lines. Section lines for various materials are shown in Fig. 4.9.

S.No.	Materials	Convention
1.	Steel, cast iron copper, aluminium and its alloys, etc.	
2.	Lead, zinc, tin, white metal, etc.	
3.	Brass, bronze, gun metal,etc.	
4.	Glass	
5.	Porcelain, stone ware, marble, slate, etc.	
6.	Asbestos, felt, paper, mica, cork, rubber, leather wax, insulating materials	
7.	Wood, plywood, etc.	
8.	Earth	
9.	Brick work, masonry fire bricks, etc.	
10.	Concrete	
11.	Water, oil, petrol, kerosene, etc.	

Fig. 4.9 Conventions for various materials

(v)   If a dimension is to be given in any sectioned area, the section lines should be interrupted around the dimension numeral. See Fig. 4.10.

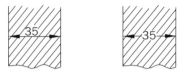

Fig. 4.10 Giving a dimension in a sectional area

4.4 Types of Sections

Following types of sections are commonly found in engineering practice:

(i)   **Full Section:** When the section plane passes entirely through the object, the view obtained is called a full section of the object. This type of section is used for both detail and assembly drawings. Fig. 4.11 is an example of a full section. The following points should be kept in mind for this type of section:
*   Invisible lines behind the cutting plane should be omitted.
*   Visible lines behind the cutting plane should be shown.
*   The parts which are actually cut by the cutting plane should be hatched.
*   The position of the cutting plane should be shown on the final drawing.

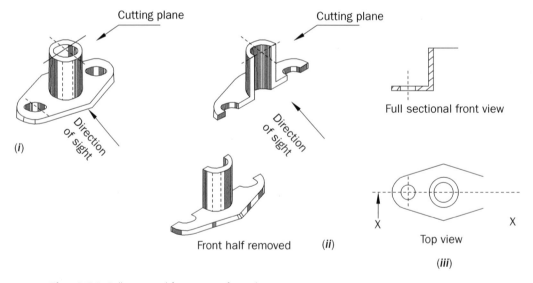

Fig. 4.11 Full sectional front view of an object

(ii)   **Half Section:** The section obtained after removing the front quarter portion of the object by two imaginary cutting planes at right angles to each other is known as half section, as shown in Fig. 4.12. The half section may be left, right, lower or upper, depending upon which quarter portion of the object is removed.

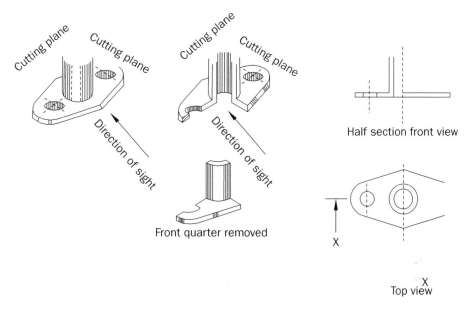

Fig. 4.12 Half sectional view of an object

(iii) **Partial or Broken Section:** In some objects, only a particular hidden detail of the object is required to be shown; a full or half section is not feasible. In such a case, only a partial section which is limited by a short break line is called a partial or broken section. See Fig. 4.13.

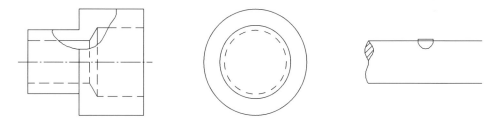

Fig. 4.13 Partial or broken section

(iv) **Offset Section:** In some objects, the full or half section does not expose the required details or it passes through features which should not be sectioned. To overcome this problem it is offset by changing the direction and is made to pass through the required details that need to be shown in the section. See Fig. 4.14.

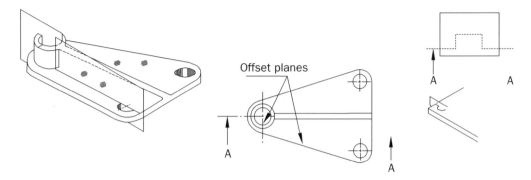

Fig. 4.14 Offset section

(v) **Revolved Section:** A revolved section is obtained when the cross section of an object is revolved through 90°. To obtain such a section, an imaginary cutting plane is made to pass perpendicular to the centre line of the object as shown in Fig. 4.15. The section shown in this manner is known as revolved section.

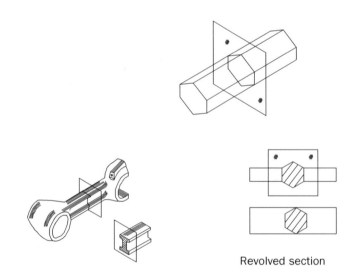

Revolved section

Revolved section of connecting rod

Fig. 4.15 Revolved section

(vi) **Removed Section:** A removed section is obtained in the same manner as the revolved section, but is drawn separately outside the view, as shown in Fig. 4.16.

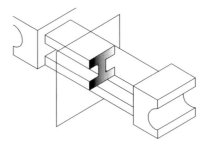

Fig. 4.16 Removed section

(vii) **Auxiliary Section:** When the cutting plane is not parallel to any of the principal planes, the section obtained is called an auxiliary section. These sections are used to show the interior details of the inclined features of the object. See Fig. 4.17.

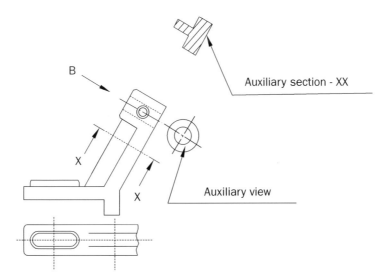

Fig. 4.17 Auxiliary section

4.5 Conventions for Various Materials

In engineering practice, different types of materials are used for manufacturing various parts of a machine. It is, therefore, desirable that different conventions should be adopted to differentiate various materials for convenience on the drawing. Fig. 4.9 illustrates the conventions for various types of materials.

4.6 Conventional Breaks

Long members of uniform cross section such as rods, shafts, pipes, etc., are generally shown by conventional breaks in the middle so as to accommodate their view of whole length, on the drawing sheet, without reducing the scale. Its true length is given by providing its overall dimension. The conventional breaks for standard members are shown in Fig. 4.18.

S.No.	Object	Convention
1.	Rectangular section	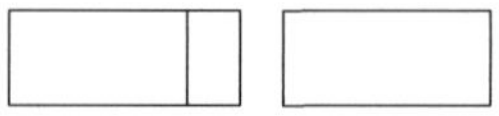
2.	Round section	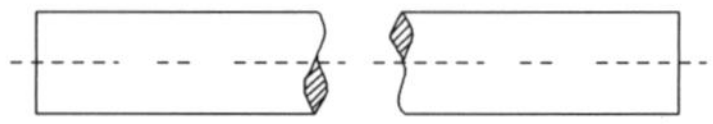
3.	Pipe or tubing	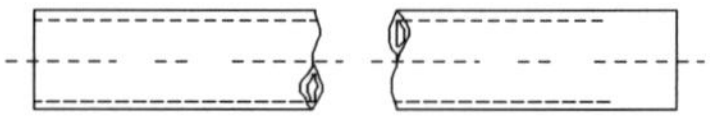
4.	Pipe or tubing	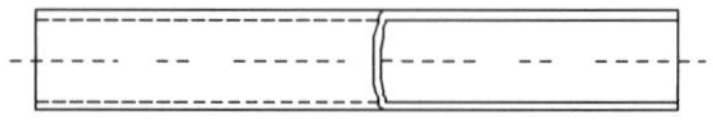
5.	Wood rectangular section	
6.	Rolled section	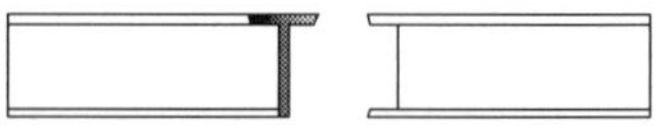
7.	Channel section	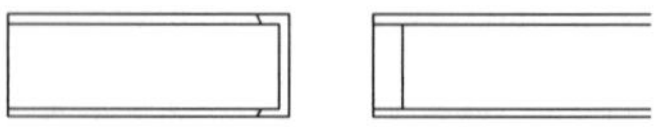

Fig. 4.18 Conventional breaks

4.7 Conventional Representation of Common Features

Conventional representation is adopted in cases where complete description of the machine component would involve unnecessary time or space on the drawing. Various examples are shown in Figs. 4.19 and 4.20.

S.No.	Title	Actual Projection/section	Convention
1.	External threads		
2.	Internal threads		
3.	Slotted head		To be drawn at 45°
4.	Square end and flat		
5.	Radial ribs		
6.	Serrated shaft		
7.	Splined shaft		
8.	Chain wheel		
9.	Ratchet and pinio		
10.	Bearings		
11.	Straight knurling		
12.	Diamond knurling		
13.	Holes on a linear pitch		

| 14. | Holes on circular pitch | 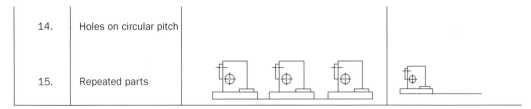 |
| 15. | Repeated parts | |

Fig. 4.19 Conventional representation of common features

S.No.	Description		Actual Projection		
			View	Section	Convention
1.	Disc springs	Spring			
		Spring assembly			
2.	Spiral springs	Spiral spring unwound			
		Spiral spring wound with barrel			
3.	Leaf springs	Without eyes			
		With eyes			
		Without eyes, with centre band			
		With eyes and centre band			

Fig. 4.20 Conventional representation of springs

Exercises

4.1   Differentiate between a full sectional view and a half sectional view with suitable diagrams.

4.2   Figure 4.21 shows a machine block. Draw the half sectional front view and top view.

4.3   Figure 4.22 shows a flange. Draw the half sectional front view and side view.

4.4   What do you mean by the following terms? Explain them with suitable diagram.
    (i)   Offset section   (ii)   Removed section
    (iii) Partial section   (iv)   Auxiliary section.

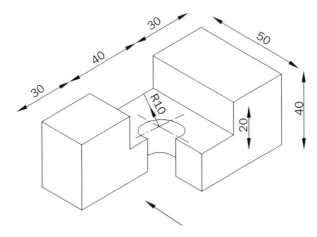

Fig. 4.21 Machine block

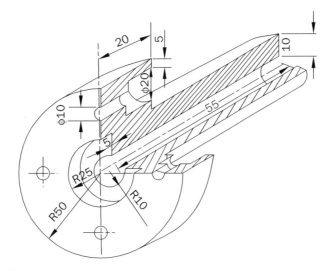

Fig. 4.22 Flange

Objective Questions

4.1   The purpose of sectional view is to show the .... shape of the object.

4.2   In sectional view, the part of the object between the cutting plane and .... is assumed to be removed.

4.3   Section lines are generally drawn at .... to the axis of the section.

4.4   For ordinary sectioning, the spacing varies from .... to .... mm.

4.5   In the sectional view, all hidden details are ....

4.6   The conventions make the drawing .... and .... to draw.

4.7   Long members are generally shown in the .... by the convention breaks.

4.8   What do you mean by convention?

4.9   Where and why is a cutting plane drawn in a drawing?

4.10   What are the principles of hatching? How do you hatch if there are more than two adjacent parts?

4.11   Give the conventional representation of the following:
(i)   Bearing   (ii)   Chain wheel   (iii)   Straight knurling.

4.12   What do you mean by revolved section?

4.13   What do you mean by sectional view?

4.14   Give the conventional breaks of the following long objects:
(i)   Rectangular section of long rod   (ii)   Round section of long rod

4.15   For a full sectional view, ... of the object is assumed to be removed.

Answers

4.1	Internal		4.5	Omitted
4.2	Observer		4.6	Simple, easy
4.3	45°		4.7	Middle
4.4	1, 3		4.15	Half

Chapter 5

Geometrical Constructions

5.1 Introduction

There are a number of simple geometrical constructions with which a draughtsman or an engineer should be familiar, as these are essential in the preparation of engineering drawing figures. Though these geometrical constructions are generally studied in the lower classes, they are being discussed here because of their importance in engineering drawing.

5.2 Bisection of a Straight Line

(i)   Draw a straight line *AB*.
(ii) With centre *A* and radius greater than half *AB*, draw arcs on either side of *AB*.
(iii) With centre *B* and same radius, draw arcs intersecting the above arcs at *C* and *D*.
(iv) Draw a line joining *C* and *D* to intersect the given line *AB* at *E*. The point *E* bisects the line *AB* and the line *CD* is called the perpendicular bisector of the line *AB*, as shown in Fig. 5.1.

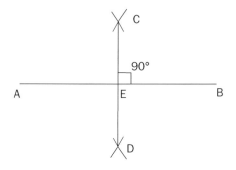

Fig. 5.1 Bisection of a straight line

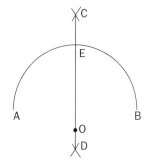

Fig. 5.2 Bisection of an arc

Note: The above procedure may be followed to bisect a given arc *AB*. See Fig. 5.2.

5.3 Dividing a Line into Equal Parts

(a) **Dividing a given straight line into a specified number of equal parts, say six:**
There are two methods to divide a given line into equal number of parts.

Method I

(i)   Draw the given line *AB*.
(ii)   From *A*, draw another line *AC* at any acute angle with respect to line *AB*.
(iii) Open the compass to suitable distance and mark six equal divisions on line *AC*, as shown in Fig. 5.3, and mark them as 1'–6'.
(iv)   Draw a line joining the mark 6' and *B*.
(v)   With the help of a mini-draughter, draw lines through 1', 2', 3', etc., parallel to 6' *B* to meet the line *AB* at 1, 2, 3..... etc., respectively.

The points 1, 2, 3, etc., divide the line *AB* into six equal parts.

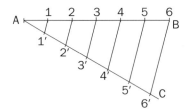

Fig. 5.3 Dividing a line into a number of equal parts (Method I)

Method II

(i)   Draw the given line *AB*.
(ii)   Draw a line *AC* that makes an angle θ with the line *AB*.
(iii) Draw *AC* and *BD* parallel to each other. Lines *AC* and *BD* make the same angle θ to *AB*, at A and B, respectively.
(iv)   Mark the required number of equal divisions (say six) of any suitable length on *AC* and *BD*.
(v)   Join 1 1', 2 2', etc., which divides the line *AB* into six equal parts. See Fig. 5.4.

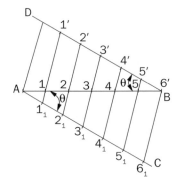

Fig. 5.4 Dividing a line into a number of equal parts (Method II)

(b) Dividing a given straight line into unequal parts:

Following are the procedural steps.

(i)   Draw the given straight line *AB*.

(ii) Draw perpendiculars *AD* and *BC* at the points *A* and *B*, respectively. Complete the square or rectangle *ABCD*.

(iii) Draw the diagonals *AC* and *BD* intersecting at *E*.

(iv) Through E, drop a perpendicular to *AB*, meeting the mid-point *F* of the line *AB*.

(v)   Join *D* and *F*. The line meets the diagonal *AC* at *G*. Then draw a perpendicular from *G* to *AB*. ($AH = 1/3\ AB$).

(vi) Similarly, for obtaining $1/4\ AB$ and $1/5\ AB$, make constructions as shown in Fig. 5.5.

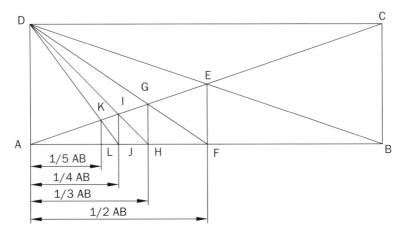

Fig. 5.5 Dividing a line into unequal parts

5.4 Drawing a Line Parallel to a Given Straight Line

(a) Drawing a line parallel to a given straight line through a given point

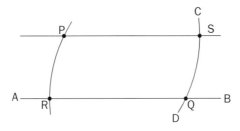

Fig. 5.6 Drawing a line parallel to a given straight line through a given point

(i)   First of all draw the given straight line *AB* and let *P* be the given point.

(ii) Considering P as the centre and with any convenient radius, draw an arc *CD* cutting *AB* at *Q*.

(iii)   Now consider Q as the centre and with the same radius, draw an arc cutting AB at R.

(iv)   Again, with Q as the centre and with the radius equal to RP, draw an arc that intersects CD at S.

(v)   Draw a straight line through P and S. The line PS is the required parallel line. See Fig. 5.6.

(b)   Drawing a line parallel to and at a given distance from a given straight line

(i)   Draw the given straight line AB and let 'x' be the given distance.

(ii)   Take two points C and D on the given line AB at a suitable distance apart.

(iii)   With C and D as centres, draw arcs on one side of AB with 'x' as radius.

(iv)   Draw a line PQ just touching the top surfaces of the two arcs. The line PQ is the required parallel line. See Fig. 5.7.

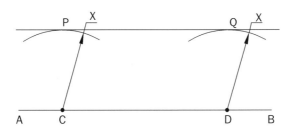

Fig. 5.7 Drawing a line parallel to and at a given distance from a given straight line

5.5 Bisecting an Angle

(a)   When the vertex of the angle is accessible

(i)   Draw the lines BC and BA, making the given angle.

(ii)   Keep B as centre and, with any convenient radius, draw an arc cutting AB at D and BC at E.

(iii)   With centres D and E and radius larger than half the chord length DE, draw arcs intersecting each other at F.

(iv)   Join B and F with a straight line. BF bisects the angle ABC, i.e., $\angle ABF = \angle FBC$. See Fig. 5.8.

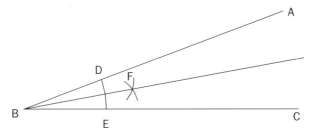

Fig. 5.8 Bisecting an angle when its vertex is accessible

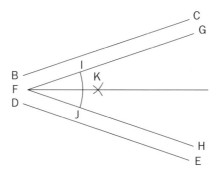

Fig. 5.9 Bisecting an angle when its vertex is inaccessible

(b) When the vertex of the angle is inaccessible
    (i)   Draw the lines *BC* and *DE*, inclined at the given angle.
    (ii)   Draw a line *FG* parallel to *BC*, at any suitable distance.
    (iii) Similarly draw a line *FH* parallel to *DE*, at any suitable distance.
    (iv) Bisect the angle *GFH*, as discussed in the previous case. See Fig. 5.9.

5.6 Finding the Centre of an Arc

(a) Find the centre of a given arc.
    (i)   Draw a given arc, say *AB*.
    (ii)   Draw any two chords *CD* and *EF* to the given arc *AB*.
    (iii) Draw perpendicular bisectors to *CD* and *EF*, intersecting each other at point *O*.
           Point *O* is the required centre of the given arc. See Fig. 5.10.

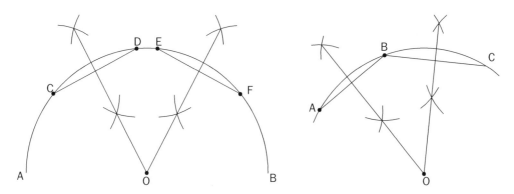

Fig. 5.10 Finding the centre of an arc   **Fig. 5.11** Drawing an arc through three points

(b) Draw an arc passing through three given points, not in a straight line.
    (i)   Locate the given points *A*, *B,* and *C*.
    (ii)   Draw straight lines joining *B* with *A* and *C*.
    (iii) Draw perpendicular bisectors of line *AB* and *BC*, intersecting each other at point *O*.
    (iv) With *O* as centre and radius equal to *OA* or *OB* or *OC*, draw an arc. See Fig. 5.11.

5.7 Constructing an Equilateral Triangle

(a) To construct an equilateral triangle, given the length of the side

Method I (Using T-square and set-square only)

     (i)   With the T-square, draw a line *AB* of given length.

     (ii)   With 30°–60° set-square and T-square, draw a line through A making 60° angle with *AB*.

     (iii) Similarly, through *B*, draw a line making the same angle with *AB* and intersecting the previous line at *C*. Then *ABC* is the required triangle. See Fig. 5.12.

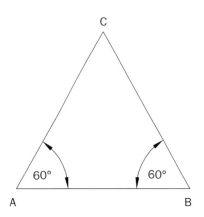

Fig. 5.12 Drawing an equilateral triangle of given length of side (Method I)

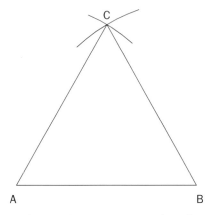

Fig. 5.13 Drawing an equilateral triangle of given length of side (Method II)

Method II (Using compass only)

     (i)   Draw a line *AB* of given length.

     (ii)   With centres *A* and *B* and radius equal to *AB*, draw arcs interescting each other at *C*.

     (iii) Draw lines joining *C* with *A* and *B*.

     Then *ABC* is the required triangle. See Fig. 5.13.

(b) To construct an equilateral triangle, when the altitude is given

     (i)   Draw any line *PQ* and select a point *D* on it.

     (ii)   Through point *D*, draw a perpendicular *DC* to the line *PQ* such that *DC* is the altitude.

     (iii) With *C* as centre and any suitable radius, draw an arc intersecting *CD* at *E*.

     (iv) With *E* as centre and same radius, draw arcs intersecting the above arc at *F* and *G*.

     (v)   Draw bisectors of the arcs *EF* and *EG* passing through *C* and meeting the line *PQ* at *A* and *B*, respectively.

Then *ABC* is the required triangle. See Fig. 5.14.

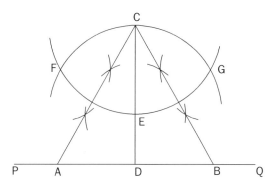

Fig. 5.14 Drawing an equilateral triangle of known altitude

5.8 Constructing Squares

(a) Construct a square when each side is given
- (i) Draw a line *AB* which is the given side.
- (ii) At *A*, draw a line *AE* perpendicular to *AB*.
- (iii) Considering A as the centre and AB as the radius, draw an arc intersecting *AE* at *D*.
- (iv) With centres *B* and *D* and the same radius, draw arcs intersecting at *C*.
- (v) Draw lines joining *C* with *B* and *D*. Consequently the square *ABCD* is drawn. See Fig. 5.15.

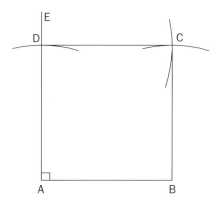

Fig. 5.15 Constructing a square of given side

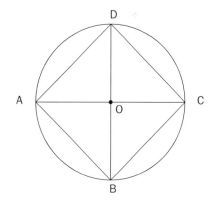

Fig. 5.16 Constructing a square of known diagonal

(b) To construct a square when the length of diagonal is given
- (i) With centre *O* and the diagonal length as the diameter, draw a circle.
- (ii) Draw two diameters *AC* and *BD* perpendicular to each other.
- (iii) Draw lines *AB*, *BC*, *CD* and *DA*, thus completing the square. So *ABCD* is the required square. See Fig. 5.16.

5.9 Constructing Regular Polygons

(a) To construct a regular pentagon when the length of side is given, any of the following three methods can be used:

Method I

   (i)   Draw a line *AB* of the given length, which is equal to the side of the polygon.
   (ii)   Bisect *AB* at point *P*.
   (iii) Draw a line *BQ* of the same length AB, perpendicular at point *B*.
   (iv) Keeping P as centre and PQ as radius, draw an arc intersecting line *AB* extended at *R*.
   (v)   After the above step, the length of the diagonal of the pentagon obtained is AR.
   (vi) Keeping A and B as centres and the radii equal to *AR* and *AB*, respectively, draw arcs intersecting at point *C*.
   (vii) Now with centres *A* and *B* and radius *AR*, draw arcs intersecting at point *D*.
   (viii) Further, taking centres A and B and the radii equal to AB and AR, respectively, draw respective arcs intersecting at point *E*.
   (ix) Draw lines *AB, BC, CD, DE* and *EA*, thus completing the regular pentagon. See Fig. 5.17.

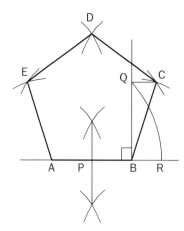

Fig. 5.17 Construction of a regular pentagon (Method I)

Method II

   (i)   Draw a line *AB* equal to the given length of the side of the pentagon.
   (ii)   With the centre *A* and the radius *AB*, draw a circle and mark it as circle
   (iii) Similarly with centre *B* and the same radius, draw a circle and mark it as circle intersecting the previous circle at points P and Q.
   (iv) With P as centre and same radius, draw another arc that intersects both circles at R and S, respectively.
   (v)   Now draw a perpendicular bisector of the line *AB* that cuts the arc *RS* at *G*.
   (vi) Draw a line *RG* and extend it to cut the circle at C.

(vii) Similarly, draw a line *SG* and extend it to cut the circle at E.

(viii) With *C* and *E* as the centres and *AB* as the radius, draw two arcs intersecting each other at point *D*.

(ix) Draw lines *AB, BC, CD, DE* and *EA*, thus completing the regular pentagon. See Fig. 5.18.

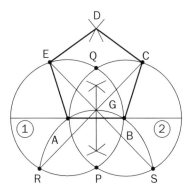

Fig. 5.18 Construction of a regular pentagon (Method II)

Method III

(i) Draw a line *AB* equal to the given length of the side.

(ii) Make an angle of 54° at each point *A* and *B*, meeting at point *O*.

(iii) Keeping O as the centre and *OA* or OB as radius, draw a circle.

(iv) Now considering B as the centre and radius *AB*, draw an arc that intersect the circle at point *C*.

(v) Similarly consider A as the centre and AB as the radius, draw an arc that intersects the circle at point *E*.

(vi) Now, considering *C* and *E* as the centres and with the radius *AB*, draw two arcs that intersect each other at point *D* on the circle.

(vii) Draw lines *AB, BC, CD*, DE and *EA*, thus completing the regular pentagon. See Fig. 5.19.

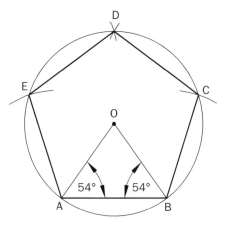

Fig. 5.19 Construction of a regular pentagon (Method III)

(b) Construction of a regular hexagon when the length of side is given

Method I

(i) Draw a line *AB* equal to the given length of the side.
(ii) Draw a semi-circle keeping A as the centre and AB as the radius.
(iii) Divide the semi-circle into the number of equal parts same as the number of sides *n* (i.e. six).
(iv) Draw radial lines through 2, 3, 4, 5, etc.
(v) Take B as the centre and with radius *AB*, draw an arc that intersects the radial line through 5 at point *C*.
(vi) Now, assume C as centre and radius *AB*, draw another arc that intersects the radial line passing through 4 at point *D*.
(vii) Repeat this procedure until the point on the radial line through 3 is obtained.
(viii) Draw lines *AB, BC, CD, DE*, etc., thus completing the regular hexagon. See Fig. 5.20.

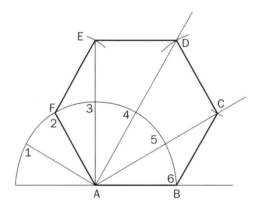

Fig. 5.20 Construction of a regular hexagon (Method I)

Method II

(i) Follow the same steps from (*i*) to (*iv*) as discussed above.
(ii) Draw perpendicular bisectors of lines 2A and *AB*, intersecting at point *O*.
(iii) Keep O as the centre and *OA* as radius and draw a circle passing through the points 2 and *B*.
(iv) Locate the corners *C, D*, etc., of the polygon where the circle meets the radial lines.
(v) Draw lines *AB, BC, CD*, etc., thus completing the regular hexagon. See Fig. 5.21.

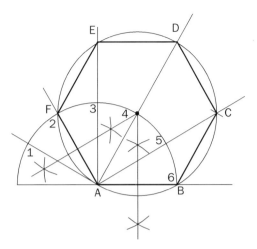

Fig. 5.21 Construction of a regular hexagon (Method II)

Method III

(i)   Draw a line *AB* of the given length as that of the side of the polygon.

(ii) Now make an angle of 60° at each point A and B, meeting at point O.

(iii) Keep O as centre and with radius *OA* or *OB*, draw a circle.

(iv) Now, considering B as the centre and *AB* as the radius, draw an arc intersecting the circle at point C.

(v)   Similarly, with A and C as the centres and *AB* as the radius, draw arcs intersecting the circle at points *F* and *D*, respectively.

(vi) Keeping D and F as centres and AB as radius, draw arcs intersecting each other at point *E* on the circle.

(vii) Draw lines *AB*, *BC*, *CD*, etc., thus completing the regular hexagon. See Fig. 5.22.

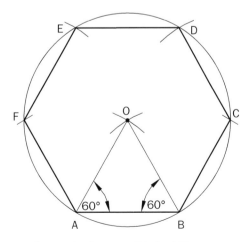

Fig. 5.22 Construction of a regular hexagon (Method III)

(c) General Method for drawing any polygon

    (i)   Draw a line *AB* equal to the length of the given side of the polygon.

    (ii) At *B*, draw a perpendicular line *BP* of same length of *AB*.

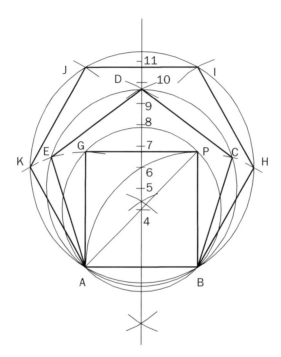

Fig. 5.23 Construction of a regular polygon (General Method)

    (iii) Draw a line joining *A* with *P*.

    (iv) Consider B as the centre and *AB* as the radius, draw the quadrant *AP*.

    (v) Next, draw a perpendicular bisector of *AB* that intersects the straight line *AP* at point 4 and the arc *AP* at 6.

    (vi) Keeping 4 as centre and A4 as the radius, draw a circle; further inscribe a square of side equal to *AB* in the circle drawn.

    (vii) Now find the mid-point (5) of the segment (4-6) of the bisector.

    (viii) Further along the bisector mark points 4-5-6-7-8, etc., at equidistant 4-5=5-6= 6-7, etc.

    (ix) A regular polygon of any number of sides can be inscribed in the circle with respective centre point number; e.g. for pentagon the centre is point 5 and radius A5.

    (x) Similarly, a regular hexagon of side equal to *AB* can be inscribed in the circle drawn with centre 6 and radius A6. A polygon of any number of sides, *N*, can be inscribed in a circle drawn with centre *N* and radius *AN*. See Fig. 5.23.

5.10 Drawing Tangents

(a) Draw a tangent to a given circle at any point on it
- (i)   With centre O, draw the given circle and mark the given point P on it.
- (ii) Join O with P and extend it.
- (iii) Draw a line TT perpendicular to the above line (as in (ii) above) at point P. The line TT is the required tangent. See Fig. 5.24.

(b) Draw a tangent to a given circle from any point outside the circle
- (i)   Keep O as the centre and draw the given circle.
- (ii)   Locate the given point P outside it.
- (iii) Join O and P and locate its mid-point A.
- (iv)   With A as the centre and AO as the radius, draw an arc that intersects the given circle at B and C.
- (v)   Join P to B and P to C and extend it.

The lines PB and PC are the two possible tangents. See Fig. 5.25.

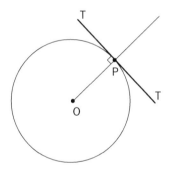

Fig. 5.24 Tangent to a circle at a point on it

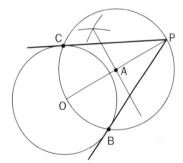

Fig. 5.25 Tangent to a circle from an outside point

(c) Draw a tangent to a given arc of inaccessible centre at any point on it
- (i)   Draw the given arc AB and locate the given point P on it.
- (ii) With the centre P and any suitable radius, draw an arc that intersects the given arc at C and D.
- (iii) Draw perpendicular bisector EF of the chord CD passing through P.
- (iv) Through P, draw a line GH perpendicular to EF. Thus GH is the required tangent. See Fig. 5.26.

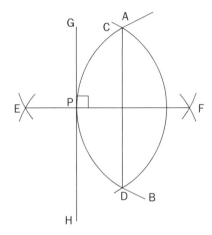

Fig. 5.26 Tangent to an arc having inaccessible centre

5.11 Inscribed Circles

(a) **Inscribe a circle in a given triangle**
   (i)   Let *ABC* be the triangle.
   (ii)   Bisect any two angles by lines intersecting each other at O.
   (iii)   Draw a perpendicular from O to any one side of the triangle, meeting at P.
   (iv)   Keeping O as the centre and OP as the radius, draw the required circle. See Fig. 5.27.

(b) **Inscribe a circle in a regular polygon of any number of sides, say a hexagon**
   (i)   Let *ABCDEF* be the hexagon.
   (ii)   Bisect any two angles by lines intersecting each other at O.
   (iii)   From O, draw a perpendicular to any one side of the hexagon cutting it at P.
   (iv)   Keeping O as the centre and OP as the radius, draw the required circle. See Fig. 5.28.

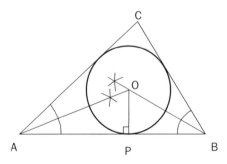

Fig. 5.27 Construction of a circle in a given triangle

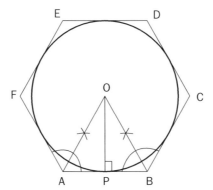

Fig. 5.28 Construction of a circle in a regular hexagon

Exercises

5.1 Divide a straight line *AB* of 100 mm length into ten equal parts.

5.2 Divide a line 120 mm long into three parts in the ratio of 2: 3: 4.

5.3 Draw an arc of 60 mm radius of convenient dimension and bisect it.

5.4 Draw a line parallel to and at a distance of 25 mm from a given straight line *AB*.

5.5 Construct a triangle *ABC* with the base *BC* equal to 35 mm and base angles as 60° and 30°, respectively.

5.6 Construct an equilateral triangle *ABC*, having its altitude 60 mm.

5.7 Construct an equilateral triangle *ABC* of 30 mm side.

5.8 Construct a right angle triangle whose hypotenuse is 150 mm and one of the sides is 70 mm.

5.9 Construct a triangle *ABC* with *AB* = 30 mm, *BC* = 50 mm and the median *AD* = 40 mm.

5.10 Construct a triangle ABC whose sides are in the ratio of 3: 4: 5. The base BC measures 50 mm.

5.11 Construct a square of side 30 mm.

5.12 Draw a regular pentagon with each side as 30 mm.

5.13 Construct a regular hexagon of 30 mm side by two different methods.

5.14 Draw a regular octagon with each side as 25 mm.

5.15 Inscribe a regular pentagon in a given circle of radius 25 mm.

5.16 Inscribe a circle in a triangle having sides 50 mm, 60 mm and 75 mm long.

5.17 Two lines converge to a point making an angle of 30° between them. Draw three circles to touch both these lines, the middle circle being of 20 mm radius and touching the other two circles.

5.18 Draw a circle to touch two converging *OA* and *OB* at 60° and passing through a point *P* between them.

5.19 Draw a circle passing through all the three points *P*, *Q* and *R* when *P* is 75 mm from *Q* and 60 mm from *R* and *R* itself is 35 mm from *Q*.

Objective Questions

5.1 What is the difference between regular and irregular polygons?

5.2 An angle that is less than a right angle is known as ....

5.3 An angle bigger than a right angle is called ....

5.4 A polygon is a plane figure having more than .... sides.

5.5 The bisector of a line meets at ....

5.6 It is .... to draw a tangent to circle from a point outside it.

5.7 The distance between any two parallel lines is ....

5.8 How many tangents can be drawn to a circle from a point outside it?

5.9 What is the difference between a parallelogram and a rhombus?

5.10 The arc is any part of the circumference of a ....

Answers

5.2	Acute angle	5.6	Possible
5.3	Obtuse angle	5.7	Constant
5.4	Four	5.8	Two
5.5	90°	5.10	Circle

Chapter 6

Scales

6.1 Introduction

It is always preferable to make the linear dimensions on a drawing the same size as the corresponding real dimensions of the object drawn. However, the drawing of a very big object, like a diesel engine, should be made considerably smaller than the object, whereas details of small precision instruments, watches, etc., are generally made larger than their real size. Objects large and small have to be drawn in such a way that the drawing can be read and handled conveniently. Thus if the linear dimensions of an object have to be enlarged or reduced, for drawing purposes, we need to use scales which enable us to enlarge (enlarging scale) or reduce (reducing scale) linear dimensions with uniformity. If the actual linear measurements of an object are shown in a drawing, the scale is called a full scale. The term 'scale' is defined as the ratio of linear dimension of an element of an object as represented in the original drawing to the real linear dimension of the same element of the object itself. All drawings should be drawn to scale and the scale used has to be stated on the drawing.

The complete designation of a scale shall consist of the word SCALE, followed by the indication of its ratio, as follows

SCALE 1 : 1   For Full Size
SCALE X : 1   For Enlarging Scales, X > 1
SCALE 1 : X   For Reducing Scales, X > 1

As discussed above, different scales are used for convenience and specific purposes. The recommended scales for use on technical drawings are specified in the Table 6.1 [SP : 46 (1988)].

Table 6.1 Recommended scales [SP : 46 (1988)]

S. No.	Category	Recommended Scales		
1.	Reducing Scales	1:2	1:5	1:10
		1:20	1:50	1:100
		1:200	1:500	1:1000
		1:2000	1:5000	1:10000
2.	Enlarging Scales	50:1	20:1	10:1
		5:1	2:1	--
3.	Full Size Scales			1:1

6.2 Representative Fraction or Scale Factor

The ratio of the drawing to the object is called the representative fraction, abbreviated as RF. In detailed words, representative fraction (RF) or scale factor (SF) is the ratio of length of a line in the drawing to actual length of the object represented.

$$\text{Representative Fraction} = \frac{Length \ of \ a \ line \ on \ the \ drawing}{Actual \ length \ of \ a \ line \ on \ the \ object}$$

The dimensions in both numerator and denominator of the fraction must be in the same units. For example, if we wish to represent a dimension 2 m by a line 4 cm long, it will be evident that when the line is divided into 4 equal parts, each part will be 1 cm long and that this 1 cm length of the line represents 0.5 m on the actual object. So we might say that we are using a scale of 1 cm to 0.5 m. This is quite a common scale and is recommended by the BIS.

$$\text{Its} \ RF = \frac{1 \ cm}{0.5 \times 100 \ cm} = \frac{1}{50}$$

We call this scale to be 1:50 without any mention of cm, m or other linear units.

6.2.1 Construction of scales

When the required scale is not available out of a set of recommended scales, it has to be constructed on the drawing sheet. For constructing a scale, the following information is needed:

- The RF of the scale
- The units it has to represent
- The maximum length, required to be measured

Length of scale = RF × maximum length to be measured by a scale; if it is not given, then we can assume the length of the scale 15 cm to 30 cm or in other words, we shall assume the maximum distance to be measured such that the length of the scale on calculation comes out to be 15 cm or so.

Scales used in engineering practice are mentioned by one of the ways given below:

1. 1:2 or ½ (RF)
2. half full size, etc.
3. 1 cm = 2 cm, etc.

6.3 Scales on Drawings

When the required scale is not available out of a set of recommended scales, it has to be constructed on the drawing sheet. To construct a scale, the following information is essential:

- The RF or scale factor of the scale
- The units which it must represent; e.g., millimetres, centimetres, metres, etc.
- The maximum distance to be measured

The length of the scale is determined by the formula

Length of the scale = RF × Maximum distance to be measured.

It may not be possible to draw as long a scale as to measure the longest length in the drawing. The scale is therefore drawn from 15 cm to 30 cm long, longer lengths being measured by marking them off in parts.

6.4 Types of Scales

Scales can be divided into the following six types:

(i)   Plain Scales   (ii)   Diagonal Scales
(iii) Comparative Scales   (iv)   Vernier Scales
(v)   Isometric Scales   (vi)   Scale of Chords

Only plain scales and diagonal scales are to be described here.

6.5 Plain Scales

A plain scale represents either two units or a unit and its sub division. A plain scale consists of a line divided into a suitable number of equal parts or units, the first of which is further sub-divided into smaller parts.

In every scale

(i)   The zero should be placed at the end of the first main division.
(ii) From zero mark, the units should be numbered to the right and its sub-division to the left.
(iii) The name of the units and sub-divisions should be written clearly below or at the appropriate places.
(iv) Always mention RF of the scale.

6.5.1 Construction of a plain scale

A plain scale represents either two units or a unit and its sub division (fractions). A plain scale consists of a line divided into a suitable number of equal parts or units, the first of which is sub-divided into small parts. For constructing a plain scale the following steps are followed:

1. Calculate RF, if not given.
2. Calculate length of the scale, using the formula

    L = RF × Maximum length (rounded off to next higher whole number)

    If the maximum length to be measured is not known, then take L = 15 or 30 cm.

3. Draw a straight line of length L and divide it into a number of equal parts as required. Each part represents the larger unit.
4. Place zero (0) at the end of the first main unit.
5. Sub-divide the division to the left zero (0) into sub-units. Number the units towards right of zero (0) and sub-units towards its left.
6. Print the names of the units and the sub-units below the corresponding length of scale.
7. Mention always the RF, or the name of the scale with the scale.

The construction of a plain scale can be best understood with the help of the following worked out examples.

Problem 6.1 Construct a scale of 1 cm = 0.5 m to show single metre and single decimetre and long enough to measure up to 8 m. Also show the distances of 7 m 5 dm and 6 m 4 dm.

Solution:

(i)   Obtain the representative fraction as:

$$1 \text{ cm} = 0.5 \text{ m} = 0.5 \times 100 \text{ cm}$$
$$1 \text{ cm} = 50 \text{ cm}$$
$$RF = \frac{1}{50}$$

(ii) Calculate the length of the scale (LOS) as:

$$LOS = RF \times \text{Maximum distance to be measured}$$
$$= \frac{1}{50} \times 8 \times 100 = 16 \text{ cm}$$

(iii) Draw a line 16 cm long. Divide it into eight equal parts to represent single metre.
(iv) Mark zero (0) after the first division and continue 1, 2, 3, etc., to the right of the scale.
(v)   Divide the first division into ten equal parts to represent single decimetre.
(vi) Mark the division points from right to left.
(vii) Write the units at the bottom of the scale in their respective positions.
(viii) Mark the distance of 7 metre 5 decimetre between the points A and B. Also show a distance of 6 metre 4 decimetre between the points C and D.
(ix) Complete the scale as shown in Fig. 6.1.

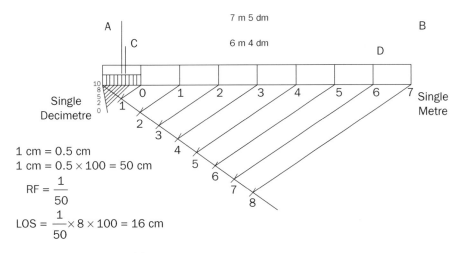

$$1 \text{ cm} = 0.5 \text{ cm}$$
$$1 \text{ cm} = 0.5 \times 100 = 50 \text{ cm}$$
$$RF = \frac{1}{50}$$
$$LOS = \frac{1}{50} \times 8 \times 100 = 16 \text{ cm}$$

Fig. 6.1 Solution to problem 6.1

Problem 6.2 Construct a plain scale of 1 cm = 1 m to show single metre and single decimetre and long enough to measure up to 10 m. Also show the distances of 6 m 4 dm, 8 m 5 dm and 9 m 8 dm.

Solution:

(i)   Obtain the representative fraction as:

$$1 \text{ cm} = 1 \text{ m} = 1 \times 100 \text{ cm}$$

$$RF = \frac{1}{100}$$

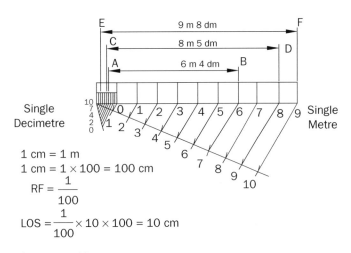

$$1 \text{ cm} = 1 \text{ m}$$
$$1 \text{ cm} = 1 \times 100 = 100 \text{ cm}$$
$$RF = \frac{1}{100}$$
$$LOS = \frac{1}{100} \times 10 \times 100 = 10 \text{ cm}$$

Fig. 6.2 Solution to problem 6.2

(ii)   Calculate the length of the scale as:

$$LOS = \frac{1}{100} \times 10 \times 100 = 10 \text{ cm}$$

(iii)   Draw a line 10 cm long. Divide it into ten equal parts to represent single metre.
(iv)   Mark zero (0) after the first division and continue 1, 2, 3, etc., to the right of the scale.
(v)   Divide the first division into ten equal parts to represent single decimetre.
(vi)   Mark the division points from right to left.
(vii)   Write the units at the bottom of the scale in their respective positions.
(viii) Mark the distances of 6 m 4 dm, 8 m 5 dm, and 9 m 8 dm between the points A to B, C to D, and E to F, respectively.
(ix)   Complete the scale as shown in Fig. 6.2.

Problem 6.3 A rectangular plot of 25 sq km is represented on a certain map by a similar rectangle of area 1 sq cm. Draw a plain scale to show ten kilometres and single kilometre on it and long enough to measure up to 80 km.

Solution:

(i)   Obtain the representative fraction as:

$$1 \text{ cm} = 25 \text{ km}$$
$$1 \text{ cm} = 5 \text{ km}$$
$$1 \text{ cm} = 5 \times 1000 \times 100 \text{ cm}$$

$$RF = \frac{1}{500000}$$

(ii)   Calculate the length of the scale as:

$$LOS = \frac{1}{500000} \times 80 \times 1000 \times 100 = 16 \text{ cm}$$

(iii)   Draw a line 16 cm long. Divide it into eight equal parts to represent ten kilometres.
(iv)   Mark zero (0) after the first division and continue 10, 20, 30, etc., to the right of the scale.
(v)   Divide the first division into ten equal parts to represent single kilometre.
(vi)   Mark the division points from right to left.
(vii)   Write the units at the bottom of the scale in their respective positions.
(viii) Complete the scale as shown in Fig. 6.3.

80 km

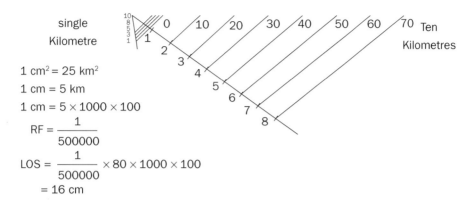

single
Kilometre

Ten
Kilometres

$1 \text{ cm}^2 = 25 \text{ km}^2$

$1 \text{ cm} = 5 \text{ km}$

$1 \text{ cm} = 5 \times 1000 \times 100$

$RF = \dfrac{1}{500000}$

$LOS = \dfrac{1}{500000} \times 80 \times 1000 \times 100$

$= 16 \text{ cm}$

Fig. 6.3 Solution to problem 6.3

Problem 6.4 Construct a plain scale of 1 cm = 0.5 km, to read single kilometre and single hectometre and long enough to measure up to 9 km. Find its RF and measure a distance of 6 km 4 hm on it.

(PTU, Jalandhar May 2009)

Solution:

(i) Obtain the representative fraction as:

1 cm = 0.5 km = 0.5 × 1000 × 100 cm

1 cm = 50000 cm

$$RF = \frac{1}{500000}$$

(ii) Calculate the length of the scale as:

$$LOS = \frac{1}{500000} \times 9 \times 1000 \times 100 = 18 \text{ cm}$$

(iii) Draw a line 18 cm long. Divide it into nine equal parts to represent single kilometre.

(iv) Divide the first part into ten equal parts to represent single hectometre and complete the scale as shown in Fig. 6.4.

The distance between points A and B is 6 km 4 hm.

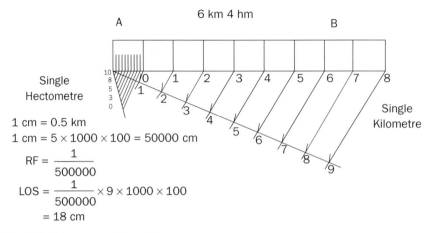

Fig. 6.4 Solution to problem 6.4

Problem 6.5 On a plan a line 20 cm long represents a distance of 400 m. Construct a plain scale to read up to 300 m. Write its RF and show a distance of 195 m on it.

Solution:

(i) Obtain the representative fraction as:
20 cm = 400 m
1 cm = 20 m = 20 × 100 = 2000 cm

$$RF = \frac{1}{2000}$$

(ii) Calculate the length of the scale as:
LOS 300 × 100 = 15 cm

(iii) Draw a line 15 cm long. Divide it into thirty equal parts to represent ten metres.

(iv) Divide the first part into two equal parts. Each part will represent 0.5 metre.

(v) Complete the scale as shown in Fig. 6.5. The distance between points A and B is 195 m.

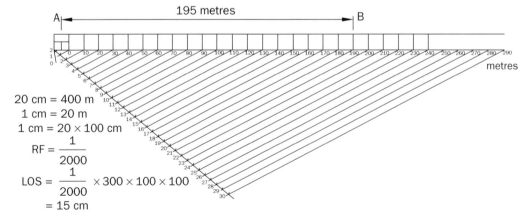

Fig. 6.5 Solution to problem 6.5

Problem 6.6 The distance between two towns is 120 km. A passenger train covers this distance in 4 hours. Construct a plain scale to measure time up to a single minute. The RF of the scale is $\frac{1}{200000}$. Indicate on this scale, the distance covered by the train is 34 minutes.

Solution:

(i) Obtain the average speed of the train $= \frac{120}{4} = 30$ km/hr

In other words, it represents that the train covers 30 km in one hour or 60 minutes.

(ii) Calculate the length of the scale, to represent 30 km as:

$$\text{LOS} = \frac{1}{200000} \times 30 \times 1000 \times 100 = 15 \text{ cm}$$

Therefore, 30 km distance is represented by 15 cm length of the scale and is covered by the train in 60 minutes.

In other words, 15 cm = 30 km = 60 minutes.

(iii) Draw a line 15 cm long and divide it into six equal parts, each representing 10 minutes or 5 km.

(iv) Divide the first part into ten equal parts, each representing the distance covered in single minute.

(v) Complete the scale as shown in Fig. 6.6.

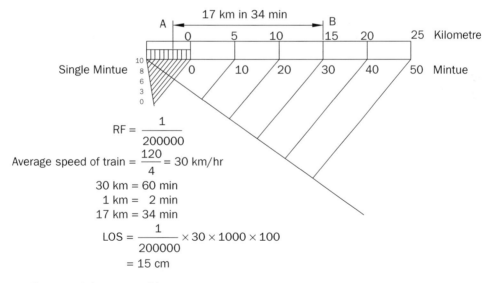

Fig. 6.6 Solution to problem 6.6

The distance covered by the train in 34 minutes is 17 km (between points A and B).

Problem 6.7 Construct a plain scale of 2 cm to 1 km to read kilometres and hectometres. Show a distance of 5.8 km on this scale.

(PTU, Jalandhar December 2004)

ENGINEERING DRAWING

76

Solution:

(i) Obtain the representative fraction as:

$$2 \text{ cm} = 1 \text{ km}$$
$$2 \text{ cm} = 1 \times 1000 \times 100 \text{ cm}$$
$$1 \text{ cm} = 50000 \text{ cm}$$
$$RF = \frac{1}{50000}$$

(ii) Calculate the length of the scale as:

$$LOS = \frac{1}{50000} \times 6 \times 1000 \times 100 = 12 \text{ cm}$$

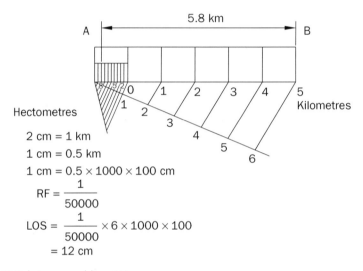

2 cm = 1 km
1 cm = 0.5 km
1 cm = 0.5 × 1000 × 100 cm
$$RF = \frac{1}{50000}$$
$$LOS = \frac{1}{50000} \times 6 \times 1000 \times 100$$
$$= 12 \text{ cm}$$

Fig. 6.7 Solution to problem 6.7

(iii) Draw a line 12 cm long. Divide it into six equal parts to represent kilometres.
(iv) Divide the first part into ten equal parts to represent hectometres.
(v) Complete the scale as shown in Fig. 6.7. The distance between the points *A* and *B* is 5.8 km.

Problem 6.8 Construct a scale having RF $= \dfrac{1}{400}$ to show metres and long enough to measure up to 60 m. Measure a distance of 44 m on the scale.

(PTU, Jalandhar December 2008)

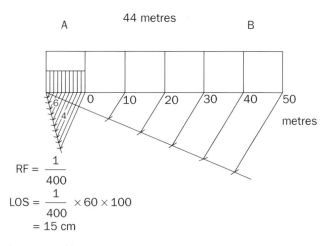

Fig. 6.8 Solution to problem 6.8

Solution:

(i) Calculate the length of the scale as:
$$\text{LOS} = \frac{1}{400} \times 60 \times 100 = 15 \text{ cm}$$

(ii) Draw a line 15 cm long. Divide it into six equal parts to represent ten metres.

(iii) Divide the first part into ten equal parts to represent single metre and complete the scale as shown in Fig. 6.8.

(iv) The distance between points *A* and *B* is 44 m.

Problem 6.9 A cube of 8 cm represents a tank of 1000 m volume. Find its RF and draw a plain scale to measure up to 80 m. Also show a distance of 65 m on it.

Solution:

(i) Obtain the representative fraction as:
$$8 \text{ cm} = 1000 \text{ m}$$
$$2 \text{ cm} = 10 \text{ m}$$
$$1 \text{ cm} = 5 \text{ m}$$
$$\text{RF} = \frac{1}{500}$$

(ii) Calculate the length of the scale as:
$$\text{LOS} = \frac{1}{500} \times 80 \times 100 = 16 \text{ cm}$$

(iii) Draw a line 16 cm long. Divide it into eight equal parts to represent ten metres.

(iv) Divide the first part into ten equal parts to represent single metre and complete the scale as shown in Fig. 6.9.

(v) The distance between points A and B is 65 m.

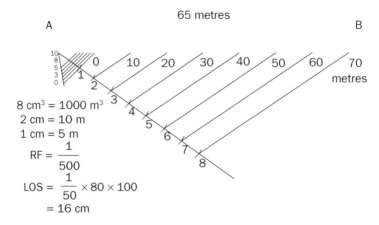

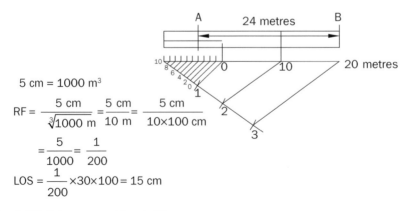

Fig. 6.9 Solution to problem 6.9

Problem 6.10 A room of 1000 m volume is shown by a cube of 5 cm side. Find its RF and draw a plain scale to measure up to 30 m. Also show a distance of 24 m on it.

Solution:

(i)   Obtain the representative fraction as:

$$5 \text{ cm} = 1000 \text{ m}^3$$

$$\text{RF} = \frac{5 \text{ cm}}{3\sqrt{1000}} = \frac{5 \text{ cm}}{10 \text{ m}} = \frac{5 \text{ cm}}{10 \times 100 \text{ cm}} = \frac{5}{1000} = \frac{5}{200}$$

(ii)   Calculate the length of the scale as:

$$\text{LOS} = \frac{1}{200} \times 30 \times 100 = 15 \text{ cm}$$

(iii) Draw a line 15 cm long. Divide it into three equal parts to represent ten metres.

(iv)   Divide the first part into ten equal parts to represent single metre and complete the scale as shown in Fig. 6.10.

(v)   The distance between points A and B is 24 m.

Fig. 6.10 Solution to problem 6.10

6.6 Diagonal Scales

A diagonal scale is used when very minute distances such as 0.1 mm, etc., are to be accurately measured or when measurements are required in three units, for example (*i*) m, dm and cm (*ii*) dm, cm and mm, etc.

Small divisions of short lines are obtained by the principle of diagonal division as explained below:

Principle of Diagonal Scales: To obtain divisions of a given short line *AB* in multiples of $\frac{1}{10}$ of its length, e.g., 0.1 *AB*, 0.2 *AB*, 0.3 *AB*, etc., as shown in Fig. 6.11.

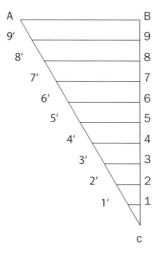

Fig. 6.11 Principle of diagonal scales

(i) At one end, say *B*, draw a line perpendicular to *AB* and along it, step off ten equal divisions of any length, starting from *B* and ending at *C*.

(ii) Number the division points 9, 8, 7 ...... 1 as shown.

(iii) Join *A* with *C*.

(iv) Through the points 1, 2, etc., draw lines parallel to *AB* and cutting *AC* at 1', 2', etc. It is evident that triangles 1'1*C*, 2'2*C* ..... . *ABC* are similar since C5 = 0.5 BC, the line 5'5 = 0.5 *AB*. Similarly, 1'1 = 0.1 *AB*, 2'2 = 0.2 *AB*, etc. Thus, each horizontal line below *AB* becomes progressively shorter in length by $\frac{1}{10}$ AB giving lengths in multiples of 0.1 AB.

Problem 6.11 Construct a diagonal scale of RF = $\frac{1}{5000}$ to show hundred metres, ten metres and single metre and long enough to measure up to 500 m. Also show the distances of 375 m and 467 m on it.

(PTU, Jalandhar December 2003, 2009)

Solution:

(i)   Calculate the length of the scale as:

$$LOS = \frac{1}{500} \times 500 \times 100 = 10 \text{ cm}$$

(ii)   Draw a line 10 cm long and divide it into five equal parts, each representing hundred metres.

(iii)   Divide the first part into ten equal parts, each representing ten metres.

(iv)   Further divide each of these small parts into ten equal parts, using the principle of diagonal scale, each representing single metre.

(v)   Mark the distances of 375 m and 467 m between the points A to B and C to D, respectively.

(vi)   Complete the scale as shown in Fig. 6.12.

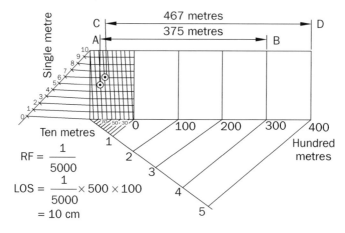

Fig. 6.12 Solution to problem 6.11

Problem 6.12 A rectangular plot of area 50,000 sq m, is represented on a map by a similar rectangle of 80 sq cm. Calculate the RF of the scale of the map. Construct a diagonal scale which can read up to single metre. Also show a distance of 234 m on it.

Solution:

(i)   Obtain the representative fraction as:

$$80 \text{ cm}^2 = 50,000 \text{ m}^2$$
$$1 \text{ cm}^2 = 625 \text{ m}^2$$
$$RF = \sqrt{\frac{1 \text{ cm}^2}{625 \text{ m}^2}} = \frac{1 \text{ cm}}{25 \text{ m}}$$
$$1 \text{ cm} = 25 \text{ m}$$
$$1 \text{ cm} = 25 \times 100 \text{ cm}$$
$$RF = \frac{1}{2500}$$

(ii)   Calculate the length of the scale as:

$$LOS = \frac{1}{2500} \times 300 \times 100 = 12 \text{ cm}$$

(iii) Draw a line 12 cm long. Divide it into three equal parts, each representing hundred metres.

(iv)   Divide the first part into ten equal parts, each representing ten metres.

(v)   Further divide each of these small parts into ten equal parts, using the principle of diagonal scale, each representing single metre.

(vi)   Complete the scale as shown in Fig. 6.13. The distance between the points *A* and *B* is 234 m.

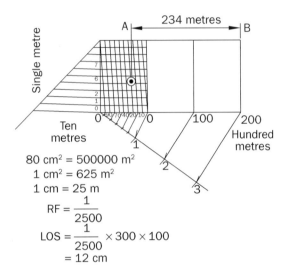

80 cm² = 500000 m²
1 cm² = 625 m²
1 cm = 25 m
$$RF = \frac{1}{2500}$$
$$LOS = \frac{1}{2500} \times 300 \times 100$$
$$= 12 \text{ cm}$$

Fig. 6.13 Solution to problem 6.12

Problem 6.13 A rectangle of 500 × 50 cm represents an area of 6250 sq km. Construct a diagonal scale to measure kilometres, hectometres and decametres. Indicate on this scale a distance of 4 km 5 hm 6 dm.

Solution:

(i)   Obtain the representative fraction as:

25000 cm² = 6250 km²
2500 cm² = 625 km²
50 cm = 25 km
2 cm = 1 km
1 cm = 0.5 km
$$RF = \frac{1}{50000}$$

(ii)   Calculate the length of the scale as:

$$LOS = \frac{1}{50000} \times 5 \times 1000 \times 100 = 10 \text{ cm}$$

(iii)   Draw a line 10 cm long. Divide it into five equal parts, each representing single kilometre.
(iv)   Divide the first part into ten equal parts, each representing single hectometre.
(v)   Further divide each of these small parts into ten equal parts, using the principle of diagonal scale, each representing single decametre.
(vi)   Complete the scale as shown in Fig. 6.14. The distance between the points A and B is 4 km 5 hm 6 dm.

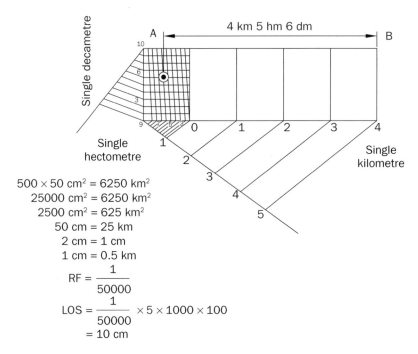

Fig. 6.14 Solution to problem 6.13

Problem 6.14 Construct a diagonal scale to measure 1/100th and 1/10th of a centimetre. Assume length of the scale to be 10 cm.

Solution:

(i)   Draw a line 10 cm long. Divide it into ten equal parts, each representing one centimetre.
(ii)   Divide the first part into ten equal parts, each representing 1/10th of a centimetre.
(iii)   Further divide each of these small parts into ten equal parts, using the principle of diagonal scale, each representing 1/100th of a centimetre.
(iv)   Complete the scale as shown in Fig. 6.15.

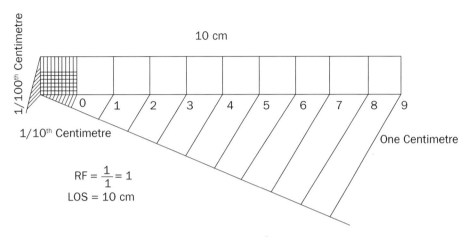

Fig. 6.15 Solution to problem 6.14

Problem 6.15 Construct a diagonal scale of 1:40 to measure a tape up to the length 5 m.

(PTU, Jalandhar June 2003, May 2004, December 2006)

Solution:

(i)   Calculate the length of the scale as:

$$LOS = \frac{1}{40} \times 5 \times 100 = 12.5 \text{ cm}$$

(ii)   Draw a line 12.5 cm long and divide it into five equal parts, each representing single metre.

(iii)   Divide the first part into ten equal parts to represent decimetres.

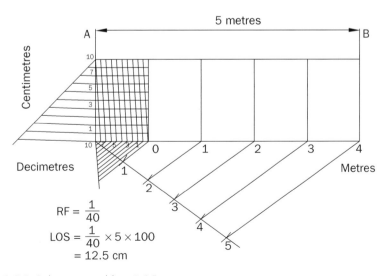

Fig. 6.16 Solution to problem 6.15

(iv)   Further divide each of these small parts into ten equal parts, using the principle of diagonal scale to represent centimetres.

(v)   Complete the scale as shown in Fig. 6.16. The distance between the points *A* and *B* is 5 m.

Problem 6.16 Construct a diagonal scale to read metres, decimetres and centimetres and long enough to measure up to 5 m when one metre is represented by 3 cm. Find RF and indicate a distance of 438 cm.

(*PTU, Jalandhar December 2004*)

Solution:

(i)   Obtain the representative fraction as:

$$3 \text{ cm} = 1 \text{ m}$$
$$3 \text{ cm} = 100 \text{ cm}$$
$$RF = \frac{3}{100}$$

(ii)   Calculate the length of the scale as:

$$LOS = \frac{3}{100} \times 5 \times 100 = 15 \text{ cm}$$

(iii) Draw a line 15 cm long. Divide it into five equal parts to represent metres.

(iv)   Divide the first part into ten equal parts to represent decimetres.

(v)   Further divide each of these small parts into ten equal parts, using the principle of diagonal scale to represent centimetres.

(vi)   Complete the scale as shown in Fig. 6.17. The distance between points *A* and *B* is 4 m 3 dcm and 8 cm.

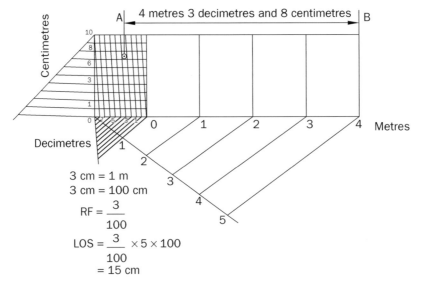

Fig. 6.17 Solution to problem 6.16

The distance between points A and B is 4 m 3 dcm and 8 cm.

Problem 6.17 Construct a diagonal scale to read metres, decimetres and centimetres for a RF of $\frac{1}{50}$ and long enough to measure up to 5 m. Show on it a length of 2.86 m and 3.48 m.

(*PTU, Jalandhar December 2003, May 2005*)

Solution:

   (i)   Calculate the length of the scale as:

$$\text{LOS} = \frac{1}{50} \times 5 \times 100 = 10 \text{ cm}$$

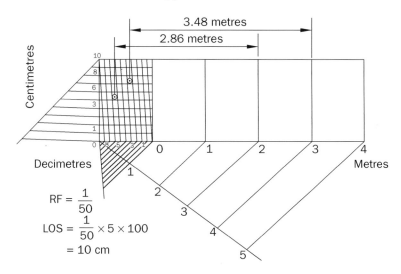

Fig. 6.18 Solution to problem 6.17

  (ii)   Draw a line 10 cm long and divide it into five equal parts to represent metres.
  (iii) Divide the first part into ten equal parts to represent decimetres.
  (iv) Further divide each of these small parts into ten equal parts, using the principle of diagonal scale to represent centimetres.
  (v)   Mark the distances of 2.86 m and 3.48 m between the points *A* to *B* and *C* to *D*, respectively.
  (vi) Complete the scale as shown in Fig. 6.18.

Problem 6.18 Draw a diagonal scale of 1:2.5 showing centimetres and millimetres and long enough to measure up to 20 cm. Show a distance of 13.4 cm on it.

(*PTU, Jalandhar May 2005*)

Solution:

   (i)   Draw a line 20 cm long and divide it into twenty equal parts to represent centimetres.
  (ii)   Divide the first part into five equal parts to represent millimetres.
  (iii) Further divide each of these small parts into two equal parts, using the principle of diagonal scale to represent half of millimetres.

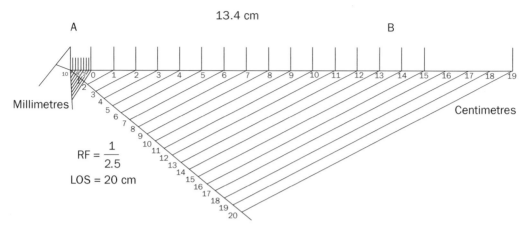

Fig. 6.19 Solution to problem 6.18

    (iv)   Mark the distance of 13.4 cm between the points *A* and *B*.

    (v)   Complete the scale as shown in Fig. 6.19.

Problem 6.19 A cube of 4 cm side represents a tank of 1728 m³ volume. Find its RF and draw a diagonal scale to measure up to 60 m. Also show a distance of 54.4 m on it.

Solution:

    (i)   Obtain the representative fraction as:

$$4 \text{ cm} = 1728 \text{ m}^3$$

$$RF = \frac{4 \text{ cm}}{\sqrt[3]{1728} \text{ m}} = \frac{4 \text{ cm}}{12 \text{ m}} = \frac{4 \text{ cm}}{12 \,100 \text{ cm}} = \frac{1 \text{ cm}}{300 \text{ cm}} = \frac{1}{300}$$

    (ii)   Calculate the length of the scale as:

$$LOS = \frac{1}{300} \times 60 \times 100 = 20 \text{ cm}$$

    (iii)   Draw a line 20 cm long. Divide it into six equal parts to represent ten metres.

    (iv)   Divide the first part into ten equal parts.

    (v)   Further divide each of these small parts into ten equal parts, using the principle of diagonal scale.

    (vi)   Mark the distance of 54.4 m between the points A and B.

    (vii)   Complete the scale as shown in Fig. 6.20.

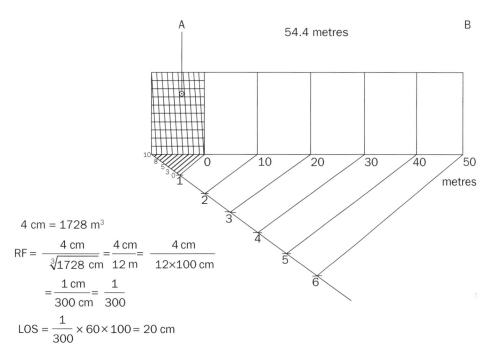

$$4 \text{ cm} = 1728 \text{ m}^3$$

$$\text{RF} = \frac{4 \text{ cm}}{\sqrt[3]{1728} \text{ cm}} = \frac{4 \text{ cm}}{12 \text{ m}} = \frac{4 \text{ cm}}{12 \times 100 \text{ cm}}$$

$$= \frac{1 \text{ cm}}{300 \text{ cm}} = \frac{1}{300}$$

$$\text{LOS} = \frac{1}{300} \times 60 \times 100 = 20 \text{ cm}$$

Fig. 6.20 Solution to problem 6.19

Problem 6.20 A rectangular field of 0.81 hectare is represented on a map by a square of 3 cm × 3 cm. Calculate its RF. Draw a diagonal scale to show hundred metres, ten metres and single metre and long enough to measure up to 600 m. Also show the distances of 475 m and 567 m on it.

Solution:

(i) Obtain the representative fraction as:

$$9 \text{ cm}^2 = 0.81 \text{ hectare}$$
$$9 \text{ cm}^2 = 0.81 \times 10000 = 8100 \text{ m}^2$$
$$3 \text{ cm} = 90 \text{ m}$$
$$1 \text{ cm} = 30 \text{ m} = 30 \times 100 = 3000 \text{ cm}$$
$$\text{RF} = \frac{1}{3000}$$

(ii) Calculate the length of the scale as:

$$\text{LOS} = \frac{1}{3000} \times 600 \times 100 = 20 \text{ cm}$$

(iii) Draw a line 20 cm long. Divide it into six equal parts to represent hundred metres.

(iv) Divide the first part into ten equal parts, each representing ten metres.

(v) Further divide each of these small parts into ten equal parts, using the principle of diagonal scale, each representing single metre.

(vi) Complete the scale as shown in Fig. 6.21.

(vii) Mark the distances of 475 m and 567 m between the points *A* to *B* and *C* to *D*, respectively.

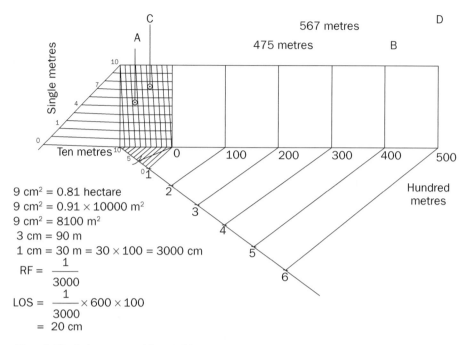

$9\ cm^2 = 0.81$ hectare
$9\ cm^2 = 0.91 \times 10000\ m^2$
$9\ cm^2 = 8100\ m^2$
$3\ cm = 90\ m$
$1\ cm = 30\ m = 30 \times 100 = 3000\ cm$
$RF = \dfrac{1}{3000}$
$LOS = \dfrac{1}{3000} \times 600 \times 100$
$= 20\ cm$

Fig. 6.21 Solution to problem 6.20

Problem 6.21 On a plan, a line 22 cm long represents a distance of 440 m. Draw a diagonal scale for the plan to read up to single metre. Measure and mark a distance of 187 m on the scale.

(*PTU, Jalandhar December 2005*)

Solution:

(i) Obtain the representative fraction as:

$22\ cm = 440\ m$
$1\ cm = 20\ m$
$1\ cm = 20 \times 100 = 2000\ cm$
$RF = \dfrac{1}{2000}$

(ii) Calculate the length of the scale as:

$LOS = \dfrac{1}{2000} \times 200 \times 100 = 10\ cm$

(iii) Draw a line 10 cm long. Divide it into two equal parts to represent hundred metres.

(iv) Divide the first part into ten equal parts, each representing ten metres.

(v) Further divide each of these small parts into ten equal parts, using the principle of diagonal scale, each representing single metre.

(vi) Complete the scale as shown in Fig. 6.22.

(vii) Mark the distance of 187 m between the points A and B.

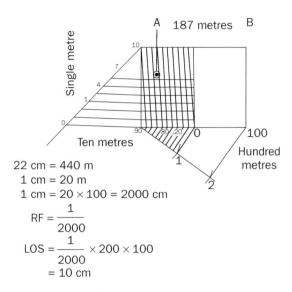

$$22 \text{ cm} = 440 \text{ m}$$
$$1 \text{ cm} = 20 \text{ m}$$
$$1 \text{ cm} = 20 \times 100 = 2000 \text{ cm}$$
$$RF = \frac{1}{2000}$$
$$LOS = \frac{1}{2000} \times 200 \times 100$$
$$= 10 \text{ cm}$$

Fig. 6.22 Solution to problem 6.21

Problem 6.22 On a plan, a line 20 cm long represents 200 m. Construct a diagonal scale for the plan to read up to 0.5 m. Measure and mark a distance of 183.5 m on the scale.

(*PTU, Jalandhar December 2005*)

Solution:

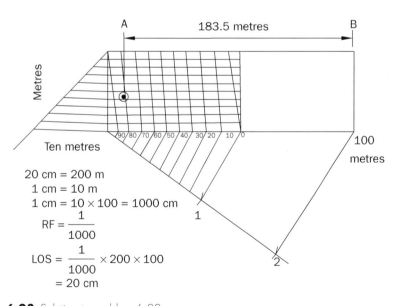

$$20 \text{ cm} = 200 \text{ m}$$
$$1 \text{ cm} = 10 \text{ m}$$
$$1 \text{ cm} = 10 \times 100 = 1000 \text{ cm}$$
$$RF = \frac{1}{1000}$$
$$LOS = \frac{1}{1000} \times 200 \times 100$$
$$= 20 \text{ cm}$$

Fig. 6.23 Solution to problem 6.22

(i)   Obtain the representative fraction as:

$$20 \text{ cm} = 200 \text{ m}$$
$$1 \text{ cm} = 10 \text{ m}$$
$$1 \text{ cm} = 10 \times 100 = 1000 \text{ cm}$$
$$RF = \frac{1}{1000}$$

(ii) Calcualate the length of the scale as:

$$LOS = \frac{1}{1000} \times 200 \times 100 = 20 \text{ cm}$$

(iii) Draw a line 20 cm long. Divide it into two equal parts to represent hundred metres.

(iv) Divide the first part into ten equal parts, each representing ten metres

(v)   Further divide each of these small parts into twenty equal parts, using the principle of diagonal scale.

(vi) Complete the scale as shown in Fig. 6.23.

(vii) Mark the distance of 183.5 m between the points A and B.

Problem 6.23 Construct a diagonal scale to measure 0.01 and 0.1 of a metre and long enough to measure up to 6 m when one metre is represented by 2.5 cm. Find RF and indicate on the scale a distance of 5.55 m.

(PTU, Jalandhar May 2011)

Solution:

(i)   Obtain the representative fraction as:

$$2.5 \text{ cm} = 1 \text{ m}$$
$$2.5 \text{ cm} = 100 \text{ cm}$$
$$1 \text{ cm} = 40 \text{ cm}$$
$$RF = \frac{1}{40}$$

(ii) Calculate the length of the scale as:

$$LOS = \frac{1}{40} \times 6 \times 100 = 15 \text{ cm}$$

(iii) Draw a line 15 cm long. Divide it into six equal parts to represent single metre.

(iv) Divide the first part into ten equal parts, each representing $\frac{1}{10}$ of a metre.

(v)   Further divide each of these small parts into ten equal parts, using the principle of diagonal scale to represent $\frac{1}{100}$ of a metre.

(vi) Complete the scale as shown in Fig. 6.24. Mark the distance of 5.55 m between the points *A* and *B*.

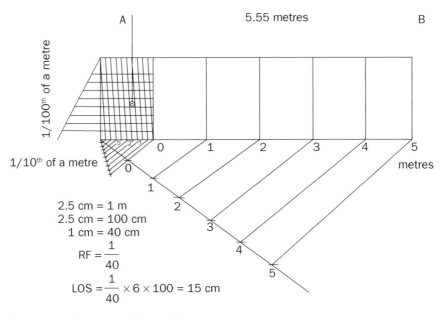

2.5 cm = 1 m
2.5 cm = 100 cm
1 cm = 40 cm

$$RF = \frac{1}{40}$$

$$LOS = \frac{1}{40} \times 6 \times 100 = 15 \text{ cm}$$

Fig. 6.24 Solution to problem 6.23

Problem 6.24 The distance between two cities is 156 km and they are shown 156 mm on a road map. Construct a diagonal scale with this RF and long enough to measure up to 200 km. Make a distance of 109 km on this scale.

(*PTU, Jalandhar December 2010*)

Solution:

(i)   Obtain the representative fraction as:

156 mm = 156 km

1 mm = 1 km

1 mm = 1 × 1000 × 100 × 10 = 1000000 mm

$$RF = \frac{1}{1000000}$$

(ii) Calculate the length of the scale as:

$$LOS = \frac{1}{1000000} \times 200 \times 1000 \times 100 \times 10$$

$$= 200 \text{ mm or } 20 \text{ cm}$$

(iii) Draw a line 20 cm long. Divide it into two equal parts to represent hundred kilometres.

(iv) Divide the first part into ten equal parts, each representing ten kilometres.

(v)   Further divide each of these small parts into ten equal parts, using the principle of diagonal scale, each representing single kilometre.

(vi) Complete the scale as shown in Fig. 6.25.

(vii) Mark the distance of 109 km between the points A and B.

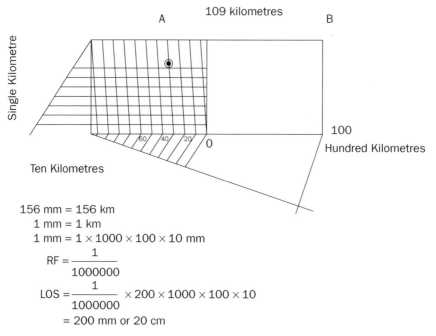

$$156 \text{ mm} = 156 \text{ km}$$
$$1 \text{ mm} = 1 \text{ km}$$
$$1 \text{ mm} = 1 \times 1000 \times 100 \times 10 \text{ mm}$$
$$RF = \frac{1}{1000000}$$
$$LOS = \frac{1}{1000000} \times 200 \times 1000 \times 100 \times 10$$
$$= 200 \text{ mm or } 20 \text{ cm}$$

Fig. 6.25 Solution to problem 6.24

Problem 6.25 The length of the Khandala tunnel on the Mumbai–Pune Expressway is 330 m. On the road map it is shown by a 16.5 cm long line. Construct a diagonal scale to show metres and to measure up to 400 m. Show the length of a 289 m long bridge on the scale.

(PTU, Jalandhar May 2011)

Solution:

(i)   Obtain the representative fraction as:
$$16.5 \text{ cm} = 330 \text{ m}$$
$$1 \text{ cm} = 20 \text{ m}$$
$$1 \text{ cm} = 2000 \text{ cm}$$
$$RF = \frac{1}{2000}$$

(ii)   Calculate the length of the scale as:
$$LOS = \frac{1}{2000} \times 400 \times 100 = 20 \text{ cm}$$

(iii) Draw a line 20 cm long. Divide it into four equal parts to represent hundred metres.

(iv)   Divide the first part into ten equal parts, each representing ten metres.

(v)   Further divide each of these small parts into ten equal parts, using principle of diagonal scale.

(vi)   Complete the scale as shown in Fig. 6.26.

(vii) Mark the distance of 289 m between the points A and B.

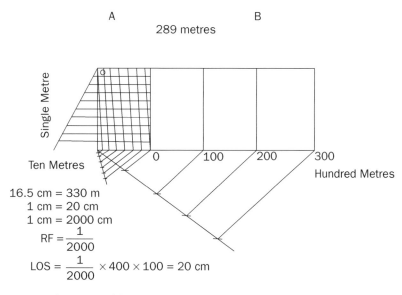

A   B

289 metres

Single Metre

Ten Metres

0   100   200   300

Hundred Metres

16.5 cm = 330 m
1 cm = 20 cm
1 cm = 2000 cm
$$RF = \frac{1}{2000}$$
$$LOS = \frac{1}{2000} \times 400 \times 100 = 20 \text{ cm}$$

Fig. 6.26 Solution to problem 6.25

Exercises

Plain Scales

6.1   Construct a plain scale of 1 cm = 1 m to show single metre and single decimetre and long enough to measure up to 14 m. Show on this scale a distance equal to 12.4 m.

6.2   Construct a plain scale to show metres and decimetres. When 5 cm distance is equal to 1 metre on the scale. The scale should be long enough to measure up to 3 m. Show a distance of 2 m 5 dm on it.

6.3   A rectangular plot 16 sq km in area is represented on a certain map by a similar rectangle of area one square centimetre. Draw a plain scale to show ten kilometres and single kilometre and long enough to measure up to 60 km. Find RF of the scale. Also show a distance of 45 km on it.

6.4   The distance between Delhi and Saharanpur is 180 km. A passenger train covers this distance in 6 hours. Construct a plain scale to measure time up to a single minute. The RF of the scale is $\frac{1}{200000}$. Indicate on this scale the distance covered by the train in 36 minutes.

6.5   Construct a plain scale of 1 cm = 1 km, to read single kilometre and single hectometre and long enough to measure up to 15 km. Find its *RF* and measure a distance of 12 km 5 hm on it.

6.6   Construct a plain scale of 5 cm to 10 km to read kilometres and hectometres. Show the distances of 16.8 km and 17.9 km on this scale.

Diagonal Scales

6.7   Distance between two railway stations is 600 km. It is represented on a railway map by a line 15 cm long. Construct a diagonal scale to show hundred kilometres, ten kilometres and single kilometre. Find its RF and indicate a distance of 345 km on it.

6.8   Construct a diagonal scale of 1:50 to show metres, decimetres and centimetres and long enough to measure up to 6 m. Also indicate on this scale a distance of 6 m 5 dm and 4 cm.

6.9   Construct a diagonal scale of RF = $\dfrac{1}{400}$. It should be long enough to measure 60 m. Show a distance of 57.6 m on it.

6.10   On a plan a line 22 cm long represents a distance of 440 m. Draw a diagonal scale for the plan to read up to single metre. Also show a distance of 234 m on it.

6.11   A rectangular plot of land area 0.45 hectare is represented on a map by a similar rectangle of 5 square cm. Calculate the RF of the scale of the map. The scale should be long enough to measure up to 400 m. Also show a distance of 345 m on it.

6.12   Construct a diagonal scale to measure 1/25th and 1/5th of a centimetre. Assume length of the scale to be 15 cm.

6.13   Construct a diagonal scale of 1:50 to measure a tape up to the length 10 m.

Objective Questions

6.1   The ratio of the length of the drawing of the object to the actual length of the object is called ....

6.2   When a drawing is made to the same size of the object, the name of the scale is ....

6.3   Drawings of buildings, maps are drawn using ....

6.4   When the measurements are desired in two units .... scale is used.

6.5   The relative values of the RF of enlarging, reducing and full size scales are ...., ... and .... respectively.

6.6   What is the function of a scale?

6.7   What are the different types of scales?

6.8   What is the difference between plain scale and diagonal scale?

6.9   What is the principle of a diagonal scale?

6.10   .... = Representative fraction × maximum distance to be measured.

6.11   Give the difference between reducing and increasing scale.

6.12   What is the representative function (RF) or scale factor (SF)?

6.13   What are the uses of diagonal scale?

Answers

6.1   Representative fraction   6.4   Plain

6.2   Full size   6.5   > 1, < 1 and 1

6.3   Reducing scale   6.10   Length of the scale

Chapter 7

Orthographic Projections

7.1 Introduction

There is a huge number of objects of different shapes which have dimensions in the three directions, i.e., length, breadth and thickness. However, these three-dimensional objects are difficult to describe on a two-dimensional sheet. Therefore, in the language of graphics, the shape of an object is described by its projection. A projection is the image of an object formed by rays of sight coming from the same direction as the object. The image is taken on a picture plane, and it appears to an observer stationed at a point from or towards which the projection is made. Therefore, in order to draw the projections of an object the following elements are considered:

(i)   An object or body (point, line, plan, solid)
(ii) The projection plane (HP, VE, PP)
(iii) The point of sight (Observer)
(iv) The lines of sight or rays of sight

7.2 Methods of Projections

In an engineering drawing, four methods of projection are generally used, which are as follows:

(i)   Orthographic Projections
(ii) Axonometric Projections
(iii) Oblique Projections
(iv) Perspective Projections

The first type, i.e., orthographic projection, is one of the most prevalent methods, in which an object can be represented by two or more than two views, on mutually perpendicular planes of projections or picture planes. However, a maximum of six views (front, top, rear, bottom and both left and right side) are used to represent an object. Each view represents two dimensions of an object. For the complete description of a three-dimensional object, at least two views are required. Whereas, in the other three methods, the objects are represented by a pictorial view only, a three-dimensional object is represented on a projection plane by single view only.

7.2.1 Perspective projection

Perspective projection represents objects as perceived by the human eye(s) (refer to Fig. 7.1). It is a pictorial drawing by the intersection of observer's visual rays (lines of sight) converging on a plane (picture plane). The observer's eye-station point or point of sight-is located at a finite distance from the picture plane (refer to Fig. 7.1). Depending on the position of the picture plane, the size of the projection may vary.

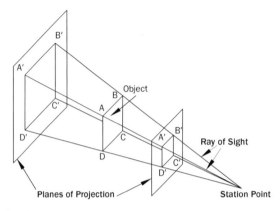

Fig. 7.1 Principle of Perspective Projection

7.2.2 Parallel projection

Parallel projection is obtained by assuming the observer at infinite distance from the object. Hence, the visual rays are considered as parallel to one another. These rays or lines of sight are used to project the object on a standard plane (refer to Fig. 7.2). The object is projected to a plane by drawing straight lines from each and every point on the object. These lines used for projecting the object are 'projectors'. The plane to which the object is projected is the 'plane of projection'. All projectors are parallel to one another and perpendicular to the plane of projection. The image or view obtained on the plane is the 'projection'.

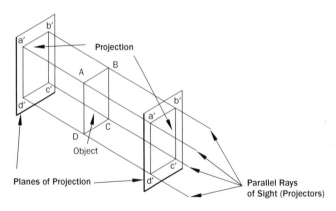

Fig. 7.2 Parallel Projection

7.3 Planes of Projection

These are the planes on which the projection or picture or shadow of an object is obtained, and hence called the projection or picture plane. Especially in orthographic projections, the different views of an object are obtained on the three planes. One plane is used for obtaining the front view or elevation, i.e., in the vertical position which is called the vertical plane (VP) or frontal plane (FP). The second one is used to obtain the top view or plan of an object; it is in horizontal direction and is therefore called the horizontal plane (HP horizontal position). The third plane, perpendicular to both HP and VP is called the profile plane (PP); the same is used to obtain the side views of objects.

Both horizontal and vertical planes, which are called the principal planes, divide the whole space on one side of the profile plane into four parts, called the four dihedral angles or quadrants, as shown in Fig. 7.3. The lines of intersection of these three planes are called coordinate axes. The line of intersection of HP and VP is commonly called as reference line and is generally denoted by the letters x, y. However, in more detail, in reality this x-y line is elevation (front view) of HP and plan (top view) of VP. The projection on VP is called the front view or elevation of the object. The projection on HP is called the top view or plan of the object. The projection on the PP is called the side view or end view or end elevation of the object. The point of intersection of the three coordinate planes is called the origin. In this book, actual points, ends of lines, corners of solids, etc., in space are denoted by capital letters A, B, C, etc. Their top views or plan are marked by corresponding small letters a, b, c, etc., and front views or elevation is mentioned as small letters with dashes i.e., a', b', c', etc., whereas the side views or end elevation is denoted by small letters with double dashes a'', b'', c'', etc. These are the standard conventions that will be used throughout this book.

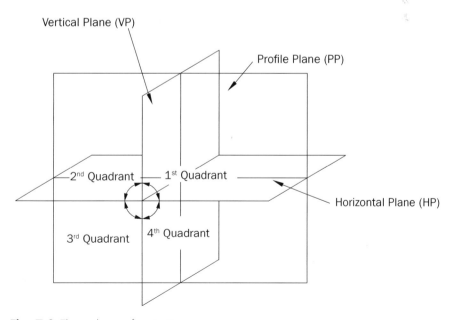

Fig. 7.3 Three planes of projections

7.4 Four Quadrants

Four quadrants or dihedral angles are formed when a projection of planes is extended to meet the line of interaction outside, which may be numbered as shown in Fig. 7.4. The object could be found in any of these quadrants. The relative position of the planes as described in 'behind or in front of VP' and 'below or above the HP'.

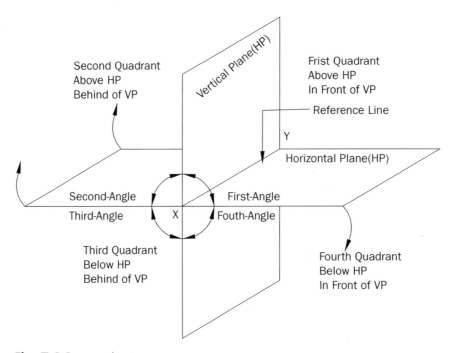

Fig. 7.4 Four quadrants

A projection is said to be the first-angle, second-angle, third-angle or fourth-angle projection when the object is imagined to be either in the first quadrant, second quadrant, third quadrant or fourth quadrant, respectively. The two principal planes are made to lie in one plane by holding VP and rotating HP through 90°, clockwise. However, only two systems, i.e., the first- and third-angle projections are being followed; this is because in the other two quadrants, i.e., in the second and fourth, there will be overlapping of projection while rotating the projection planes. However, in first- and third-angle projections there is no such overlapping while rotating the planes, always opening out first or third quadrant.

7.5 First-Angle Projection

In first-angle projection, an object is imagined to be positioned in the first quadrant. The front view is obtained by looking at the object from the right side of the quadrant. In this case, the object will be in between the observer and the plane of projection.

Similarly, by observing the front surface of the object, the side and top views can be easily obtained. After observing this, there are two sides of the object—one is left side and other one is right side. From the side of the object there are two possible views that may be obtained for any object, i.e., the left side view and the right side view.

By placing the right side of the object on the profile plane, a left side view is obtained in the first-angle of projection. It represents the side view of the object in any view of first-angle projection that is away from it, as shown in Fig.7.5.

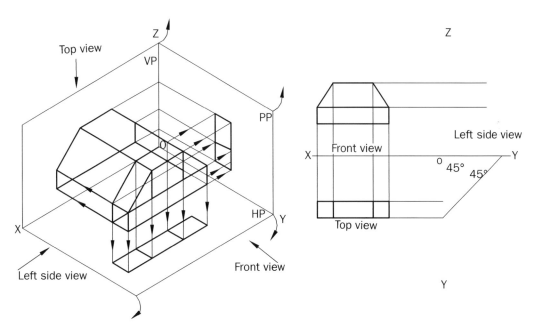

Fig. 7.5 Presentation of orthographic views in first-angle projection

7.6 Third-Angle Projection

In third-angle projection, an object is considered located in the third quadrant, the object is behind the vertical plane (VP) and below the horizontal plane (HP). The front view is obtained on the vertical plane by projecting the ray of sight passing through VP. In this case, the plane of projection is in between the object and the observer. The plane of projection is imagined to be transparent and rays of sight pass through it and then reach the object.

Similarly, its side view and top view are obtained by looking at the normal surface of the object. It may be noted that in the third-angle projection, a left side view is obtained by placing the profile plane to the left side of the object. Thus it represents the side of object that is nearer to it; if any view takes place in the third-angle projection as shown in Fig. 7.6.

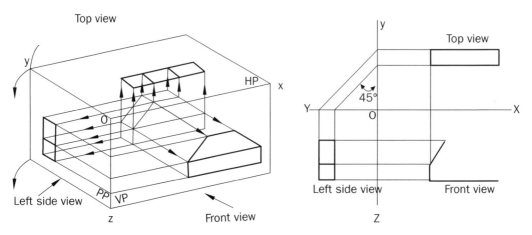

Fig. 7.6 Presentation of orthographic views in third-angle projection

Table 7.1 Difference between first-angle projection and third-angle projection methods

S. No.	Points of Comparison	In First-angle Projection Method	In Third-angle Projection Method
1.	Quadrant	The object or body considered to lie in the first quadrant.	The object is considered to lie in the third quadrant.
2.	Octant	The object is considered in the first octant.	The object lies in the seventh octant.
3.	Position of object	The object lies in between the observer and the plane of projection.	The plane of projection lies in between the object and the observer.
4.	Plane of projection	The plane of projection is considered to be non-transparent.	The plane of projection is assumed to be transparent
5.	Position of view	In this method, the front view or elevation lies above the x-y, the top view or plane below the x-y and the left side view is drawn from the right of elevation.	In this method, the front view or elevation lies below the x-y, top view or plane above the x-y and the left side view is drawn from the left of elevation.
6.	Commonly used countries	This method is commonly used by the Bureau of Indians Standards (BIS) and European countries.	This method is commonly used in the USA and other countries.

7.7 Symbols Used for First-Angle Projection and Third-Angle Projection Methods

For every drawing, it is absolutely essential to indicate the method of projection adopted. This is done by means of a symbolic figure drawn on a title block. The symbol drawn for first-angle projection method is shown in Fig. 7.7, while that for the third-angle projection method is shown in Fig. 7.8. These symbols are actually obtained from the projections of frustum of a cone as shown in Fig. 7.9.

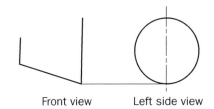

Front view   Left side view

Fig. 7.7 Symbol for first-angle projection method

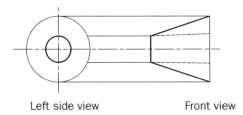

Left side view   Front view

Fig. 7.8 Symbol for third-angle projection method

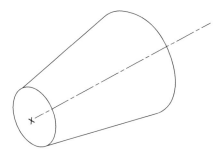

Fig. 7.9 Frustum of a cone

Exercises

7.1   What do you mean by orthographic projection? Describe briefly the method of obtaining the orthographic projection of an object.

7.2   Explain with the help of neat sketches the difference between the first-angle and the third-angle projection methods.

7.3   Why are the projections of an object not drawn in the second and the fourth quadrants?

7.4   Write short notes on
(i) Principal planes	(ii) Reference line	(iii) Projection
(iv) Front view	(v) Top view	(vi) Side view

Objective Questions

7.1   In .... projection, the object is positioned in between the observer and the plane of projection.

7.2   In .... angle projection, the plane is positioned between the object and observer.

7.3   Draw the symbols for the first-angle and the third-angle projection methods.

7.4   A surface of an object appears in its ...., when it is parallel to the plane of projection.

7.5   What is a plane of projection?

7.6   In orthographic projection the lines of sight are .... to the plane of projection.

7.7   The three planes of projection are ...., ... and ....

7.8   What does the following symbol represent—the first- or the third-angle projection?

7.9   What are the full forms of 'HP' and 'VP'?

7.10 Distinguish between the terms 'Projector' and 'Projection' by means of suitable sketches.

7.11 What is a profile plane and what is its use?

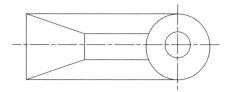

Answers

7.1   First-angle

7.2   Third-angle

7.4   True shape

7.6   Perpendicular

7.7   Horizontal, vertical, profile planes

7.8   Third–angle projection

7.9   HP → Horizontal plane, VP → Vertical plane

Chapter 8

Projections of Points

8.1 Introduction

In this chapter, an attempt is being made to introduce the readers to the projection of points. Following the treatment here it will be easy for them to understand the projection of lines, planes and solids in the subsequent chapters. For the projection of points, the quadrant system is considered and a point lying in space is assumed in any one of the four quadrants that are obtained by the intersection of two principal planes. A point can lie with reference to both the reference planes, i.e., HP and VP. Its projections are obtained by extending projectors perpendicular to the planes.

In order to obtain the projection of a point lying in three-dimensional space on a two-dimensional plane (drawing sheet), the principal plane HP is rotated clockwise through 90° and made co-planner with the VP. This process coverts the three-dimensional quadrant system into two-dimensional front and top views, i.e., the front view VP is obtained above x-y line where this x-y line represents the elevation or front view of HP; similarly the top view of HP is obtained below the x-y line, here the x-y line represents the top view of VP.

8.2 Projection of a Point Lying in the First Quadrant

The pictorial view Fig. 8.1(*a*) shows a point A lying in the first quadrant, i.e., above the HP and in front of the VP. When the point is viewed in the direction of *l*, the view from front *a*′ is obtained as the intersection point between the ray of sight through *A* and the VP. When the point is viewed in the direction *m*, the top view (a) of point 'A' is obtained as the ray of sight intersect with HP at a. Similarly front view (a′) of 'A' is obtained when the ray of sight coming from l direction intersects VP at a′. Hold VP and rotate HP 90° in the clockwise direction; these projections are seen in the Fig. 8.1(*b*). The front view *a*′ is above x-y and top view below it.

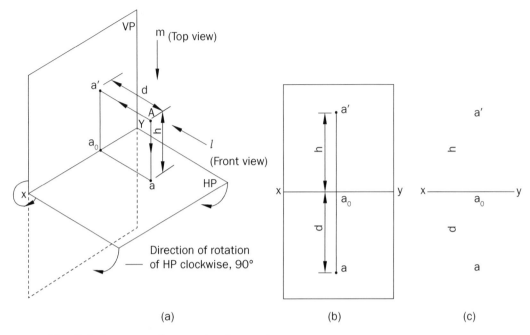

Fig. 8.1 Projections of a point A in first quadrant

The line joining a' and a (which is called as projector) intersects x-y at right angle (90°) at a point a. It is quite evident from pictorial view that $a'a_0 = Aa$, i.e., distance of front view from x-y line = distance of A from the HP *viz.*, h. Similarly, $aa_0 = Aa'$, i.e., distance of the top view from x-y line = distance of A from the VP *viz.*, d. Fig. 8.1(*c*) shows only the relative positions of the views, as it is customary not to show the planes of projection.

The following points may be noted from the study of Fig. 8.1.

(i)   The line x-y is the intersection line between HP and VP, as shown in Fig. 8.1(*a*). In Figs. 8.1(*b*) and 8.1(*c*), the line is represented by x-y, which is known as the reference line or common axis. Actually, x-y is the line about which the rotation of the plane is made.

(ii)   It is customary to use capital letters to specify the position of the points in space and small letters for their projections. As an example, for the point A in space, the views from front, top and side are represented by a', a and a''.

8.3 Projection of a Point Lying in the Second Quadrant

As shown in Fig. 8.2 (*a*), when a point B is lying above the HP and behind the VP, it is in the second quadrant. Here b' and b are the front and top views obtained on VP and HP, respectively, by viewing the points in the direction l and m, respectively. Fig. 8.2 (*b*) shows the relative positions of the views these are obtained by rotating the HP clockwise until it is co-planer with the VP. It may be noted that both the views are seen above reference line x-y. Also $b'b_0 = Bb$ and $bb_0 = Bb'$.

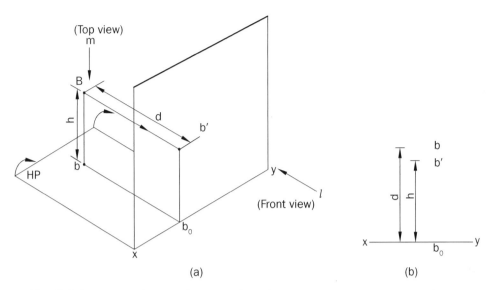

Fig. 8.2 Projections of a point B in second quadrant

8.4 Projection of a Point Lying in the Third Quadrant

As shown in Fig. 8.3(*a*), a point *C* is below the HP and behind the VP, i.e., in the third quadrant. *c'* and *c* are the front and top views obtained on VP and HP by viewing the points in the direction *l* and *m*, respectively. Here, it is assumed that both HP and VP are transparent. Fig. 8.3 (*b*) shows the relative positions of the views. These are obtained by rotating the HP clockwise until it coincides with the VP. Here, it may be noted that front view *c'* is below x-y and top view *c* above the x-y line. Also $c'c_0 = Cc$ and $cc_0 = Cc'$.

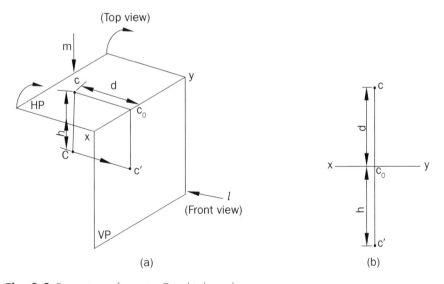

Fig. 8.3 Projections of a point C in third quadrant

8.5 Projection of a Point Lying in the Fourth Quadrant

When a point, say 'D', as shown in Fig. 8.4(a), is lying below the HP and in front of the VP, it is in the fourth quadrant. Here d' and d are the front and top views obtained on VP and HP by viewing the points in the direction l and m, respectively. Fig. 8.4 (b) shows the relative positions of the views. These are obtained by rotating the HP in the clockwise direction until it coincides with the VP. Here, it can be noted that both the views are below x-y. Also $d'd_o = Dd$ and $dd_o = Dd'$.

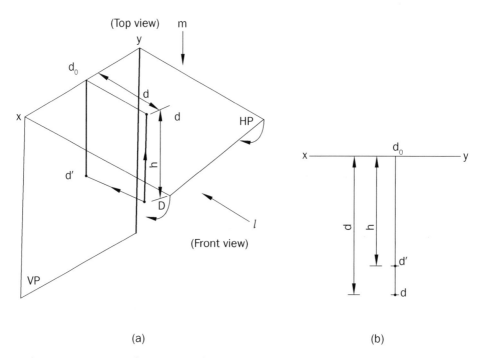

(a)   (b)

Fig. 8.4 Projections of a point D in fourth quadrant

8.6 Special Cases

    (i)   A point E is situated in the HP and in front of the VP.
    (ii)   A point F is situated at height 'h' above HP and is in the VP.
    (iii) A point G is in both HP and VP.

All the three cases are shown in Fig. 8.5 (a). Fig. 8.5 (b) shows the relative positions of the views for each case.

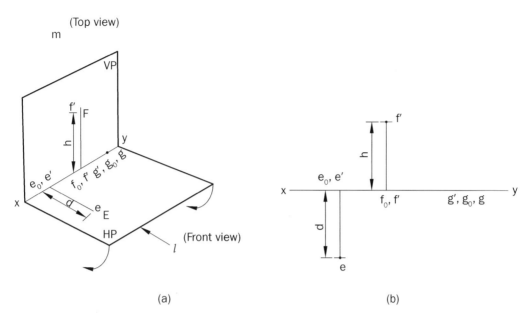

Fig. 8.5 Special cases of points E, F and G in the first quadrant

The following points may be noted from Fig. 8.5 (*b*).

(i)   When a point lies in the HP, its front view will lie on x-y.
(ii) When a point lies in the VP, its top view will lie on x-y.
(iii) When a point lies on both HP and VP, its front and top views will lie on x-y.

8.7 A Point is Situated in the Three Planes of Projection

Sometimes two views of an object are not sufficient to describe the shape and size completely. It is then necessary to draw a third view, preferably a side view or an end view on a profile plane. As already described, HP and VP divide the space into four quadrants. Introduction of PP will divide each quadrant further into two parts, i.e., four to the left and four to the right of the PP. Thus we have eight spaces, called octants, as shown in Fig. 8.6. Only anticlockwise or left hand system will be used throughout the book.

   A point may be situated in either of the eight octants. As discussed in Chapter 7, rotate the HP through 90° clockwise and PP through 90° anticlockwise, so as to be coplanar with the VP, such that the first octant opens out. On the same pattern, it can be noted that the planes after rabatment do not overlap each other in the seventh octant. Except in the first and seventh octants, there is overlapping of planes in all other octants, which results in overlapping of views too. So only the first and seventh octants will be discussed here.

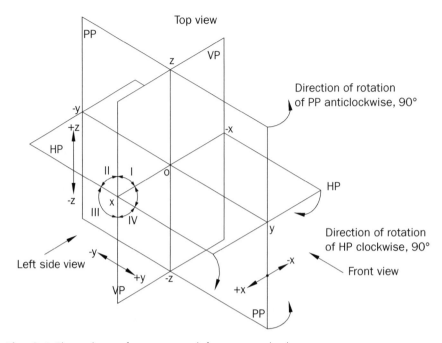

Fig. 8.6 Three planes of projection with front, top and side views

(a) **A point is situated in the first octant**

As shown in Fig. 8.7 (*a*), a point *E* is placed in the first quadrant along with the profile plane (PP). The projection of a point *E* on the PP is obtained by viewing the point in the direction *n*. The view obtained on the PP, i.e., *e″* is known as left side view. Fig. 8.7 (*b*) shows the relative positions of the views along with the three planes of projections. Fig. 8.7 (*c*) shows the relative positions of the views without showing the planes of projection.

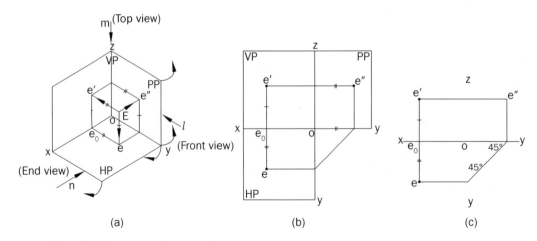

Fig. 8.7 Projections of a point E in the first octant

(b) **A point is situated in the seventh octant**

As shown in Fig. 8.8 (*a*), a point F is placed in the third quadrant along with the profile plane (PP). The projection of a point F on the PP is obtained by viewing the point in the direction *n*. Here it is assumed that all the three planes are transparent. Fig. 8.8(*b*) shows the relative positions of the views.

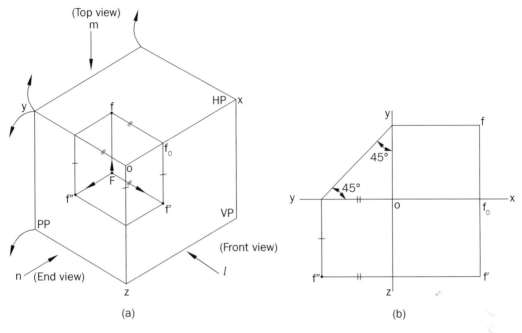

Fig. 8.8 Projections of a point F in seventh octant

Problem 8.1 Consider the following points and draw their projections on a common x-y line. Keep the distance between two consecutive projectors as 20 mm.

(a) Point 'A' is 30 mm above the HP and 30 mm in-front of the VP
(b) Point 'B' is 30 mm above the HP and 30 mm behind the VP
(c) Point 'C' is 30 mm below the HP and 30 mm behind the VP
(d) Point 'D' is 30 mm below the HP and 30 mm in-front the VP
(e) Point 'E' is in HP and 30 mm in front of the VP
(f) Point 'F' is in VP and 40 mm above the HP
(g) Point 'G' is in VP and 40 mm below the HP
(h) Point 'H' is in both in HP and VP.

Solution:
The solution to this problem is shown in Fig. 8.9, and the same is self-explanatory.

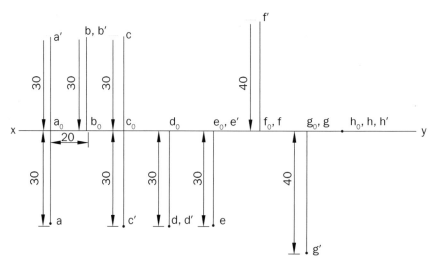

Fig. 8.9 Solution to problem 8.1

Problem 8.2 Consider a point 'A' which is 20 mm in front of VP and 30 mm above HP. On the other hand point 'B' is 25 mm behind the VP and 40 mm below the HP. The distance is 40 mm between the end projectors. Draw the joining line of their top and front views and also draw the points of the projections.

Solution:

  (i)   Draw two projectors 40 mm apart on reference line x-y.

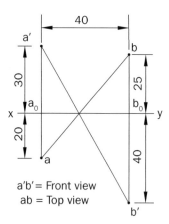

a'b' = Front view
ab = Top view

Fig. 8.10 Solution to problem 8.2

  (ii)   Locate front and top views for points A and B on the projectors.
  (iii) Join a'b' and ab with straight lines.
  (iv)   a' b' and ab are the required front and top views respectively. See Fig. 8.10.

Problem 8.3 Determine the least distance from x-y lines between point 'A' which is 25 mm in front of VP and 30 mm above the HP.

(PTU, Jalandhar December 2008)

Solution:

- (i) Draw reference line x-y.
- (ii) Locate front and top views for the point A.
- (iii) Draw a parallel line through a point y-z which is perpendicular to x-y and intersect point o.
- (iv) Through point a, draw a line parallel to x-y line to intersect y-z at o_2
- (v) With oo_2 as radius and centre o draw an arc that meets x-y at o_3.
  or
  At point o_2, draw an angle of 45° to meet x-y line at o_3.
- (vi) Locate the side view a'' as the intersection point between the projectors drawn from o_3 and a'.
- (vii) Join o with a'', which will represent the least distance from x-y line. See Fig. 8.11.

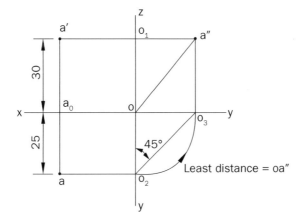

Fig. 8.11 Solution to problem 8.3

Problem 8.4 Point P is 30 mm behind the VP and 25 mm below the HP. Determine the least distance from x-y line.

(PTU, Jalandhar May 2009)

Solution:

- (i) Draw reference line x-y.
- (ii) Locate front and top views for the point P.
- (iii) Draw another reference line y-z perpendicular to x-y, which will intersect at point o.
- (iv) Draw a line over point P that intersects y-z at o_2.
- (v) At point o_2, draw an angle of 45° to meet x-y at o_3.
- (vi) Locate the side view p'' as the intersection point between the projectors drawn from o_3 and p'.
- (vii) Join o with p'', which will represent the least distance from x-y line. See Fig. 8.12.

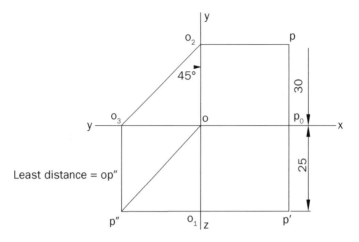

Fig. 8.12 Solution to problem 8.4

Problem 8.5 A point "C" is located on HP and 40 mm in front of VP. There is another point D which is 30 mm below HP and 25 mm behind the VP. If the distance between the end projectors is 40 mm, draw the projections of the points and draw straight lines joining their front and top views.

Solution:

    (i)   Draw two projectors 40 mm apart from x-y reference line.
    (ii)   Locate front and top views for the C and D points on the projectors.
    (iii) Join $c'd'$ and cd with straight lines.
    (iv) $c'd'$ and cd are the required front and top views, respectively See Fig. 8.13.

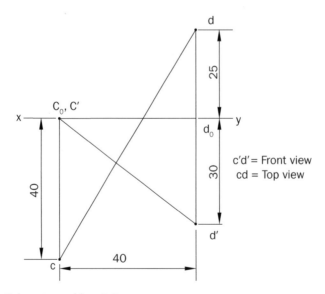

Fig. 8.13 Solution to problem 8.5

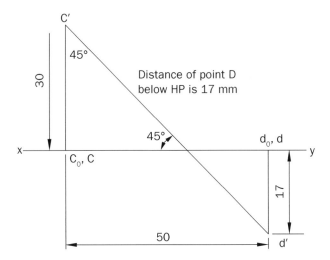

Fig. 8.14 Solution to problem 8.6

Problem 8.6 Two points C and D are in VP. Point C is 30 mm above HP while point D is below the HP. The distance is 50 mm between the end projectors. The line joining the front views of the two points makes angles of 45° with the reference line. Draw the projections line of CD and find out the distance of the point D below the HP.

Solution:
The solution to this problem is self-explanatory. See Fig. 8.14.

Exercises

8.1   Draw the projection lines of the following points on a common x-y line, keeping the distance between two consecutive projectors as 20 mm.
   (a)   40 mm in front of VP and 30 mm above the HP
   (b)   40 mm behind of VP and 30 mm above the HP
   (c)   40 mm behind of VP and 30 mm below the HP
   (d)   40 mm in front of VP and 30 mm below the HP
   (e)   in HP and 40 mm in front of VP
   (f)   in VP and 30 mm above the HP
   (g)   in VP and 30 mm below the HP
   (h)   in HP and 40 mm behind the VP
   (i)   both in HP and VP.

8.2   State the quadrants in which the given points are located:
   (a)   A point A, front view is 40 mm below x-y line and top view 30 mm above x-y line.
   (b)   A point B, its top view and front view coincide with each other 40 mm above x-y line.
   (c)   A point C, its top view on x-y line and front view 30 mm above the x-y line.
   (d)   A point D, its top view is 35 mm below x-y line and front view on x-y line.

8.3   A point A is located on 25 mm in front of VP and 40 mm above HP. Location of point B is 30 mm behind VP and 20 mm below HP. The distance between end projectors is 50 mm. Draw the straight line joining their front and top views and also projections of the points. Also draw straight lines joining their front and top views.

8.4   A point P is located 40 mm in front of VP and 20 mm above HP. Location of point Q lies in the first quadrant and the distance between the vertical projections of P and Q be equal to 60 mm and the distance between the projectors through P and Q be 45 mm and the point Q is 35 mm in front of VP. Draw its projections.

8.5   A point A is located 30 mm above HP and is in the first quadrant. Its shortest distance from x-y line is 55 mm. Draw its front and top views.

8.6   A point P is located 40 mm below HP and 35 mm behind VP. Determine its least distance from x-y.

8.7   A point P is located 20 mm below HP and is in the third quadrant. Its shortest distance from x-y line is 50 mm. Draw its projections.

Objective Questions

8.1   Draw the projection of a point 'P' when it is lying on the x-y line, i.e., where HP and VP meet each other.

8.2   To represent the projections on a sheet, the planes must be rotated such that .... or .... quadrant always opens out.

8.3   When a point is below the HP, its front view is .... x-y line.

8.4   When a point lies on both HP and VP, its front view and top view .... x-y line.

8.5   How is the side view of a point obtained from its front and top views?

8.6   The .... is obtained as the intersection point between the ray of the sight and the VP.

8.7   The top view is obtained as the intersection point between the ray of the sight and the ....

8.8   When a point lies in the HP, its front view will lie ....

8.9   The line joining the projections of a point intersects at an angle of .... to the x-y line.

8.10   When a point lies in the VP, its top view will lie ....

8.11   If the front view is above x-y line and the top view is below x-y line, then the point is in .... quadrant.

8.12   When a point is situated in the first quadrant it will be located above the HP and in front of VP. (True/False)

8.13   When a point lies on both HP and VP, its front and top views will lie ....

8.14   The projector of a point is always .... to the reference line.

8.15   What is a projector?

Answers

8.2	First, third	8.7	HP	8.11	First
8.3	Below	8.8	On x-y line	8.12	True
8.4	Lie on	8.9	90°	8.13	On x-y line
8.6	Front view	8.10	On x-y line	8.14	Perpendicular

Chapter 9

Projections of Lines

9.1 Introduction

A line can be defined as the route or pathway of a point displaced from one position to another. A line can be a curved one or a straight one. A straight line is the shortest track or distance between two points. The projections of a straight line can be drawn by joining the respective projections of its end points. The position of a straight line in space may be defined by

- the location of its end points or extremities from principal planes and its true length (TL).
- the location of one of its end points and inclination of the line with the principal planes.
- the location of its end points and distance between the end projectors, etc.

9.2 Position of a Straight Line

A line in space can be in various positions w.r.t. to the principal planes of projection as shown in Table 9.1:

- (i) Line parallel to both HP and VP.
- (ii) Line inclined to one plane and parallel to the other
  - Line inclined to the HP and parallel to the VP.
  - Line inclined to the VP and parallel to the HP.
- (iii) Line perpendicular to one of the planes.
  - Line perpendicular to the HP and parallel to the VP.
  - Line perpendicular to the VP and parallel to the HP.
- (iv) Line contained by one or both of the principal planes
  - Line contained by the HP.
  - Line contained by the VP.
  - Line contained by both HP and VP, i.e., in x-y line.
- (v) Line inclined to both HP and VP.
- (vi) Line contained by a profile plane (PP) or line contained by a plane, perpendicular to both HP and VP, i.e., in x-y line.

Table 9.1 Summary of positions of line w.r.t. reference planes

S. No	Position of Line w.r.t. Horizontal Plane (HP)	Position of Line w.r.t. Vertical Plane (VP)	The Projection Will Be
1	Parallel	Parallel	True length in both views
2	Inclined	Parallel	Apparent length in plan and true length in elevation
3	Parallel	Inclined	Apparent length in elevation and true length in plan
4	Perpendicular	Parallel	Single point in plan and true length in elevation
5	Parallel	Perpendicular	Single point in elevation and true length in plan
6	Contained by	--	True length in plan
7	--	Contained by	True length in elevation
8	Contained by	Contained by	True length in both views (on X-Y line)
9	Inclined	Inclined	Apparent length in both the views
10	Perpendicular or parallel when the line is contained by the profile plane	Perpendicular or parallel when the line is contained by the profile plane	True length in either of the views

9.3 Line Parallel to Both HP and VP

Line *AB*, parallel to both HP and VP, is shown in Fig. 9.1 (*a*). Here ends *A* and *B* of the line *AB* are at equal distance from the HP and VP. When a line is parallel to any plane, its projections on that plane is a straight line of the same length. The elevation (front view) *a'b'* and plan (top view) *ab* are both parallel to x-y line and their lengths represent the true length (*AB* of a line).

Fig. 9.1 (*b*) shows the orthographic projection of the line *AB*.

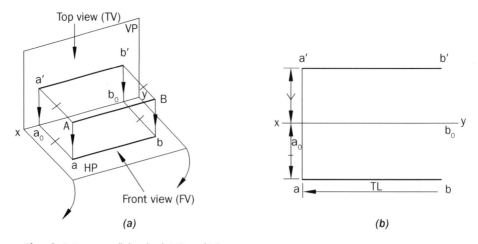

(a)   (b)

Fig. 9.1 Line parallel to both HP and VP

Problem 9.1 Draw the projections in all the four quadrants of a given line *AB* of length 50 mm, which is parallel to both HP and VP; it has its end A 30 mm away from HP and 20 mm away from the VP.

Solution:

(i) Draw the x-y line and take a convenient point a_0 on it and draw the end projector for end point A.

(ii) On *this vertical projector*, mark front and top views of A as *a'* *and a* at a distance equal to the distance of *A* from HP and VP respectively.

(iii) Through *a'* and *a*, draw horizontal lines parallel to x-y and mark each 50 mm true length, to locate *b'* and *b*.

(iv) Join *a'* with *b'* and *a* with *b*.

(v) Now *a'b'* and *ab* are the required elevation and plan, respectively.

Follow the above mentioned steps to draw the projections of a line in all the four quadrants. See Fig 9.2.

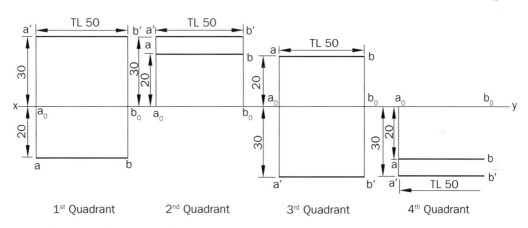

Fig. 9.2 Solution to problem 9.1

9.4 Line Inclined to One Plane and Parallel to the Other

When a line is inclined to one plane and parallel to the other, its projections on the plane to which it is parallel gives the true length of a line and its projections on the plane to which it is inclined is a straight line, shorter than its true length but parallel to the x-y.

(a) **Projection of a line inclined to the HP and parallel to the VP**

The given line *AB* is inclined to the HP at an angle θ and parallel to the VP; the pictorial view of the same is exhibited in Fig. 9.3 (*a*). The front view (elevation) *a'b'* of the line is inclined to the x-y at an angle θ and the length of the elevation is equal to its true length. Its top view (or plan) *ab* is shorter than the line *AB* and is parallel to the x-y.

The orthographic projections of a line *AB* are shown in Fig. 9.3 (*b*)

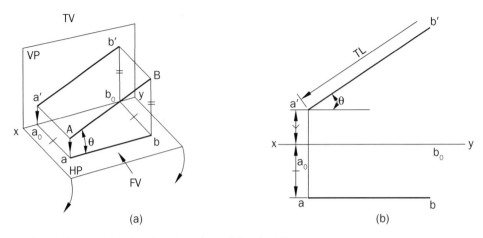

Fig. 9.3 Line inclined to the HP and parallel to the VP

Problem 9.2 The end *A* of a line AB of length 50 mm, is 25 mm away from HP and 15 mm in front of the VP. The line is parallel to the VP and inclined at 30° w.r.t. the HP. Draw its projections in all the four quadrants. Assume that the whole of the line lies in the same quadrant.

Solution:

(i)   Draw the x-y, and take a convenient point on it and draw the end projector for end point A.
(ii) On this projector, draw front view *a'* and top view *a*.
(iii) Draw elevation (front view) *a'b'* of the line at an angle 30° to x-y. Cut *a'b'* = 50 mm (TL).
(iv) Project *a'* to a and *b'* to b on top view (plan) *ab*, which is a line parallel to x-y at a distance of 15 mm.

Follow the above mentioned steps to draw the projections of a line in all the four quadrants. See Fig. 9.4.

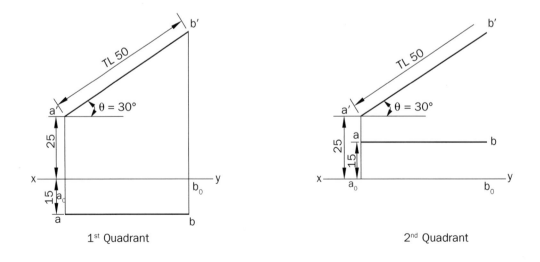

1ˢᵗ Quadrant   2ⁿᵈ Quadrant

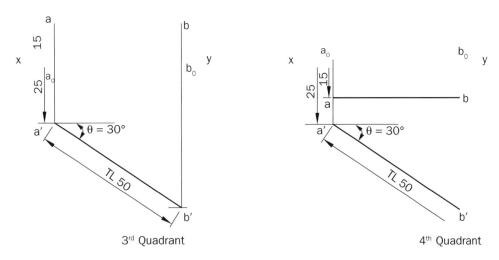

Fig. 9.4 Solution to problem 9.2

Problem 9.3 A line *AB* of 50 mm length is inclined at 30° to the HP. The end A of the line is at 12 mm height from the HP and 15 mm away from VP. If the line AB is parallel to VP, draw its plan and elevation.

(PTU, Jalandhar December 2007)

Solution: It is given here that the line is inclined to the HP and parallel to the VP.

(i)   Draw the x-y line. At a convenient distance on x-y, draw the projector of end point *A*.

(ii) On this projector, draw front view *a'* 12 mm above x-y and top view *a* 15 mm below x-y line.

(iii) Draw a line at 30° to x-y and cut *a' b'* = 50 mm (TL), here line is parallel to VP therefore its elevation (front view) *a' b'* of the line will provides the true length of the line.

(iv) Now project *a'* to *a* and *b'* to *b* to produce plan (top view) *ab*, which is parallel to x-y line.

(Note: The x-y line represent the top view of VP and front view of HP, because here the line AB is parallel to VP; therefore, its top view is parallel to x-y line, and the front view is inclined to x-y line because line is inclined to HP.)

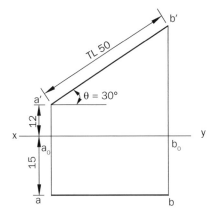

Fig. 9.5 Solution to problem 9.3

(b) The given line is inclined to the VP and parallel to the HP

The pictorial view of the line AB, inclined to the VP at an angle φ and parallel to the HP, is shown in Fig. 9.6 (*a*). The top view *ab* of the line is inclined to the x-y at angle φ and the length of the top view (or plan) is equal to its true length. Its front view (or elevation) $a'b'$ is shorter than the line AB and is parallel to the x-y line. The orthographic projections of a line AB is shown in Fig. 9.6 (*b*).

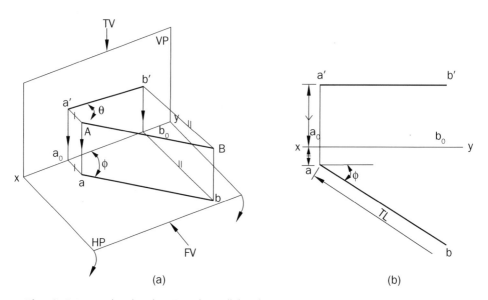

Fig. 9.6 Line inclined to the VP and parallel to the HP

Problem 9.4 The end A of a line AB, 50 mm long, is 25 mm away from the HP and 15 mm away from the VP. The line is inclined to the VP at 30° and is parallel to the HP. Draw its projections in all the four quadrants. Assume that the whole of the line lies in the same quadrant.

Solution:

(i) Draw a x-y line. At a convenient distance on x-y, draw the projector of end point A.
(ii) On this projector, draw front view a' at 25 mm above x-y and top view a 15 mm below x-y line.
(iii) Draw top view *ab* of the line at an angle of 30° to x-y. Cut *ab* = 50 mm (TL).
(iv) Now project a to a' and b to b' to produce plan (front view) $a'b'$, which is parallel to x-y line.

(Note: The x-y line represents the top view of VP and front view of HP, because here the line AB is parallel to HP; therefore, its front view is parallel to x-y line, and the top view is inclined to x-y line because the line is inclined to VP.)

Follow the above mentioned steps to draw the projections of a line in all the four quadrants. See Fig. 9.7.

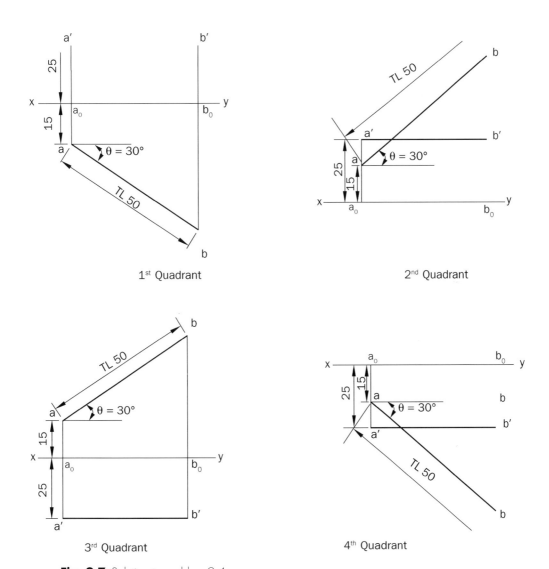

Fig. 9.7 Solution to problem 9.4

Problem 9.5 The end A of a line AB, 50 mm long, is in HP and 15 mm in front of the VP. The line is inclined to the VP at 30 and is parallel to the HP. Draw its projections.

Solution:

(i)   Draw a x-y line. At a convenient distance on x-y, draw the projector of end point *A*.

(ii)   On this projector, draw front view *a'* and top view *a* on it.

(iii)   Draw top view *ab* of the line at an angle of 30° to x-y. Cut *ab* = 50 mm (TL).

(iv)   Project *a* to *a'* and *b* to *b'* on front view *a'b'*, which is a line coincides with x-y as shown in Fig. 9.8.

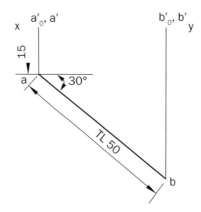

Fig. 9.8 Solution to problem 9.5

Problem 9.6 The front view of a line AB, 60 mm long, measures 40 mm. The line is parallel to the HP and the end A is 15 mm above HP and 20 mm in front of the VP. Draw the projections of the line and determine its inclination (φ) with the VP.

Solution: As the line is parallel to the HP, its front view will be parallel to x-y.

(i)   Draw front view a' and top view a of the end point A.

(ii)   Draw the front view $a'b'$ 40 mm long and parallel to x-y. With a as centre and radius equal to 60 mm, draw an arc cutting the projector through b' to b. Join a with b. Thus ab is the top view of the line and φ is the inclination of the line AB with the VP as shown in Fig. 9.9.

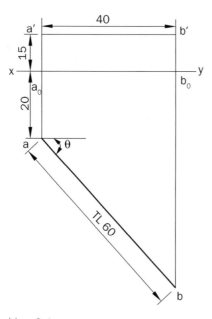

Fig. 9.9 Solution to problem 9.6

9.5 Line Perpendicular to One of the Planes

When a line is perpendicular to one of the principal planes, it will be parallel to the other.

(a) **Line perpendicular to the HP**

Figure 9.10 (*a*) shows the pictorial view of a line *AB*, when it is perpendicular to the HP. Its front view *a'b'*, is true length of a line *AB* and perpendicular to x-y. Its top view *ab* is a point where the projections of the two end points coincide with each other

(Note: here in Fig. 9.10, the top view is indicated as *ba*; this is because when line *BA* is seen from the top, the nearest end to the observer is '*B*' and farthest end is '*A*'; therefore in all such cases notation of point will be given in same manner).

Figure 9.10 (*b*) shows the orthographic projections of a line *AB*.

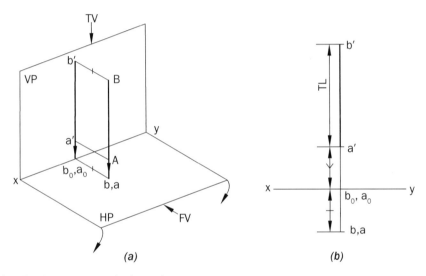

Fig. 9.10 Line perpendicular to the HP

Problem 9.7 A line AB, 40 mm long, is perpendicular to the HP and its end A is 15 mm away from the HP and 10 mm away from the VP. Draw its projections in all the four quadrants. Assume that the whole of the line lies in the same quadrant.

Solution:

(i)   Draw a x-y line and at a convenient distance on x-y, draw the projector of end point *A*.

(ii)   On this projector, draw front view *a'* and top view *a* on it.

(iii) As the line is parallel to the VP, its projection on the VP is true length and perpendicular to x-y.

(iv)   Its top view *ab* is a point, which coincides with each other.

Follow the above mentioned steps to draw the projections of a line in all the four quadrants. See Fig. 9.11.

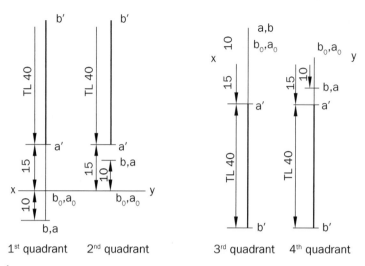

Fig. 9.11 Solution to problem 9.7

(b) **Line perpendicular to the VP**

Figure 9.12 (*a*) shows the line *AB*, perpendicular to the VP. Its top view, *ab*, is true length of a line *AB* and perpendicular to x-y. Its front view *a′b′* is a point where the projections of the two end points coincide with each other.

Figure 9.12 (*b*) shows the orthographic projections of a line *AB*.

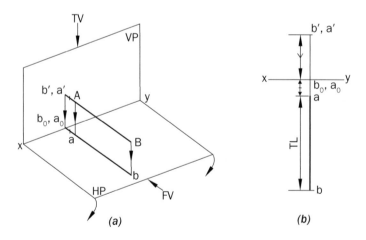

Fig. 9.12 Line perpendicular to the VP

Problem 9.8 A line AB, 40 mm long, is perpendicular to the VP and its end A is 15 mm away from the HP and 10 mm away from the VP. Draw its projections in all the four quadrants. Assume that the whole of the line lies in the same quadrant.

Solution:

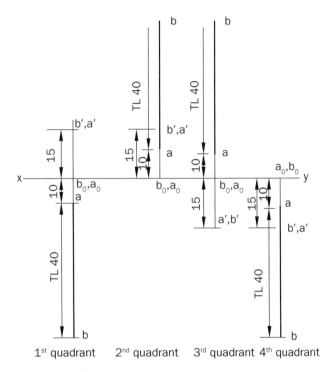

Fig. 9.13 Solution to problem 9.8

(i) Draw a x-y line and at a suitable distance on x-y, draw the projector of end point *A*.
(ii) On this projector, draw front view *a′* and top view *a* on it.
(iii) As the line is perpendicular to the VP, its projection on the VP is a point, i.e., *a′* and *b′* coincide with each other.
(iv) Its projection on HP, is a true length i.e., *ab* = 40 mm and perpendicular to the x-y.

Follow the above mentioned steps to draw the projections of a line in all the four quadrants. See Fig. 9.13.

9.6 Line Contained by One or Both of the Principal Planes

When a line is contained by a plane, its distances of the end points from the plane are zero.

(a) **Line contained by the HP**

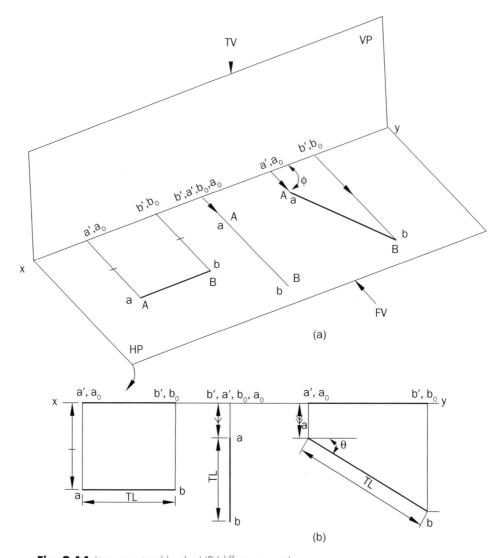

Fig. 9.14 Line contained by the HP (different cases)

Figure 9.14 (*a*) shows the different cases of a line *AB* lying in the HP. Here, in the third part the top view *ab* is inclined to the x-y line at an angle φ, which is the same angle the line *AB* makes with the VP. Here top view *ab* is the line *AB* itself (true length of the line). The front view *ab* lies on x-y line in all the cases.

Figure 9.14 (*b*) shows the orthographic projections of a line *AB* in different cases.

Problem 9.9 A line *EF*, 50 mm long, is contained by HP and is inclined at 45° to the VP, and has its end E 10 mm in front of the VP. Draw its projections.

Solution:

　(i)　Draw a x-y line and at a convenient distance on x-y, draw the projector of end point E.
　(ii)　On this projector, draw front view *e'* and top view *e* on it.

(iii) As the line is contained by the HP, so it will give the true length in the top view. Therefore, through *e* draw a line inclined at 45° to the x-y and along it mark a point *f* at a distance of 50 mm. Then *ef* is the top view of the given line *EF* (TL of the given line *EF*).

(iv) Project the front view e'f'; the same will fall on the x-y. See Fig 9.15.

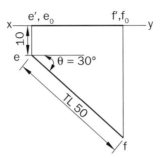

Fig. 9.15 Solution to problem 9.9

(b) **Line contained by the VP**

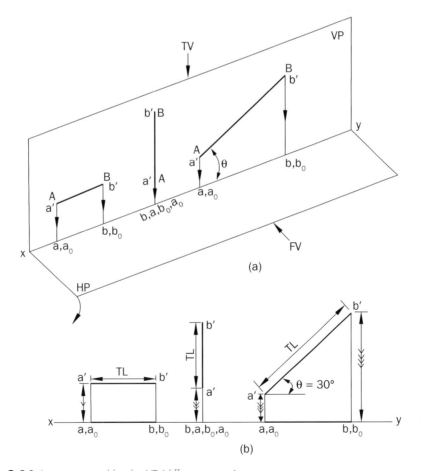

Fig. 9.16 Line contained by the VP (different cases)

Figure 9.16 (*a*) shows the different cases of a line AB lying in the VP. Here, in third part the elevation *a'b'* is inclined at an angle θ with respect to the x-y line, which is the same angle the line *AB* makes with the HP. Here the elevation *a'b'* is the line *AB* itself (TL of the line). The plan *ab* lies on x-y line in all the cases.

Figure 9.16 (*b*) shows the orthographic projections of a line *AB* in different cases.

Problem 9.10 A line EF, 40 mm long, is contained by VP and is perpendicular to the HP, has its end E is at a height of 10 mm from the HP. Draw its projections in first and third quadrants only.

Solution:

(i)   Draw a x-y line and at a convenient distance on x-y, draw the projector of end point *E*.
(ii) On this projector, draw front view *e'* and top view *e* on it.
(iii) As the line is contained by the VP, so its projections on the VP is the true length and perpendicular to x-y.
(iv) In top view *ef* is a point view on x-y. See Fig. 9.17.

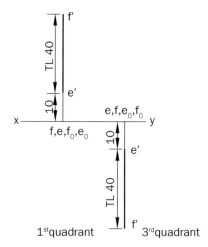

Fig. 9.17 Solution to problem 9.10

Problem 9.11 A line AB, 50 mm long, is contained by VP and is inclined to the HP at 45°, and has its end A 10 mm below the HP. Draw its projections when whole of the line lies in the same quadrant.

Solution:

(i)   Draw the x-y line. At a suitable distance on x-y, mark the projector of end point *A*.
(ii) On this projector, draw front view *a'* and top view *a* on it.
(iii) As the line is contained by the VP, so it will give the true length in the front view. Therefore, through *a'* draw the line inclined at 45° to the x-y and along it mark a point *b'* at a distance of 50 mm. Then *a'b'* is the front view of the given line *AB* (True Length).
(iv) Project the top view *ab*, will be a line on the x-y. See Fig. 9.18.

(c) **Line contained by both HP and VP**

Figure 9.19 (*a*) shows when a line *AB* is contained by both HP and VP. Its front view *a'b'* and top view *ab* of a line *AB* both lie on x-y line.

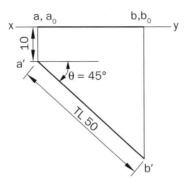

Fig. 9.18 Solution to problem 9.11

Both front view *a'b'* and top view *ab* represent the true length of the line *AB*.

Figure 9.19 (*b*) illustrates the orthographic projections of the line *AB*.

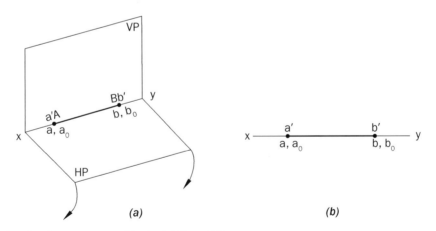

Fig. 9.19 Line contained by both HP and VP

Problem 9.12 A line *AB*, 50 mm long, is contained by both HP and VP. Draw its projections.

Solution:
As the line is contained by both HP and VP, its front view *a'b'* and top view *ab* will show its true length. See Fig. 9.20.

TL 50

a′　　　　　　　　　　　　　b′

x　　　　　　　　　　　　　　　　　　　　　y

　　a, a₀　　　　　　　　b, b₀

Fig. 9.20 Solution to problem 9.12

9.7 Line Inclined to Both HP and VP

Figure 9.21 (*a*) shows, when a line *AB* is inclined to both HP and VP, its projections are shorter than the true lengths and inclined to x-y line at angles greater than the true inclinations. The angles θ' and φ' are known as apparent angles.

Figure 9.21 (*b*) illustrates the orthographic projections of the line *AB*. To determine the true length (*TL*) and inclinations of a line with HP and VP, either of the following three methods is used here.

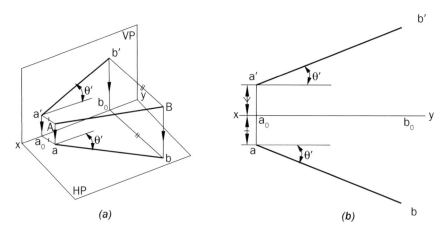

(*a*)　　　　　　　　　　　　　　　(*b*)

Fig. 9.21 Line inclined to both HP and VP

(a) Rotation Method

When a line is inclined to both HP and VP, its true lengths and true inclinations can neither be obtained in front view or top view. In Fig. 9.21 (*a*), the line AB is inclined at θ' to the HP and φ' to the VP.

(i) Fix the locus of the end *B* in front view and top view by drawing the lines *pp'* and *qq'* through *b'* and *b* parallel to the x-y line, respectively.

(ii) Draw the projections of the line *AB*, assuming it to be parallel to the VP and inclined to the HP at an angle θ. Here, for obtaining the true length of the line in front view, the top view *ab* is to be rotated and made parallel to x-y line. Keeping the end *A* fixed and *ab* radius, draw an arc that intersects it at b_1 (by drawing the line ab_1 parallel to x-y). Then draw the vertical projector $b_1 b_1'$ so that it intersect the horizontal locus (pp') of *b'* at b_1'. Now, join $a'b_1'$; it gives the true length of the line *AB* in the front view and the angle made with $a'b_1'$ is known as the true angle of inclination (θ) with the HP. Generally, it is denoted by angle θ. See Figs. 9.22 (*a*) and 9.22 (*b*).

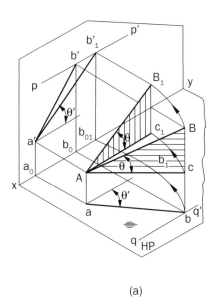

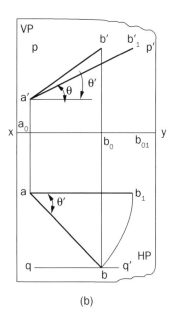

(a)   (b)

Fig. 9.22 Determination of TL on HP and true inclination θ with HP by rotation method

(iii) Similarly, draw the projections of the line *AB*, assuming it to be parallel to the HP and inclined to the VP at angle φ. Here the front view *a'b'* is to be made parallel to x-y. Again with centre *a'* and *a'b'* radius, draw an arc that intersects it at b_2' (by drawing the line $a'b_2'$ parallel to x-y line). Drop the perpendicular from b_2' to b_2. Join ab_2, Which shows the true length of the line *AB* in the top view and the angle made with ab_2 is known as true angle of inclination with the VP. It is denoted by angle φ. See Figs. 9.23 (*a*) and 9.23 (*b*).

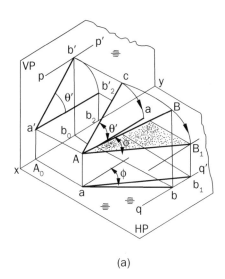

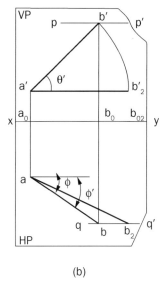

(a)   (b)

Fig. 9.23 Determination of TL on VP and true inclination φ with VP by rotation method

(iv)   Figure 9.24 shows the true length and true inclinations, when the line *AB* is inclined to both HP and VP.

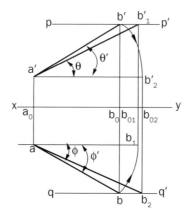

Fig. 9.24 Determination of TL on HP and VP and true inclination θ and φ with HP and VP, respectively, by rotation method

(b) Trapezoid Method

Figure 9.25 (*a*) shows when a line *AB* is inclined to both HP and VP. It is seen here that by rotating AB about the front view *a'b'* into the VP, $A_1 B_1$ is obtained, i.e., the trapezoid a'*ABb*' takes up the new position $a'A_1 B_1 b'$ in the VP. Since it shows the true shape, so $A_1 B_1$ is the true length of the line *AB*.

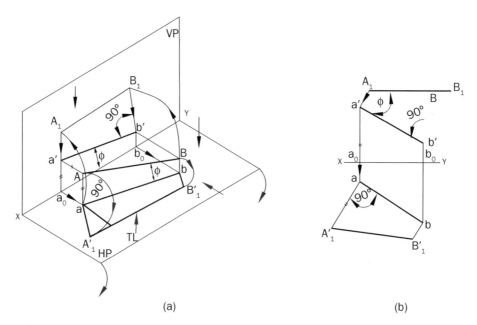

(a)   (b)

Fig. 9.25 Determination of TL on HP and VP and true inclination θ and φ with HP and VP, respectively, by trapezoid method

Similarly, the trapezoid $aABb$ takes up another new position $aA_1'B_1'b$ in the HP. Since it shows the true shape, so $A_1'B_1'$ is the true length of the line AB.

Figure 9.25 (b) illustrates the orthographic projections of the line AB. Here $a'A_1$ and $b_1'B_1$ are drawn perpendicular to $a'b'$ and are equal to a_0a and b_0b, respectively, giving the trapezoid $a'A_1 B_1 b'$. The true length of the line AB is given by $A_1 B_1$ and makes angle φ between $a'b'$ and $A_1 B_1$. (True inclination with VP.)

Similarly, aA_1' and bB_1 are drawn perpendicular to the top view ab and are made equal to $a_0 a'$ and $b_0 b_0'$ respectively, giving the trapezoid $aA_1'B_1'b$. Here $A_1'B_1'$ gives the true length of the line AB and makes angle θ between ab and $A_1'B_1'$. (True inclination with HP.)

The basic differences between the rotation method and trapezoid method are as follows:

1. In rotation method one end of the line is rotated about the other end, which is kept fixed. For example in first quadrant, the true length is obtained on the same reference plane to which the line is parallel, and true inclination with HP and VP is obtained is obtained in elevation and plan respectively.

2. In trapezoid method the whole line is rotated about one of its view (either front view or top view) and made to lie in the respective reference plane. For example, in first quadrant, if the line is rotated about the front view (elevation) it will be made to lie on the VP; similarly if line is rotated about the top view (plan), then the line will be made to lie in the HP. This is the reason that true inclination θ and φ are obtained with the top view and front view unlike the rotation method with front view and top view respectively.

(c) **Auxiliary Plane Method**

This method will be dealt in chapter 11 on Auxiliary Projections.

Problem 9.13 A line AB has its end A 15 mm above the HP and 20 mm in front of the VP, end B 40 mm above the HP and 50 mm in front of the VP. The distance between the end projectors is 45 mm. Draw the projections of the line and find out true length and true inclinations with HP and VP by using (i) Rotation Method and (ii) Trapezoid Method.

Solution:

Rotation Method

(i)   Draw the x-y line and locate the projections of two ends of the line on two projectors drawn perpendicular to x-y and 45 mm apart.

(ii) Join $a'b'$ and ab to obtain the front view and top view of the line, respectively.

(iii) Since the angles θ and φ made by the line with the HP and VP, respectively, are to be found out, so find out true length (TL) of the line in front view (elevation) and top view (plan). Fix the locus of the end B in front view and top view by drawing lines pp' and qq' through b' and b parallel to the x-y, respectively.

    (a) To do so, with centre a and radius ab, draw an arc that intersects it at b_1 (by drawing the line ab_1 parallel to x-y). Then drop the perpendicular from b to b_1'. Join $a_1'b_1'$, gives the true length of the line AB in the front view and the angle made with the $a'b_1'$ is known as the true angle of inclination with the HP.

    (b) Similarly, find out the true length of the line AB in the top view and the angle made with the ab is known as the true angle of inclination with the VP. See Fig. 9.26.

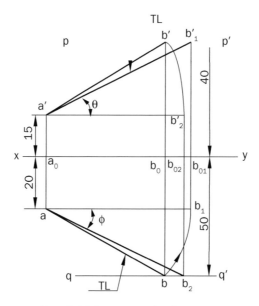

Fig. 9.26 Solution to problem 9.13 (rotation method)

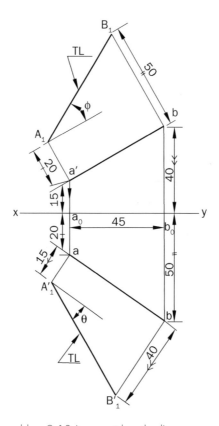

Fig. 9.27 Solution to problem 9.13 (trapezoid method)

Trapezoid Method

(i)   Draw the x-y line and locate the projections of two ends of the line on two projectors drawn perpendicular to x-y and 45 mm apart.

(ii) Join $a'b'$ and ab to obtain the front view and top view of the line, respectively.

(iii) Draw perpendiculars to the front view (elevation) $a'b'$ at the each end a' and b'. Cut along the perpendicular drawn through a', distance equal to $a\,a_0$ to locate the point A_1. Similarly cut along the perpendicular drawn through b', distance equal to $b\,b_0$ to locate the point B_1. Join $A_1\,B_1$, gives the true length of the line AB and makes angle φ with the $a'b'$.

(iv) Repeat the above procedure by drawing perpendiculars at a and b and mark along, the distances equal to $a\,a_0'$ and $b\,b_0'$, respectively, to obtain the points A_1' and B_1'. Join $A_1'B_1'$, gives the true length of the line AB and makes angle θ with the ab. See Fig. 9.27.

Problem 9.14 A line AB has its end A 15 mm above the HP and 20 mm in front of the VP, end B 40 mm below the HP and 30 mm behind the VP. The distance between the end projectors is 45 mm. Draw the projections of the line and find out true length and true inclinations with HP and VP by using (i) Rotation Method and (ii) Trapezoid Method.

Solution:

Rotation Method
Repeat the same procedure, as described in the previous problem 9.13. See Fig. 9.28.

Trapezoid Method

(i)   Draw the x-y line and locate the projections of two ends of the line on two projectors drawn perpendicular to x-y and 45 mm apart.

(ii) Join $a'b'$ and ab to obtain the front view and top view of the line, respectively.

(iii) Since the points a' and b' are on the opposite sides of the x-y, the distances from the x-y are marked in the opposite directions along the perpendicular drawn at a' and b'. Cut along the perpendicular drawn through a', distance equal to $a\,a_0$ to locate the point A. Similarly, cut along the perpendicular drawn through b', distance equal to $b\,b_0$ to locate the point B_1. Join $A_1\,B_1$ and it gives the TL of the line AB and makes angle φ with the $a'b'$.

(iv) Repeat the above procedure by drawing perpendiculars at a and b and cut off the distances equal to $a_1\,a'$ and $b_0\,b'$, respectively, to obtain the points A_1' and B_1'. Join $A_1'B_1'$ and it gives the TL of the line AB and makes an angle θ with the ab. See Fig. 9.29.

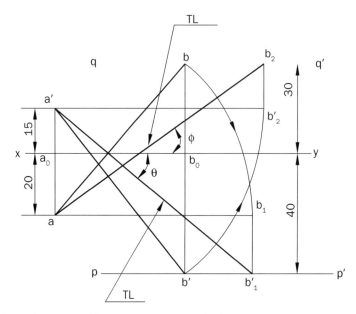

Fig. 9.28 Solution to problem 9.14 (rotation method)

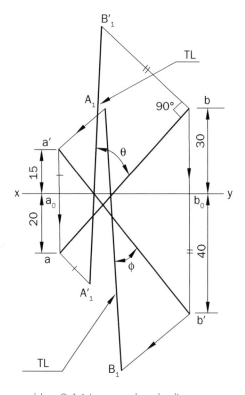

Fig. 9.29 Solution to problem 9.14 (trapezoid method)

Problem 9.15 A line *AB*, 50 mm long has its end *A* 15 mm away from HP and 20 mm away from the VP. It is inclined at 30° to the HP and 45° to the VP. Draw its projections in all the four quadrants. Assume that the whole of the line lies in the same quadrant.

Solution:

First quadrant

(i) Draw the x-y line and locate the front view (elevation) a' and top view (plan) a of the given point *A* on a projector drawn perpendicular to x-y.

(ii) Assuming the line *AB* to be parallel to the VP and inclined to the HP at an angle 30°, draw its front view $a'b_1'$ (equal to *AB*) and project the top view ab_1.

(iii) Again assuming the line *AB* to be parallel to the HP and inclined at 45° to the VP, draw its top view ab_2 (equal to *AB*) and project the front view $a'b_2'$.

(iv) ab_1 and $a'b_2'$ are the lengths of *AB* in the top view and the front view, respectively. Fix the locus of the end *B* in front view and top view by drawing pp' and qq' through b' and b parallel to the x-y line, respectively.

(v) With a' as centre and $a'b_2'$ as radius, draw an arc that intersects the locus pp' at b'. Join $a'b'$.

(vi) With a as centre and ab_1 as radius, draw an arc that intersects the locus qq' at b. Join ab.

(vii) Then $a'b'$ and ab are the required front view and top view of the line *AB*, respectively.

(viii) If the construction of the given problem is correct, join b' to b, which should be straight vertical line. See Fig. 9.30.

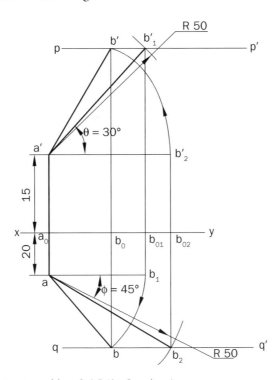

Fig. 9.30 Solution to problem 9.15 (1st Quadrant)

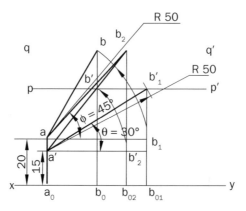

Fig. 9.31 Solution to problem 9.15 (2nd Quadrant)

Second Quadrant

If the line AB is located in the second quadrant, the front view and top view of the line will remain above the x-y line. Using the same procedure as in problem 9.15 (first quadrant), the projections can be drawn. Fig. 9.31 shows the projections of a line AB in the second quadrant.

Third Quadrant

If the line AB is placed in the third quadrant, the front view will be below the x-y line and the top view will be above the x-y line. Fig. 9.32 shows the projections of a line AB in the third quadrant.

Fourth Quadrant

If the line AB is located in the fourth quadrant, the front view and top view will be below the x-y line. Fig. 9.33 shows the projections of a line AB in the fourth quadrant.

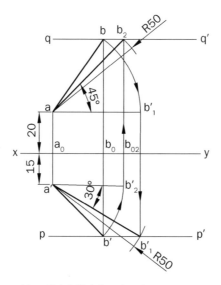

Fig. 9.32 Solution to problem 9.15 (3rd Quadrant)

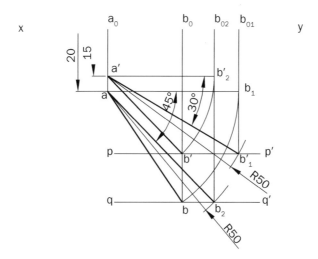

Fig. 9.33 Solution to problem 9.15 (4ᵗʰ Quadrant)

Problem 9.16 A straight line AB, 60 mm long, makes an angle of 45° to HP and 30° to the VP. The end *A* is 15 mm in front of VP and 25 mm above HP. Draw the projections of the line AB.

(*PTU, Jalandhar December 2004*)

Solution: The procedure to draw this problem has already been explained in previous problems. See Fig. 9.34.

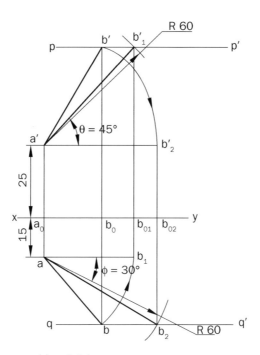

Fig. 9.34 Solution to problem 9.16

Problem 9.17 The end point A of a straight line AB = 36 mm long is 12 mm away from HP and VP and another point B is 24 mm away from HP and VP, respectively. Draw the top view and front view of the straight line AB and determine the true inclination with HP and VP, respectively.

(PTU, Jalandhar December 2003)

Solution:

When whole of the line *AB*, lies in the first quadrant

(i)   Draw the x-y line and locate the front view (elevation) a' and top view (plan) a of the given point A on a projector drawn perpendicular to x-y.

(ii)   Draw two lines pp' and qq' lines 24 mm above and 24 mm below, respectively, parallel to x-y. These lines represent the locus of end B in front view and top view, respectively.

(iii)   With a' as centre and ab_1' as radius (36 mm), draw an arc that intersects the locus pp' at b_1'. Similarly, with a as centre and ab_2 as radius (36 mm *TL*), draw an arc that intersects the locus qq' at b_2.

(iv)   With a as centre and ab_1 as radius, draw an arc that intersects the locus qq' at b. Join ab.

(v)   With a' as centre and ab_2' as radius, draw an arc that intersects the locus pp' at b'. Join $a'b'$.

(vi)   Then $a'b'$ and ab are the required front view and top view of the line AB, respectively.

Here $a'b_1'$ makes true angle of inclination (θ) with HP and ab makes true angle of inclination (φ) with VP. See Fig. 9.35.

When the whole of the line *AB* lies in the third quadrant
Repeat the same procedure, as described above. See Fig. 9.36.

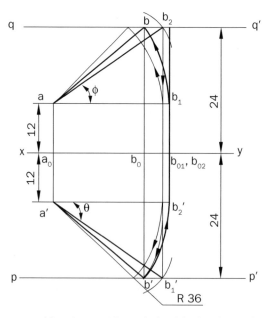

Fig. 9.35 Solution to problem 9.17 (When whole of the line lies in the 1ˢᵗ Quadrant)

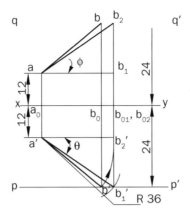

Fig. 9.36 Solution to problem 9.17 (When whole of the line lies in the 3rd Quadrant)

Problem 9.18 A straight line AB, 70 mm long, makes an angle of 45° to the HP and 30° to the VP. The end A is 15 mm in front of the VP and 20 mm above HP. Draw the plan and elevation of the line AB.

(PTU, Jalandhar May 2005, May 2008, May 2009, December 2010)

Solution: The procedure to draw this problem has already been explained in previous problems. See Fig. 9.37.

Problem 9.19. A line *AB*, 60 mm long, has its end *A* both in HP and VP. It is inclined at 45° to the VP and 30° to the HP. Draw its projections in first quadrant only.

Solution: The solution to this problem is self-explanatory. See Fig. 9.38.

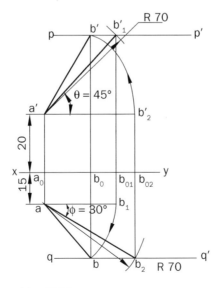

Fig. 9.37 Solution to problem 9.18

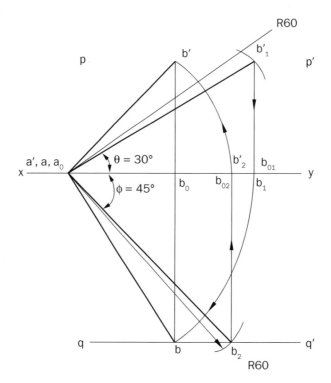

Fig. 9.38 Solution to problem 9.19

Problem 9.20 Plan and elevation of a line AB, 60 mm long, measure 50 mm and 40 mm, respectively. End A is 15 mm above HP and 20 mm in front of the VP. Draw its projections and determine the true inclination with HP and VP, respectively.

Solution:

(i)   Draw the x-y line and locate the projections of end A on a projector drawn perpendicular to x-y.

(ii) Draw ab_1 equal to 50 mm parallel to x-y. From b_1, draw a vertical projector. With a' as centre and radius 60 mm, draw an arc cutting the projector at b_1'. Join $a'b_1'$ and measure the inclination of the line $a'b_1'$ with x-y to find out the true inclination with HP (angle θ).

(iii) Draw $a'b_2'$ equal to 40 mm parallel to x-y. From b_2', draw a vertical projector. With a as centre and radius 60 mm, draw an arc cutting the projector at b_2. Join ab_2 and measure the inclination of the line ab_2 with x-y to find out the true inclination with VP (angle φ).

(iv) Fix the locus of the end B in elevation and plan by drawing lines pp' and qq' through b_1' and b_2 parallel to the x-y line respectively.

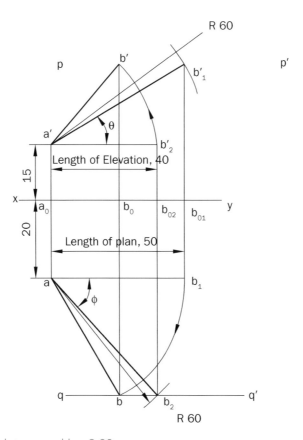

Fig. 9.39 Solution to problem 9.20

(v)   With a' as centre and radius $a'b'_2$, draw an arc that intersects the locus pp' at b'. Join $a'b'$.

(vi)   With a as centre and radius ab_1, draw an arc that intersects the locus qq' at b. Join ab.

(vii) Then $a'b'$ and ab are the required elevation (front view) and plan (top view) of the line AB respectively. As a check, join b' to b, which should be a straight vertical line. See Fig. 9.39.

Problem 9.21 A line AB, 60 mm long, is inclined at 30° to the HP and 45° to the VP. Its mid-point C is 30 mm above the HP and 40 mm in front of the VP. Draw its projections.

Solution:

(i)   Draw the x-y line and locate the projections of mid-point C on a projector drawn perpendicular to x-y.

(ii)   Assume the line to be divided into two equal parts AC and CB. Take part CB (30 mm long), draw the projections of the line $c'b'$ and cb, as described in the previous problems. Then, these projections can be extended to the other side of c and c' to a and a'. (i.e., $cb = ca$ and $c'b' = c'a'$) to obtain the projections of the line AB. See Fig. 9.40.

ENGINEERING DRAWING

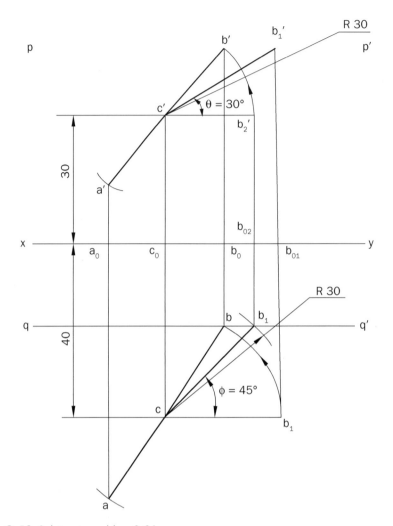

Fig. 9.40 Solution to problem 9.21

Problem 9.22 The auditorium of the college is 50 m long, 30 m wide and 15 m high. A light point is fitted at the centre of the roof and its switch is kept on one of the side walls of the auditorium, 2 m above the floor and 10 m from one of the adjacent walls. Find the actual distance between the light point and its switch.

Solution:

(i) Draw the front view and top view of the auditorium with suitable scale.

(ii) Locate the position of the light point in top view and front view as l and l', respectively.

(iii) Now locate the position of the switch in top view and front view as s and s', respectively. Join $s'l'$ and sl.

(iv) With s as centre and sl as radius, rotate sl to sl_1 such that it is parallel to x-y. Project l to l_1' in the front view. Join $s'l_1'$ to obtain the required actual distance between the light point and the switch. See Fig. 9.41.

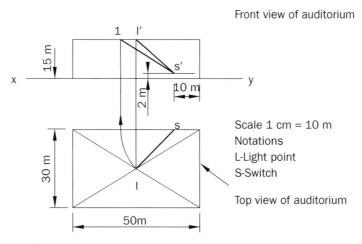

Fig. 9.41 Solution to problem 9.22

Problem 9.23 A room is 10 m long, 6 m wide and 4 m high. An electric bulb hangs in the centre of the ceiling and 1 m below it. A thin straight wire connects the bulb to a switch kept in one of the corners of the room and 1.5 m above the floor. Draw the projections of the wire and determine its TL and slope with the floor.

Solution:

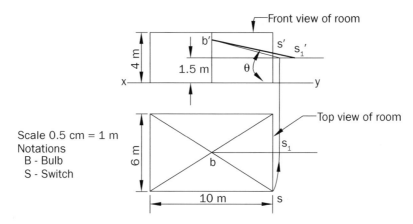

Fig. 9.42 Solution to problem 9.23

(i)   Draw the front view and top view of the room with suitable scale.
(ii)   Locate the position of the bulb in top view and front view as b and b', respectively.
(iii)   Similarly, locate the position of the switch in top view and front view as s and s', respectively. Join $b's'$ and bs.
(iv)   With b as centre and bs as radius, rotate bs to bs_1, such that it is parallel to x-y. Project s to s_1' in the front view. Join $b's_1'$ to obtain the true length of the wire and measure the true angle of inclination of $b's_1'$ line with x-y. See Fig. 9.42.

Problem 9.24 A line *AB*, 60 mm long, is inclined at 45° to the HP and its top view makes an angle of 60° with the VP. The end A is in the HP and 15 mm in front of the VP. Draw its front view and find its true inclination with the VP.

Solution: The solution to this problem is self-explanatory. See Fig. 9.43.

Problem 9.25 A line *AB*, 60 mm long, is inclined to the HP at 30°. Its end A is 10 mm above the HP and 15 mm in front of the VP. Its front view measures 40 mm. Draw the top view of AB and determine its inclination with the VP.

Solution: All the construction lines are retained to make the solution self-explanatory. See Fig. 9.44.

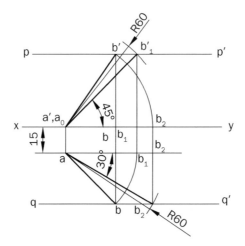

Fig. 9.43 Solution to problem 9.24

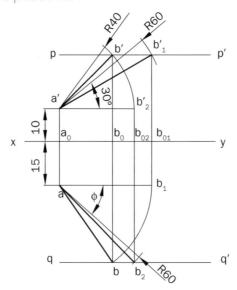

Fig. 9.44 Solution to problem 9.25

Problem 9.26 A line *AB* has its end *A* 15 mm above the HP and 20 mm in front of the VP. The front view of the line is 45 mm long and is inclined at 30° to the x-y line. The top view of the line is inclined at 45° to the x-y line. Draw its projections and find out the true length and true inclination with HP and VP.

Solution:

(i)   Draw the x-y line and locate the projections of end *A* on a projector drawn perpendicular to x-y.

(ii) Through *a'*, draw a line at an angle of 30° to x-y and mark the distance 45 mm to locate the point *b'*.

(iii) Since *b'* is being known, *b'b* can be drawn. Through *a* draw a line at an angle of 45° to intersect the second end projector at *b*. Then *ab* is the required top view.

(iv) Since the angles θ and φ made by the line with the *HP* and *VP*, respectively, are to be found out. Fix the locus of the end *B* in front view and top view by drawing lines *pp'* and *qq'* through *b'* and *b* parallel to the x-y, respectively.

(a) Now to make the line parallel with the HP, rotate the front view *a'b'*, keep *a'* as the centre and radius *a'b'*, draw an arc that intersects the horizontal locus of *a'* at b'_2. Then drop the perpendicular from b'_2 to b_2. Join ab_2, gives the true length and inclination with the VP. (Note: here we have drawn the line length of the line AB in the top view and the angle made with ab_2 is known as true angle.)

(b) Similarly, find out the true length of the line *AB* is the front view and the angle made with the *a'b'* is known as the true angle of inclination with the HP. See Fig. 9.45.

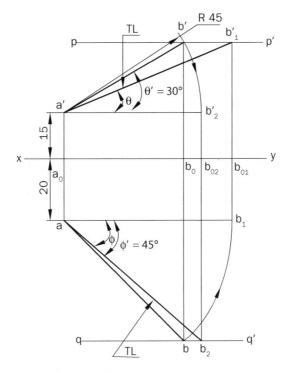

Fig. 9.45 Solution to problem 9.26

Problem 9.27 A line *AB* has its end *A* 20 mm in front of the VP and end *B* 55 mm above the HP. The line is inclined at 30° to the HP while its front view makes an angle of 45° to the x-y line. Draw its projections, when its top view is 50 mm long. Find the true length and true angle of inclination with the VP.

Solution: The procedure to draw this problem has already been explained in previous problems. See Fig. 9.46.

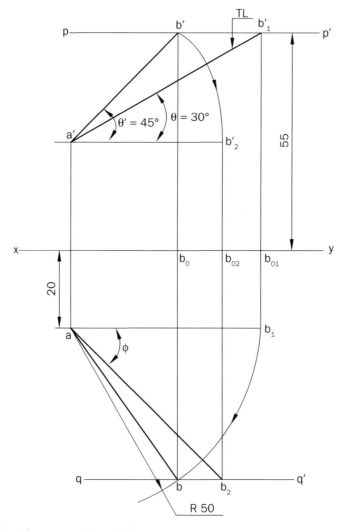

Fig. 9.46 Solution to problem 9.27

Problem 9.28 A line *AB*, 60 mm long, has its end *A* 15 mm above the HP and 20 mm in front of the VP. The front and top views of the line AB are inclined 45° and 30°, respectively. Draw its projections and find the true angles of inclination with HP and VP.

Solution:

(i) Draw the x-y line and locate the projections of end *A* on a projector drawn perpendicular to x-y.

(ii) Through *a'*, draw a line at an angle of 45° to x-y. Similarly through *a*, draw a line at angle of 30° to x-y.

(iii) Since the location of the end *B* cannot be found out directly. For this take any point *C* on the line *AB*. Then *a'c'* and *ac* can be drawn.

(iv) Since the angles θ and φ made by the part line or full length line with the *HP* and *VP*, respectively will remain same, so either of θ or φ for AC line should be found. To do so, with centre *a'* and radius *a'c'*, draw an *arc* that intersects its horizontal locus at c_2'. From c_2', draw a vertical projector. Through *c* draw a line parallel to x-y, which will intersect at a point c_2. Join *ac* and extend ac_2 to b_2 (True length of the line). Measure the inclination of the line ab_2 with x-y to find the true inclination with *VP* (angle φ).

(v) Repeat the same procedure as described in the previous problems. See Fig. 9.47.

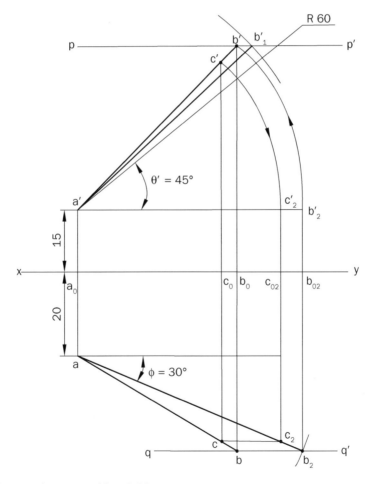

Fig. 9.47 Solution to problem 9.28

Problem 9.29 A straight line PQ, 65 mm long, makes an angle of 50° to the HP and 30° to the VP. The end P of the straight line PQ lies in VP and is 20 mm above HP. Draw the projections of the line PQ.

(PTU, Jalandhar December 2005, May 2011)

Solution: All the construction lines are retained to make the solution self-explanatory. See Fig. 9.48.

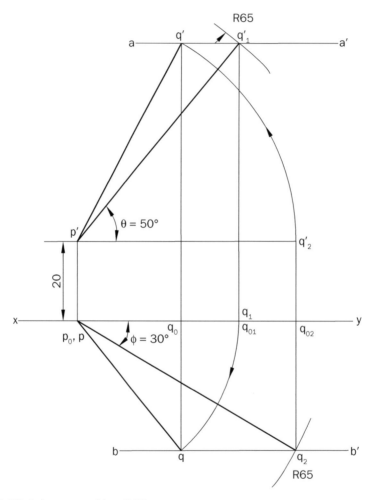

Fig. 9.48 Solution to problem 9.29

Problem 9.30 Three lines, *oa, ob* and *oc*, are 25, 45 and 60 mm long, respectively. Each makes an angle of 120° with the other two. The shortest line being vertical, the figure represents the top view of the three rods OA, OB and OC where each A, B and C are on the ground, while O is 100 mm above it. Draw the front view and determine the true length of each rod and its inclination with the ground.

(PTU, Jalandhar December 2005)

Solution:

(i) Draw lines *oa*, *ob* and *oc* meeting at point *o*, an angle of 120° each. Keep the line *oa* to be vertical. Mark the distances of 25 mm, 45 mm and 60 mm long for *oa*, *ob* and *oc*, respectively.

(ii) Draw the projections from points *a, b* and *c* to meet x-y line at a', b' and c', respectively.

(iii) Project o to o' and set-off a distance of 100 mm above the x-y line.

(iv) Join a', b' and c' with o'.

(v) Since none of the three lines of the rod represents true lengths, so true lengths can be found out by rotation method as described in the previous problems. See. Fig. 9.49.

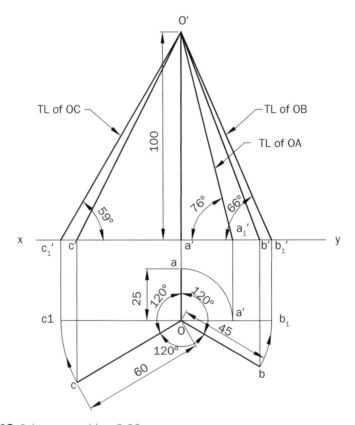

Fig. 9.49 Solution to problem 9.30

Problem 9.31 The top view of a 75 mm long line AB measures 65 mm, while the length of its front view is 50 mm. Its one end *A* is in the HP and 20 mm in front of the VP. Draw the projections of *AB* and determine its inclinations with the HP and the VP.

(PTU, Jalandhar May 2006)

Solution: The solution to this problem has already been explained in previous problem 9.20. See Fig. 9.50.

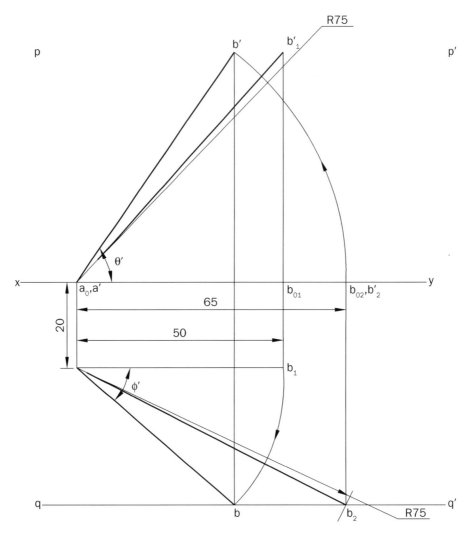

Fig. 9.50 Solution to problem 9.31

9.8　Line Contained by a Profile Plane (PP) or Line Contained by a Plane, Perpendicular to Both HP and VP

As shown in Fig. 9.51 (*a*), two principal planes are at right angles to each other and profile plane (PP) is perpendicular to both of them. For a line contained by the profile plane, the sum of the inclinations of a line with two planes, i.e., θ and φ can never be more than 90°, but $\theta + \varphi = 90°$.

A line *AB* is inclined at θ to the HP and φ (equal to $90° - \theta$) to the VP. The front view *a'b'* and the top view *ab* are both collinear and perpendicular to x-y and are shorter than *AB*. However, the true length of the line *AB* may be obtained on the profile plane (PP). Fig. 9.51 (*b*) shows the orthographic projections of the line *AB*.

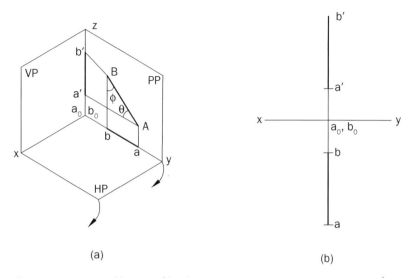

(a)   (b)

Fig. 9.51 Line contained by a profile plane (PP)

Problem 9.32 A line AB has its end A 20 mm away from the HP and 40 mm away from the VP, end B 50 mm away from the HP and 15 mm away from the VP. The line AB is contained by the profile plane. Draw its projections in first and third quadrants only.

Solution:

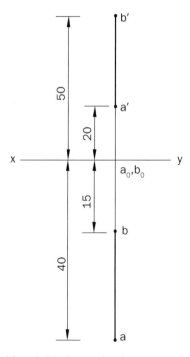

Fig. 9.52 Solution to problem 9.32 (1st quadrant)

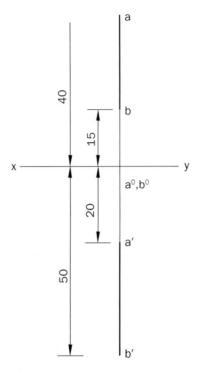

Fig. 9.53 Solution to problem 9.32 (3rd Quadrant)

As the line *AB* is contained by the profile plane, both front view (elevation) *a'b'* and top view (plan) *ab* are collinear and perpendicular to the x-y.

   (i)   Draw the x-y line and locate the projections of two ends of the line on the projectors perpendicular to x-y.

   (ii)   Mark along the projector the front view *a'b'* and top view *ab* of the two ends of the line AB.

  (iii) Join *a'b'* and *ab*. Then *a'b'* and *ab* are the required elevation (front view) and plan (top view) of the line AB, respectively as shown in Fig. 9.52.

   See Fig. 9.53, follow the above mentioned steps to draw the projections of a line in third quadrant.

Problem 9.33 Line *AB* lies in profile plane. *A* is 40 mm in front of VP and 20 mm above HP. End B is 20 mm in front of VP and 40 mm above HP. Find its true length and inclinations.

(PTU, Jalandhar December 2002)

Solution:

Method I

(i) Obtain the projections by repeating steps (i) to (iii) of previous problem 9.32. Once the projections are obtained, then true length and true inclinations can be obtained using trapezoid method in the following steps.

(ii) Draw perpendiculars to the front view (elevation) $a'b'$ at the ends a' and b'. Cut along the perpendicular drawn through a', distance equal to $a_0\,a$ to locate the point A_1. Similarly, cut along the perpendicular drawn through b', distance equal to $b_0\,b$ to locate the point B_1. Join $A_1\,B_1$ gives the true length of the line AB and it makes angle φ with the $a'b'$.

(iii) Repeat the above procedure by drawing perpendiculars at a and b and mark along the distances equal to $a_0\,a'$ and $b_0\,b'$, respectively, to obtain the points $A_1{}'$ and $B_1{}'$. Join $A_1{}'\,B_1{}'$ gives the true length of the line AB and makes angle θ with the ab as shown in Fig. 9.54.

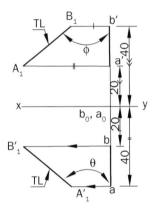

Fig. 9.54 Solution to problem 9.33 (Method I)

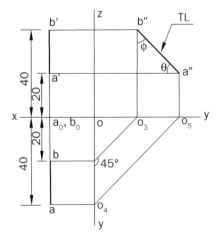

Fig. 9.55 Solution to problem 9.33 (Method II)

Method II

(i)   Repeat steps (*i*) to (*iii*) of previous problem 9.32.

(ii) Draw another reference line *yz* perpendicular to x-y, which will intersect at *o*.

(iii) Through the point *b*, draw a line parallel to x-y to intersect *yz* at o_2. From o_2, draw an angle of 45°, which will intersect x-y at o_3. Locate the side view *b″* as the intersection point between the projectors drawn from o_3 and b′.

(iv) Repeat the above procedure for the end point *A*.

(v)   Join *a″b″*, gives the true length of the line *AB* and makes true inclination θ and φ with *oy* and *oz* reference lines, respectively. See Fig. 9.55.

9.9 Traces of a Line

The point of intersection of a given line, produced or extended if necessary, where it meets one of the principal planes is called its trace. The point of intersection of a given line if extended with the HP is called the horizontal trace. It is usually denoted by HT or H. The point of intersection of a given line if extended with the VP is called the vertical trace and is usually denoted by VT or V′. The different positions of a line in space are as follows:

(a) **Line parallel to both HP and VP**

When a straight line *AB* is parallel to both HP and VP, the line will neither meet the HP nor VP, even when it is extended. Therefore, there are no traces for the line *AB*. See Figs. 9.56 (*a*) and 9.56 (*b*).

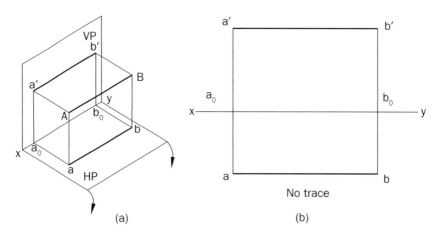

Fig. 9.56 Line parallel to both HP and VP

(b) **Line inclined to one plane and parallel to the other**

(i)   *Line inclined to the HP and parallel to the VP*

Figure 9.57 (*a*) shows a line *AB* is inclined to the HP and parallel to the VP. The line *BA* when extended will meet the HP at the HT; so it has got only HT, but no VT. Fig. 9.57 (*b*) shows the projection of a line *AB* with the horizontal trace (HT) only.

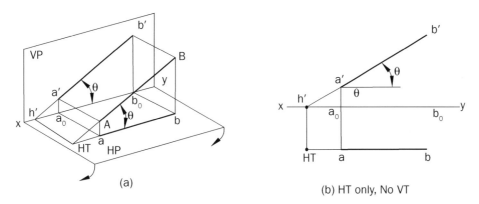

Fig. 9.57 Line inclined to the HP and parallel to the VP

(a)

(b) HT only, No VT

(ii) *Line inclined to the VP and parallel to the HP*

Figure 9.58 (*a*) shows a line *AB* is inclined to the VP and parallel to the HP. The line *BA* when produced meets the VP at the VT; so it has got only VT, but no HT. Fig. 9.58 (*b*) shows the projections of a line *AB* with the vertical trace (VT) only.

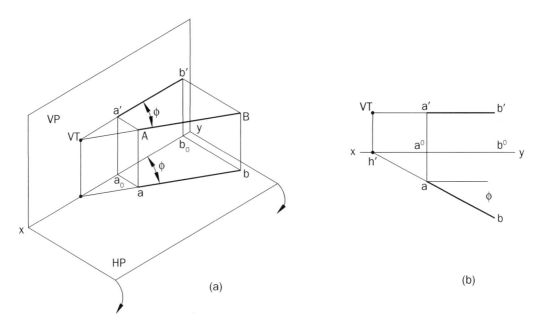

(a)

(b)

Fig. 9.58 Line inclined to the VP and parallel to the HP

(c) **Line perpendicular to one of the planes**

(i) *Line perpendicular to the HP*

Figure 9.59 (*a*) shows the pictorial view of a line *AB*, perpendicular to the HP. If extend the line *AB*, it will meet the HP; so it has got only HT but no VT. Fig. 9.59 (*b*) shows the orthographic projections of a line AB with the horizontal trace (HT) only.

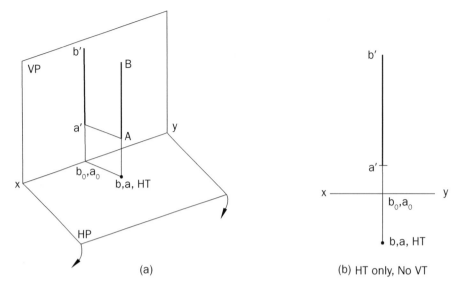

Fig. 9.59 Line perpendicular to the HP

(ii)   *Line perpendicular to the VP*

Figure 9.60 (*a*) illustrates the pictorial view of a line *AB*, perpendicular to the VP. The line *BA* when extended meets the VP; so it has got only VT, but no HT. Fig. 9.60 (*b*) shows the orthographic projections of a line *AB* with the vertical trace (VT).

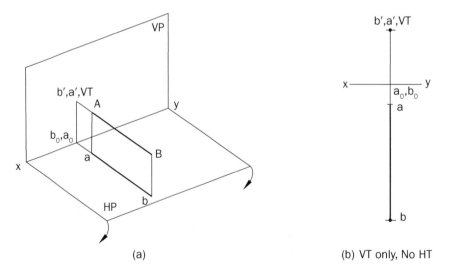

Fig. 9.60 Line perpendicular to the VP

(d) Line contained by one or both of the principal planes

(i) *Line contained by the HP*

Figure 9.61 (*a*) illustrates the pictorial view of a line AB, lying in the HP with their different cases. Fig. 9.61 (*b*) shows the orthographic projections of a line AB with their traces.

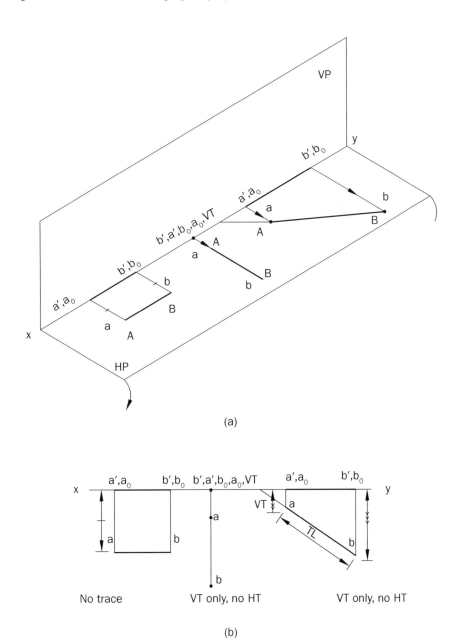

(a)

No trace   VT only, no HT   VT only, no HT

(b)

Fig. 9.61 Line contained by the HP (different cases)

(ii) *Line contained by the VP*

Figure 9.62 (*a*) shows the different cases of a line *AB*, lying in the VP. In these cases, it may have only one trace or no trace. Fig. 9.62 (*b*) shows the projections of a line *AB* with their traces.

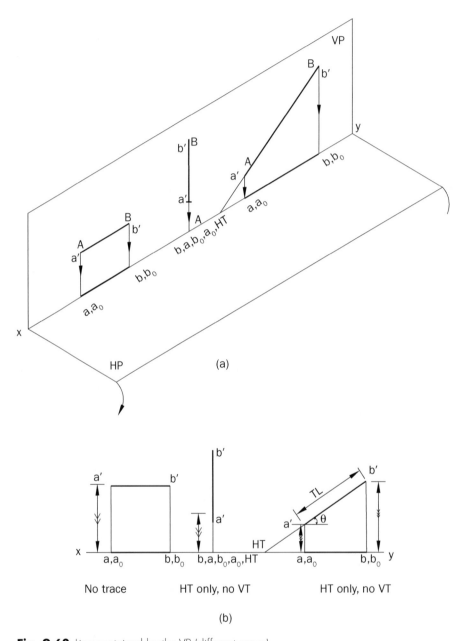

Fig. 9.62 Line contained by the VP (different cases)

(iii) *Line contained by both HP and VP*

See Fig. 9.63 (*a*). Since the line *AB* is contained by both HP and VP, so it will not have any trace. Fig. 9.63 (*b*) shows the projections of a line *AB* with no trace.

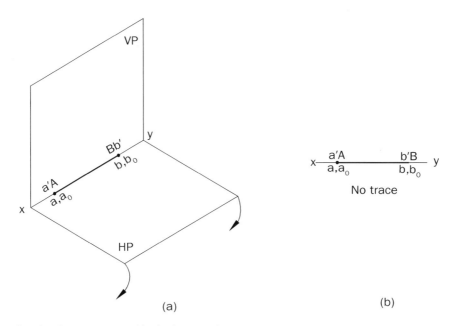

(a)　　　　　　　　　　　　　　　　(b)

Fig. 9.63 Line contained by both HP and VP

(e) **Line inclined to both HP and VP**

There are two methods to find the traces of given line AB, which is inclined to both HP and VP.

Method I

Figure 9.64 (*a*) shows a line *AB*, inclined to both the principal planes and its traces marked clearly. Its traces may be determined as described below:

(i)　　Draw the projections *a'b'* and *ab*.

(ii)　　Extend the front view *b'a'* to meet x-y at a point *h'*.

(iii)　　Draw a projector from *h'* to intersect the line b-a extended at HT or H of the line.

(iv)　　Similarly, extend the line b-a to meet x-y at a point v.

(v)　　Draw a projector from v to intersect the line b'-a' extended, at VT or V' of the line. See Fig. 9.64 (*b*).

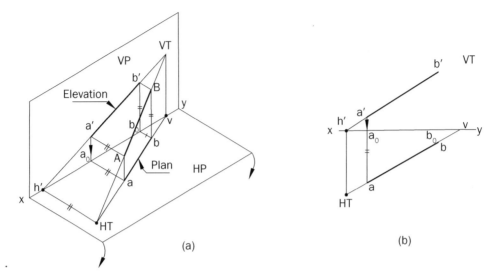

Fig. 9.64 To determine the traces of a line inclined to both HP and VP (Method I)

Method II

Figure 9.65 (*a*) shows a line *CD*, inclined to both the principal planes. For its traces by this method,

(i)　Determine its true length CD from the front view $c'd'$ by trapezoid method. The point of intersection between $c'd'$ produced and CD produced will represent the VT of the line.

(ii)　Similarly, determine the length $C_1'D_1'$ from the top view cd. Produce these lines to intersect at the HT of the line. See Fig. 9.65 (*b*).

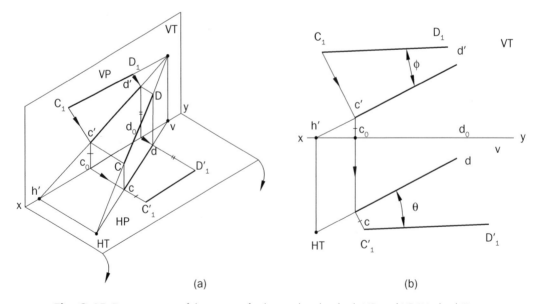

Fig. 9.65 Determination of the traces of a line inclined to both HP and VP (Method II)

(f) Line contained by a profile plane

As shown in Fig. 9.66 (*a*), when a line *AB* is placed in such a way that the sum of its inclinations with both HP and VP is 90°, then it is not possible to find the traces by the method I. Method II must be employed to find its traces as shown in Fig. 9.66 (*b*).

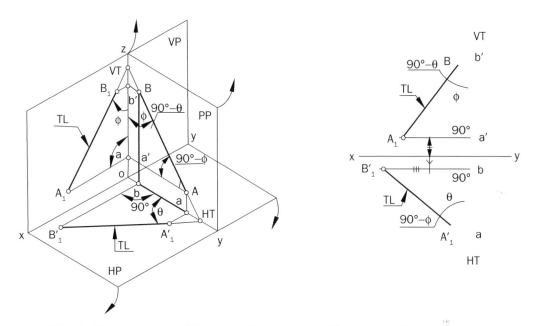

Fig. 9.66 Determination of the traces of a line contained by a profile plane

Note: To have the complete knowledge of projections of lines, the student must acquaint himself or herself with TL, θ, φ, HT and VT. All these five parameters may or may not exist in the particular position of a line in the space.

Problem 9.34 A line *AB*, 50 mm long, has its end *A* is 25 mm away from the HP and 15 mm away from the VP. The line is inclined to the HP at 30° and is parallel to the VP. Draw its projections and determine its traces in first and second quadrant only.

Solution:

(i)   For drawing the projections of a line, repeat the same steps as described in the problem 9.2.

(ii) To determine its trace, produce the line *b'a'* so as to meet the x-y at *h'*. Locate the HT as the point of intersection between a projector drawn from *h'* and top view *ba* (or its extension). See Fig. 9.67.

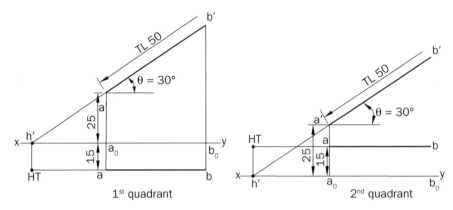

Fig. 9.67 Solution to problem 9.34

Problem 9.35 End *A* of a line *AB* is 30 mm above the HP and 10 mm in front of the VP and end *B* is 15 mm above the HP and 40 mm in front of the VP. The distance between the end projectors is 40 mm. Draw its projections and determine its traces.

Solution:

  (i)   Draw the projections of the line *a'b'* and *ab*.
  (ii) To determine its traces, produce the line *ba*, so as to meet the x-y at *v*. Locate the VT as the point of intersection between a projector drawn from *v* and front view *b'a'* (or its extension).
  (iii) Similarly, produce the line *a'b'*, so as to meet the x-y line at *h'*. Locate the HT as the point of intersection between a projector drawn from *h'* and top view *ab* (or its extension). See Fig. 9.68.

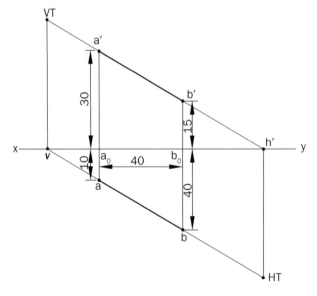

Fig. 9.68 Solution to problem 9.35

Problem 9.36 End *A* of a line *AB* is 40 mm below the HP and 15 mm behind the VP. End *B* is 10 mm above the HP and 30 mm in front of the VP. The distance between the end projectors is 45 mm. Draw its projections and determine its traces.

(PTU, Jalandhar May 2010)

Solution: The solution to this problem is self-explanatory. See Fig. 9.69 for Method I and Fig. 9.70 for Method II.

Problem 9.37 A line *AB* has its end *A* 15 mm above the HP and 20 mm in front of the VP. End *B* is 30 mm above the HP and 50 mm in front of the VP. The distance between the end projectors is 45 mm. Draw the projections of the line and determine its TL, θ, φ, HT, VT by using (i) Rotation Method and (ii) Trapezoid Method.

Solution:

Rotation Method

(i)   For drawing the projections of a line, true length, true angle of inclination with HP and VP, follow the same procedure as described in the problem 9.13 for the Rotation Method.

(ii) To determine its traces, produce the line *ba*, so as to meet x-y at *v*. Locate the VT as the point of intersection between a projector drawn from *v* and front view *b'a'* (or its extension).

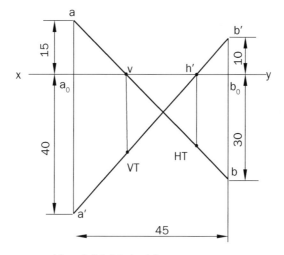

Fig. 9.69 Solution to problem 9.36 (Method I)

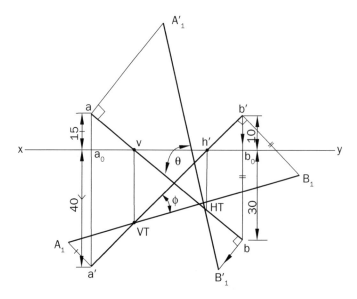

Fig. 9.70 Solution to problem 9.36 (Method II)

(iii) Similarly, produce the line $b'a'$, so as to meet x-y at h'. Locate the HT as the point of intersection between a projector drawn from h' and top view ba (or its extension) as shown in Fig. 9.71.

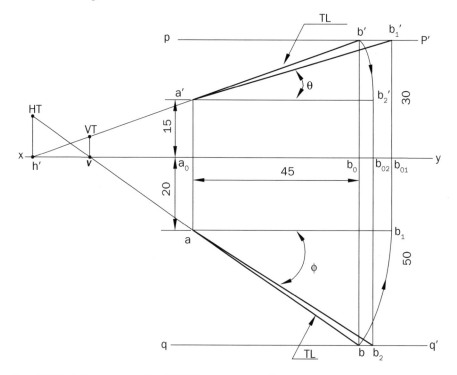

Fig. 9.71 Solution to problem 9.37 (rotation method)

Trapezoid Method

All the construction lines are retained to make the solution self-explanatory. See Fig. 9.72.

Problem 9.38 A line *EF* is contained by a profile plane. Its end *E* is 45 mm behind the VP and 10 mm below the HP and end *F* is 10 mm behind the VP and 50 mm below the HP. Draw its projections and determine its TL, θ, φ, HT and VT.

Solution: All the construction lines are retained to make the solution self-explanatory. See Fig. 9.73.

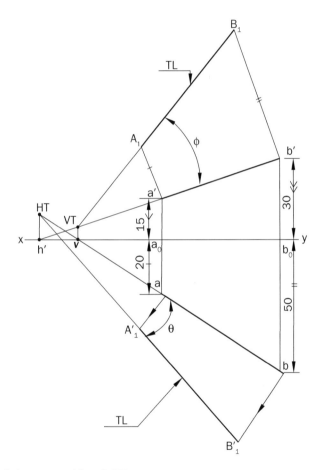

Fig. 9.72 Solution to problem 9.37

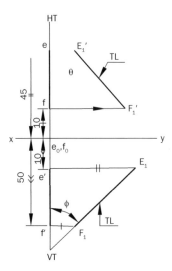

Fig. 9.73 Solution to problem 9.38

Problem 9.39 A line *AB* has its end *A* 30 mm above the HP and 10 mm in front of the VP. End *B* is 15 mm above the HP and 40 mm in front of the VP. The distance between end projectors is 40 mm. Draw projections of the line and determine its TL, θ, φ, HT, VT by using (i) Rotation Method and (ii) Trapezoid Method.

Solution:

Rotation Method
Repeat the same procedure as described in the previous problems. See Fig 9.74.

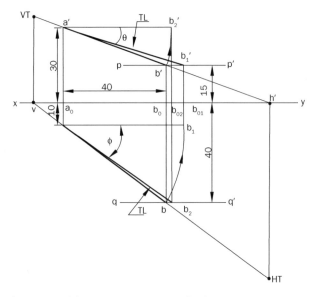

Fig. 9.74 Solution to problem 9.39 (Rotation Method)

Trapezoid Method

All the construction lines are retained to make the solution self-explanatory. See Fig. 9.75.

Problem 9.40 A line *AB*, 60 mm long, has its end *A* 10 mm above the HP and 15 mm in front of the VP. Whereas end B is 40 mm above the HP and 60 mm in front of the VP. Draw its projections and determine the true inclination with HP and VP, respectively.

Solution: Repeat the same procedure as described in the previous problem 9.17. See Fig. 9.76.

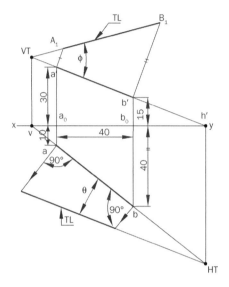

Fig. 9.75 Solution to problem 9.39 (Trapezoid Method)

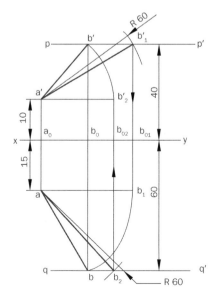

Fig. 9.76 Solution to problem 9.40

Problem 9.41 A line *AB* has its end *A* 15 mm above the HP and end *B* is 60 mm in front of the VP. The front view of the line is inclined at 45° to the x-y line. The horizontal trace (HT) of the line is 12 mm in front of the VP and its vertical trace (VT) is 15 mm below the HP. Draw its projection and determine its TL and true angle of inclination with HP and VP, respectively.

Solution:

(i)   Draw the x-y line and locate the horizontal trace (*HT*) by drawing perpendicular to the x-y line at a distance of 12 mm. Through the point *h'*, draw a line inclined at angle of 45° to the x-y line.

(ii)   Mark vertical trace (*VT*) of the line such that its perpendicular distance from *v* is 15 mm of the inclined line passing through *h'*.

(iii)   Locate the projection *a'* (15 mm above *HP*) of end *A* on a projector drawn perpendicular to x-y and similarly locate the projection *a* of end *A* on a projector passing through *HT* and *v*.

(iv)   Locate the projection *b* (60 mm in front of *VP*) of end *B* on a projector drawn perpendicular to x-y. Join *a* with *b*, which gives the required top view. Also mark the point *b'*. Then *a'b'* is the required front view.

(v)   Repeat the same procedure as described in the previous problems. See Fig. 9.77.

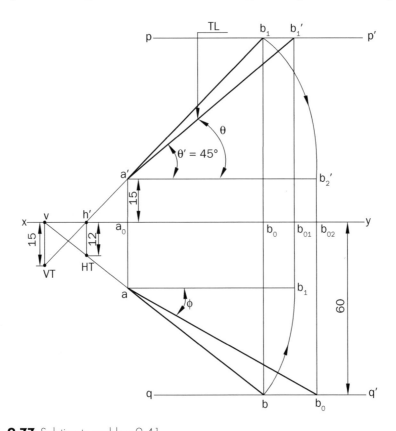

Fig. 9.77 Solution to problem 9.41

Problem 9.42 A line *AB* has its end *A* 15 mm above the HP and 20 mm in front of the VP. End *B* is 50 mm in front of the VP. The vertical trace (VT) is 7 mm above the HP. Draw the projections of the line if the distance between end projectors is 45 mm and find its TL, θ, ϕ, HT.

Solution: The solution to this problem is self-explanatory. Value of TL, θ, ϕ, HT should be given in answer.

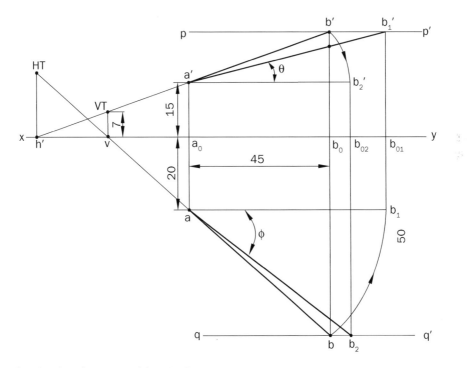

Fig. 9.78 Solution to problem 9.42

Problem 9.43 A line *AB* has its end *A* 20 mm in front of the VP and end *B* is 60 mm in front of the VP. The HT of the line is 10 mm in front of the VP. The length of its front view is 45 mm and makes an angle of 45° to x-y line. The length of its top view is 55 mm. Draw the projections of the line and find it TL, θ, ϕ, VT.

Solution: All the construction lines are retained to make the solution self-explanatory. See Fig. 9.79

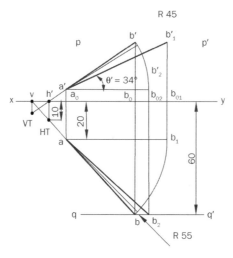

Fig. 9.79 Solution to problem 9.43

Problem 9.44 The front view of a line *AB* measures 65 mm and makes an angle of 45° with x-y. *A* is in the HP and the VT of the line is 15 mm below the HP. The line is inclined at 30° to the VP. Draw the projections of *AB* and find its true length and inclination with the HP. Also locate the HT.

(PTU, Jalandhar May 2006)

Solution: All the construction lines are retained to make the solution self-explanatory. See Fig. 9.80.

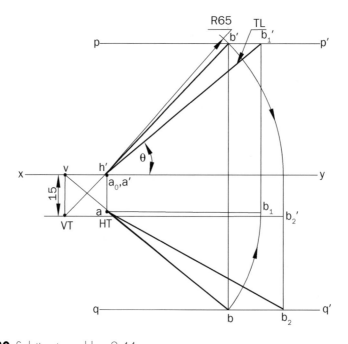

Fig. 9.80 Solution to problem 9.44

Problem 9.45 A room is 6 m × 5 m × 4 m high. Find the actual distance between a top corner and the bottom corner diagonally opposite to it.

Solution:

(i)   Draw the front view and top view of the room with suitable scale.
(ii) Locate the positions of the corner points in the font and top view as shown is Fig. 9.81. Complete the projections as being done in the previous problems.

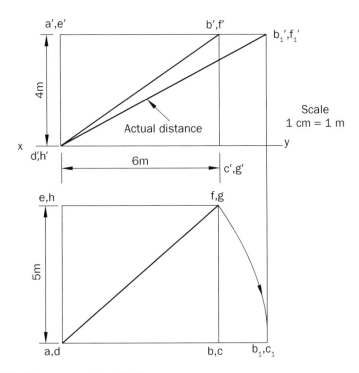

Fig. 9.81 Solution to problem 9.45

Problem 9.46 A line *CD*, 90 mm long, measures 72 mm in elevation and 65 mm in plan. Draw the projections of the line where point C is 20 mm above HP and 15 mm in front of VP. Also find out its traces.

(PTU, Jalandhar May 2011)

Solution: The solution to this problem has already been explained in previous problems. See Fig. 9.82.

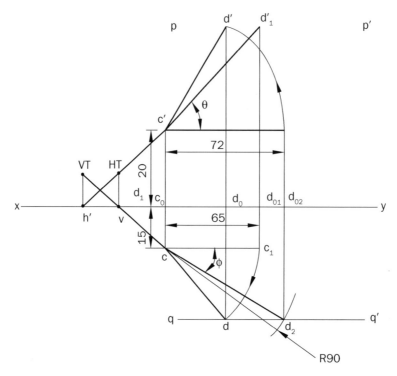

Fig. 9.82 Solution to problem 9.46

Problem 9.47 A line AB appears as 60 mm in front view. The end A is 15 mm above HP while end B is 25 mm in front of VP. The line is inclined at 30° to the HP and 45° to the VP. Draw the projections of line and find out it TL and traces.

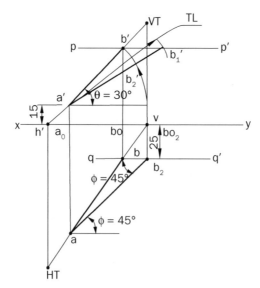

Fig. 9.83 Solution to problem 9.47

Solution: The solution to this problem has already been explained in previous problems. See Fig. 9.83.

Problem 9.48 A line *AB*, measuring 70 mm long, has one of its ends 50 mm in front of VP and 15 mm above HP. The top view of the line is 50 mm long. The other end is 15 mm in front of VP and is in the first quadrant. Draw the projections of line and find out its true inclination with HP and VP and traces.

Solution: All the construction lines are retained to make the solution self-explanatory. See Fig. 9.84.

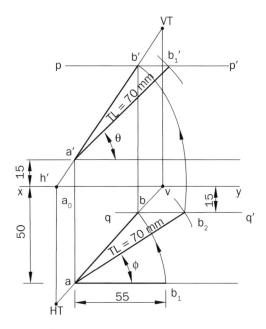

Fig. 9.84 Solution to problem 9.48

Problem 9.49 A line *AB* has its end *A* in HP and 50 mm in front of VP. Its front view is inclined at 55° to x-y line and has a length of 70 mm. The other end *B* is in VP. Draw the projections of a line and find out its true inclinations with HP and VP.

Solution: The solution to this problem is self-explanatory. See Fig. 9.85.

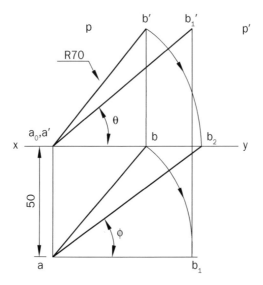

Fig. 9.85 Solution to problem 9.49

Exercises

Parallel or Perpendicular Lines

9.1   A line AB, 60 mm long, has its end A 25 mm away from the HP and 20 m from the VP.
      Draw its projections in all the four quadrants, when the line is parallel to both HP and VP.

9.2   A line AB, 40 mm long, is perpendicular to the HP and its end A is 20 mm away from the
      HP and 15 mm in-front of/behind the VP. Draw its projections in all the four quadrants.

9.3   A line AB, 40 mm long, is perpendicular to the VP and its end A is 20 mm away from the
      HP and 15 mm away from the VP. Draw its projection in all the four quadrants.

Line Inclined to One of the Planes

9.4   A line AB of length 60 mm has its end A 25 mm away from the HP and 15 mm away from
      the VP. The line is inclined to the HP at 30° and parallel to the VP. Draw its projections in
      all the four quadrants. Assume that the whole of the line lies in the same quadrant.

9.5   A line AB, 60 mm long, has its end A 20 mm away from the HP and 15 mm away from the
      VP. The line is inclined to the VP at 30° and parallel to the HP. Draw its projections in all
      the four quadrants. Assume that the whole of the line lies in the same quadrant.

9.6   A line AB, 50 mm long, has its end A is in VP and 15 mm above the HP. The line is
      inclined to the HP at 30° and parallel to the VP. Draw its projections.

9.7   A line AB, 50 mm long, is contained by HP and is inclined to the VP at 45°, and has its
      end A 10 mm behind the VP. Draw its projections when whole of the line lies in the same
      quadrant.

Line Inclined to Both HP and VP

9.8 A line AB has its end A at 20 mm height from the HP and 25 mm in front of VP; the other end B is 40 mm above the HP and 50 mm in front of the VP. The distance between end projectors is 45 mm. Draw the projections of the line and find out TL, θ, φ, HT and VT by using (i) Rotation Method and (ii) Trapezoid Method.

9.9 A line AB has its end A 10 mm above the HP and 15 mm in front of the VP. End B is 35 mm below the HP and 30 mm behind the VP. The distance between the end projectors is 40 mm. Draw the projections of the line and find out TL, θ, φ, HT and VT by using (i) Rotation Method and (ii) Trapezoid Method.

9.10 A line CD, 60 mm long, has its end C 15 mm away from HP and 25 mm away from the VP. It is inclined at 45° to the HP and 30° to the VP. Draw its projections in all the four quadrants. Assume that the whole of the line lies in the same quadrant.

9.11 A straight line AB, 50 mm long, makes an angle of 30° to HP and 45° to the VP. The end A is in both HP and VP. Draw the projections of the line AB in first quadrant only.

9.12 Plan and elevation of a line AB, 60 mm long, measure 54 mm and 43 mm, respectively. End A is 10 mm above the HP and 15 mm in front of the VP. Draw its projections and determine true inclinations with HP and VP.

9.13 A line AB, 70 mm long, is inclined at 45° to the HP and 30° to the VP. Its mid-point C is 30 mm above the HP and 40 mm in front of the VP. Draw its projections.

9.14 A room is 8 m long, 6 m wide and 5 m high. An electric bulb hangs in the centre of the ceiling and 1 m below it. A thin straight wire connects the bulb to a switch kept in one of the corners of the room and 1.5 m above the floor. Draw the projections of the wire and determine its TL and slope with the floor.

9.15 A line AB, 65 mm long, is inclined at 30° to the HP and its top view makes an angle of 45° with the VP. The end A is in the HP and 15 mm in front of the VP. Draw its front view and find out its true inclination with the VP.

9.16 A line AB has its end A 20 mm away from the HP and 40 mm away from the VP. End B is 50 mm away from the HP and 15 mm away from the VP. The line AB is contained by profile plane. Draw its projections in first quadrant and third quadrant only. Also find out its TL, θ, φ, HT and VT.

9.17 End A of a line AB is 25 mm above the HP and 15 mm in front of the VP. End B is 10 mm above the HP and 40 mm in front of the VP. The distance between the end projectors is 40 mm. Draw its projections and determine its traces.

9.18 A line AB, 50 mm long, has its end A 15 mm above the HP and 10 mm in front of the VP. End B is 40 mm above the HP and 45 mm in front of the VP. Draw its projections and determine θ and φ.

9.19 A line AB, 60 mm long, has its end A 40 mm below the HP and 10 mm behind the VP. End B is 10 mm below the HP and 45 mm behind the VP. Draw its projections.

9.20 A line AB has its end A 10 mm in front of the VP and 15 mm above the HP. End B is 35 mm behind the VP and 40 mm below the HP. The distance between the end projectors is 40 mm. Draw its projections and determine its traces too.

9.21 A line AB has its end A 20 mm above the HP and 25 mm in front of the VP. The front view of the line is 50 mm long and is inclined at 30° to the x-y line. The top view of the line is inclined at 45° to the x-y line. Draw its projections and find out the true length and true inclination with HP and VP.

9.22 A line AB has its end A 20 mm in front of the VP and end B 55 mm above the HP. The
      line is inclined at 45° to the HP while its front view makes an angle of 30° to the x-y line.
      Draw its projections when its top view is 50 mm long. Find the true length and true angle
      of inclination with the VP.

Projection of Lines Using Traces

9.23 A line AB has its end A 15 mm above the HP and end B is 60 mm in front of the VP.
      The front view of the line is inclined at 30° to the x-y line. The horizontal trace (HT) of
      the line is 10 mm in front of the VP and its vertical trace (VT) is 15 mm below the HP.
      Draw its projection and determine its TL and true angle of inclination with HP and VP,
      respectively.

9.24 A line AB has its end A 15 mm above the HP and 20 mm in front of the VP. End B is
      50 mm in front of the VP. The vertical trace (VT) is 10 mm above the HP. Draw the
      projections of the line if the distance between end projectors is 45 mm and find its TL, θ,
      φ, HT.

9.25 A line AB has its end A 20 mm in front of the VP and end B is 60 mm in front of the VP.
      The HT of the line is 10 mm in front of the VP. The length of its front view is 50 mm and
      makes an angle of 30° to x-y line. The length of its top view is 55 mm. Draw the projections
      of the line and find it TL, θ, φ, VT.

Objective Questions

9.1   A straight line is defined as the .... distance between two points or
      extremities.

9.2   When a line is perpendicular to one of the planes, it is .... to the other plane.

9.3   When a line is inclined to .... and parallel to ...., its top view
      represents the true length of the line.

9.4   When a line is inclined to HP and parallel to VP, the inclination of the front view with x-y
      represents its ....

9.5   When a line is perpendicular to the VP, its .... trace will coincide with
      .... of the line.

9.6   When a line is contained by a profile plane, the sum of the angles of the inclination with
      the HP and VP is equal to ....

9.7   The trace of a line is a ....

9.8   A straight line will represent its true length in that plane to which it is ....

9.9   Define a straight line.

9.10 What are the apparent angles of inclinations?

9.11 What do you mean by the trace of a line?

9.12 Draw the projections of two parallel lines.

9.13 A line lies in a profile plane with equal elevation and plan length. Draw its projections and
      give the magnitude of angle θ and φ.

9.14 What is the true length of a line?

9.15 When both the views of a line coincide with x-y line, the line is lying on ....

9.16 When a line is parallel to both HP and VP, it has .... trace.

9.17 The side view of a line is a ....

9.18 A straight line is generated as the .... of a moving point.

9.19 Draw free hand the trace of a line when it is parallel to VP and inclined to HP. Name the trace.

Answers

9.1	Shortest	9.6	90°	9.17	Point
9.2	Parallel	9.7	Point	9.18	Locus
9.3	VP, HP	9.8	Parallel	9.19	Horizontal trace (HT)
9.4	True inclination	9.15	Both on HP and VP		
9.5	Vertical, Front view	9.16	No		

Chapter 10

Projections of Planes

10.1 Introduction

Planes or surfaces are objects that have two dimensions, i.e., length and breadth; they have negligible thickness. Plane surfaces may be considered of infinite sizes. However, for convenience, segments of planes are only considered in the solutions. Planes are represented in space by either of the following:

- Three non-collinear points, Fig. 10.1(*a*)
- A line and a point, Fig. 10.1(*b*)
- Two intersecting lines, Fig. 10.1(*c*)
- Two parallel lines, Fig. 10.1(*d*)
- A plane, Fig. 10.1(*e*)

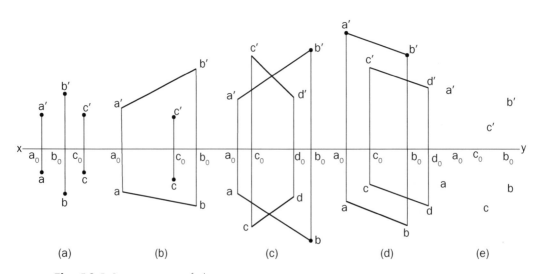

Fig. 10.1 Representation of planes

10.2 Types of Planes

Planes are mainly of two types:

- Principal Planes
- Secondary Planes

Principal planes: The planes on which the projections are obtained are called the principal planes. Examples of principal planes are horizontal and vertical planes.

Secondary planes: Secondary planes are of two types:

    (i)   Perpendicular planes
    (ii)   Oblique planes

  (i)   *Perpendicular planes*: These planes can be divided into the following sub-types:

1. Perpendicular to both the principal planes
2. Perpendicular to one of the principal planes and parallel to the other plane
3. Perpendicular to one of the principal planes and inclined to the other plane

1. **A plane perpendicular to both the principal planes.** A square plane *ABCD* is perpendicular to both the principal planes. Its horizontal trace (HT) and vertical trace (VT) are in a straight line perpendicular to the reference line x-y, as shown in Fig. 10.2. The elevation, *b′c′*, and plan, *ab*, of the square are both straight lines coinciding with VT and HT, respectively, i.e., VT and elevation, HT and plan overlapping.

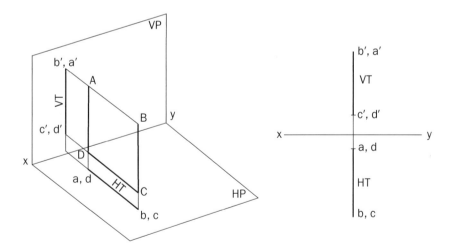

Fig. 10.2 A plane perpendicular to both HP and VP

2. **Perpendicular to one of the principal planes and parallel to the other plane.**

(a) A *plane* perpendicular *to HP and parallel to the VP*. A square lamina *ABCD* is perpendicular to the HP and parallel to the VP. Its HT, is parallel to x-y and it has no VT. The front view *a′b′c′d′* shows the true shape and size of the square object. The top view *ab* is a line, parallel to x-y, coinciding with HT, as shown in Fig. 10.3.

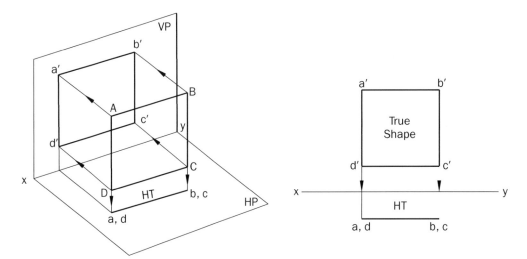

Fig. 10.3 Plane perpendicular to HP and parallel to VP

(b) *Plane, perpendicular to VP and parallel to the HP.* A square *ABCD* is perpendicular to the VP and parallel to the HP. Its VT is parallel to x-y and it has no HT. The top view *abcd* shows the true shape and size of the square object. The front view *d′c′* is a line, parallel to x-y, coinciding with VT, as shown in Fig. 10.4.

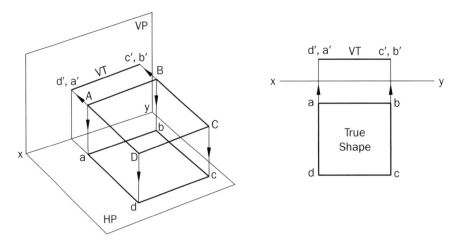

Fig. 10.4 Plane perpendicular to VP and parallel to HP

3. **Perpendicular to one of the principal planes and inclined to the other plane**

(a) *Plane, perpendicular to the HP and inclined to the VP.* A square *ABCD* is perpendicular to the HP and inclined at an angle φ to the VP. Its VT is perpendicular to x-y, and HT is inclined at angle φ to x-y. Its top view *ab* is a line inclined at an angle φ to x-y and the front view *a′b′c′d′* is smaller than square *ABCD*, as shown in Fig. 10.5.

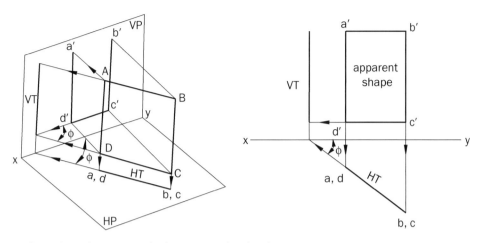

Fig. 10.5 Plane perpendicular to HP and inclined to VP

(b) *Plane, perpendicular to the VP and inclined to the HP.* A square *ABCD* is perpendicular to the VP and inclined at an angle θ to the HP. Its HT is perpendicular to x-y and VT is inclined at an angle θ to the x-y. Its front view $d'c'$ is a line inclined at an angle θ to x-y and the top view *abcd* is a rectangle which is smaller (apparent shape) than square *ABCD*, as shown in Fig. 10.6.

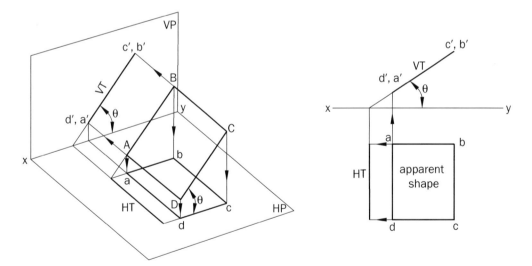

Fig. 10.6 A plane perpendicular to VP and inclined to HP

(ii) *Oblique planes.* These are the planes that are inclined to both the principal planes. However, representation of oblique planes by their traces is beyond the scope of this book.

10.3 Traces of Planes

When a plane is extended, it meets the principal planes in lines. These lines are termed as the traces of the planes. The line which exhibits the plane meeting the HP is called the horizontal trace or HT of the plane. Similarly the line which shows the plane meet the VP is called the vertical trace or VT of the plane. A plane is usually represented by its traces.

Problem 10.1 Draw the traces for the following planes in the first quadrant.

(a) Perpendicular to both HP and VP
(b) Parallel to and 40 mm from VP
(c) Parallel to and 40 mm from HP
(d) Perpendicular to HP and inclined at 45° to VP
(e) Perpendicular to VP and inclined at 45° to HP.

Solution: Figure 10.7 illustrates the various traces.

(a) The HT and VT are in a line perpendicular to x-y.
(b) The HT is parallel to and 40 mm below x-y. It has no VT.
(c) The VT is parallel to and 40 mm above x-y. It has no HT.
(d) The HT is inclined at 45° to x-y, the VT is normal to x-y, both the traces intersect in x-y.
(e) The VT is inclined at 45° to x-y, the HT is normal to x-y, both the traces intersect in x-y.

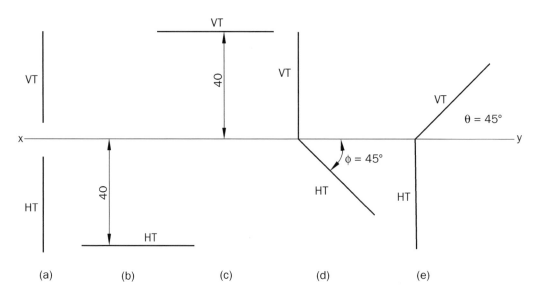

Fig. 10.7 Solution to problem 10.1

10.4 A Secondary Plane in Different Positions with Respect to the Principal Planes

Various positions, which a secondary plane in space can take, with reference to the principal planes of projection and the corresponding projections, are described in the subsequent paragraphs.

10.5 Projections of Plane Parallel to One of the Principal Planes

Projections of such plane can be drawn in one step. First draw the projection of the plane on that principal plane to which it is parallel, as that view will represent true shape and size. The other view which will be a line, should then be projected from it.

(a) **When the plane is parallel to the HP.** The top view should be drawn first and front view will be projected from it.

Problem 10.2 A square lamina ABCD of 30 mm side is parallel to HP and is 15 mm from it. Draw its projections in first quadrant only and locate its traces.

Solution: Figure 10.8, as the square lamina *ABCD* is parallel to HP, its projection on HP provides true shape and size. It has no HT, its front view will be an edge view, i.e,. a line and will be its VT also.

(i)   Draw a square *abcd* of 30 mm side, keeping one side, say *ab*, parallel to x-y.
(ii)   Keep 15 mm distance above x-y.
(iii) Project the points *a'*, *b'*, *c'*, and *d'* in the front view.
(iv)   In the front view *c'* overlaps point *b'* and point *d'* overlaps *a'*.
(v)   Therefore *d'*, *c'* will be written before the front view *a'*, *b'*, respectively, e.g., in front view write *d'*, *a'* and *c'*, *b'*.

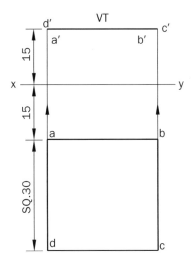

Fig. 10.8 Solution to problem 10.2

Problem 10.3 A regular hexagonal lamina, of side 25 mm side, rests on HP such that one of its sides is perpendicular to VP. Draw its projections in first quadrant only and locate its traces.

Solution: As the hexagonal lamina lies on HP, its projection on HP provides true shape and size. It has no HT, its front view will be an edge view and will be its VT also.

(i)   Draw a regular hexagon *abcdef* of 25 mm side, keeping one of its sides, *cb* or *ef*, perpendicular to x-y.
(ii)   Project the points *a*', *b*', *c*', *d*', *e*', and *f* in the front view.
(iii) Complete the projections as shown in Fig. 10.9.

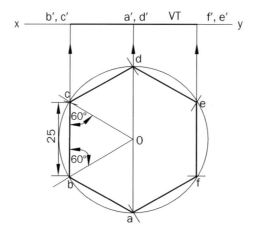

Fig. 10.9 Solution to problem 10.3

Problem 10.4 A square lamina *ABCD* of 30 mm side is parallel to HP and is 15 mm from it. Draw its projections when (a) a side is parallel to the VP; (b) a side is inclined at 30° to the VP. Locate its traces too.

Solution: Here, the given square lamina, *ABCD*, is parallel to *HP*; its projection on *HP* provides true shape and size. It has no *HT*, its front view will be an edge view, i.e., a line, and the same will be its *VT* also.

(a)   (i)   Make a square *abcd* of 30 mm side, and keep one side say *ab* parallel to x-y.
       (ii)   Keep 15 mm distance above x-y.
       (iii) Project the points *a*,' *b*,' *c*,' and *d*' in the front view.
       (iv)   In the front view *c*' overlaps point *b*' and *d*' overlaps *a*'.
       (v)   Therefore *d*', *c*' will be written before the front view *a*', *b*' respectively. e.g., in front view writes *d*', *a*' and *c*', *b*'. See Fig. 10.10 (*a*).

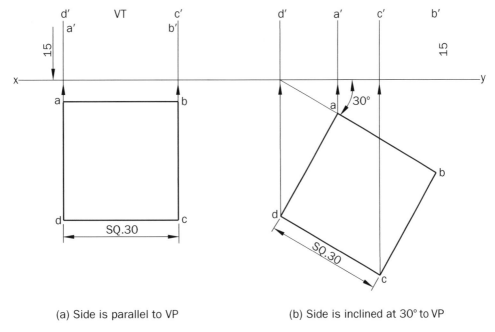

(a) Side is parallel to VP   (b) Side is inclined at 30° to VP

Fig. 10.10 Solution to problem 10.4

(b) (i)   Make a square *abcd* of 30 mm side, keeping one side *ab* to be inclined at 30° to the x-y.

(ii)   Keep 15 mm distance above x-y.

(iii)   Project the points *a'* *b'* *c'* *d'* in the front view as shown in Fig. 10.33 (*b*)

Problem 10.5 A regular hexagonal lamina of side 25 mm is parallel to HP and 15 mm from it. Draw its projections when (a) one side is perpendicular to the VP; (b) one side is parallel to the VP; (c) one side is inclined at 45° to the VP. Locate its traces too.

Solution: As the hexagonal lamina is parallel to the *HP*, therefore, its projection on *HP* provides true shape and size. It has no *HT*, its front view will be an edge view, i.e., a line, and the same will be its *VT* also.

(a) (i)   Draw a regular hexagon *abcdef* of 25 mm side, keeping one of its sides, *ab* or *de*, perpendicular to x-y.

(ii)   Keep 15 mm distance above x-y.

(iii)   Project the points a', b', c', d', e', and f' in the front view as shown in Fig. 10.11 (*a*).

(b) (i)   Draw a regular hexagon *abcdef* of 25 mm side, keeping one of its sides say *ab* or *de* parallel to x-y.

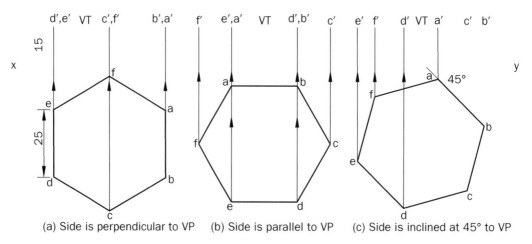

(a) Side is perpendicular to VP   (b) Side is parallel to VP   (c) Side is inclined at 45° to VP

Fig. 10.11 Solution to problem 10.5

      (ii)   Keep 15 mm distance above x-y.

      (iii)   Project the points a', b', c', d', e', and f' in the front view as shown in Fig. 10.11 (*b*)

(c)   (i)   Draw a regular hexagon *abcdef* of 25 mm side, keeping one side say *ab* to be inclined at 45° to the x-y.

      (ii)   Keep the distance of 15 mm above x-y.

      (iii)   Project the points a', b', c', d', e', *and f* in the front view. See Fig. 10.11 (*c*)

Problem 10.6 A regular hexagonal lamina *ABCDEF* 25 mm side rests on HP such that one of its sides (say AF) inclined to VP at 45°. Draw its projections and locate its traces.

Solution: As the hexagonal lamina lies on HP, so its projection on HP provides true shape and size. It has no HT, its front view will be an edge view and will be its VT also.

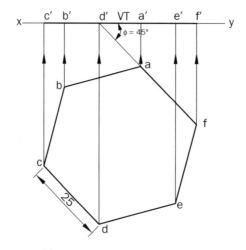

Fig. 10.12 Solution to problem 10.6

(i) Draw a regular hexagon *abcdef* of 25 mm side, keeping one of its sides (say, *af*) inclined at 45° to x-y.
(ii) Project the points *a'*, *b'*, *c'*, *d'*, *e'*, and *f* in the front view.
(iii) Complete the projections as shown in Fig. 10.12.

Problem 10.7 An equilateral triangle of side 30 mm is lying on the HP. It is placed in such a way that one of the corners of the triangle is 15 mm in front of the VP and the side containing that corner is inclined at 45° to the VP. Draw its projections and locate its traces too.

(PTU, Jalandhar May 2008)

Solution: As the triangular lamina is lying on the HP, so its projection on HP provides true shape and size. It has no HT, its front view will be an edge view and will be its VT also.

(i) Fix the position of one of corners (say *a*) of the equilateral triangle at 15 mm below x-y line.
(ii) Draw an equilateral triangle *abc* of 30 mm side, keeping side *ab* to be inclined at 45° to x-y.
(iii) Project the points *a'*, *b'*, and *c'* in the front view.
(iv) Complete the projections of top view as shown in Fig. 10.13.

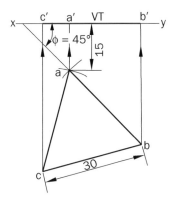

Fig. 10.13 Solution to problem 10.7

Problem 10.8 A regular pentagonal lamina *ABCDE* of 25 mm side, having one of its sides parallel to VP and 10 mm above HP. A corner opposite to this side is 12 mm in front of the VP. Draw the projections when the lamina is parallel to HP and locate its traces too.

Solution: As the pentagonal lamina is parallel to the HP, its projections will start from the top view, where it will provide true shape and size. It has no HT, its front view will be an edge view and will be its VT also.

(i) Fix the position of one of corners (say a) of the regular pentagon at 12 mm below the x-y line.
(ii) Draw a regular pentagon *abcde* of 25 mm side, keeping side *cd* to be parallel to x-y line.
(iii) Project the points a', b', c', d', and e' in the front view at a distance of 10 mm above x-y line.
(iv) Complete the projections as shown in Fig. 10.14

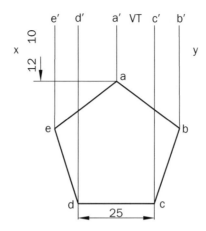

Fig. 10.14 Solution to problem 10.8

(b) **When the plane is parallel to the VP.** The front view should be drawn first and top view will be projected from it.

Problem 10.9 A regular pentagonal lamina *ABCDE* with 25 mm side, has its side *CD* lying on HP. Draw its projections when its plane is parallel to and 15 mm in front of VP. Also locate its traces.

Solution: As the regular pentagon *ABCDE* is parallel to VP, its projection on VP provides true shape and size. It has no VT, its top view will be an edge view and will be its HT also.

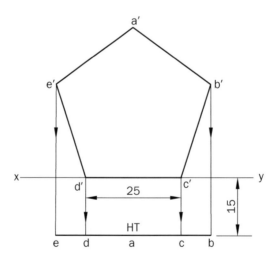

Fig. 10.15 Solution to problem 10.9

(i)   Draw a regular pentagon *a'b'c'd'e'* of 25 mm side, keeping its side *c'd'* on x-y.
(ii)   Keep a distance of 15 mm below x-y.
(iii) Project the points *a, b, c, d,* and *e* in the top view.
(iv)   Complete the projections as shown in Fig. 10.15.

Problem 10.10 A regular hexagonal lamina ABCDEF 25 mm side, has its side DE lies on HP. Draw its projections when its plane is parallel to and 10 mm in front of VP. Also locate its traces.

Solution: As the regular hexagon *ABCDEF* is parallel to VP, its projection will start from the front view, where it will provide true shape and size. It has no VT, its top view will be an edge view or line and will be its HT also.

    (i)   Draw regular hexagon *a'b'c'd'e'f'* of 25 mm side, keeping its side *d'e'* on x-y.
    (ii)   Keep a distance of 10 mm below x-y.
    (iii) Project the points *a, b, c, d, e,* and *f* in the top view.
    (iv) Complete the projections as shown in Fig. 10.16.

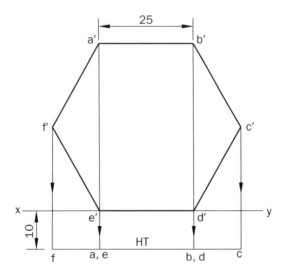

Fig. 10.16 Solution to problem 10.10

Problem 10.11 A regular hexagonal lamina ABCDEF 20 mm side, has its corner D in HP and the side EF perpendicular to HP. Draw its projections when its plane is parallel to and 15 mm from VP. Locate it traces too.

Solution: As the regular hexagon *ABCDEF* is parallel to VP, its projection on VP provides true shape and size and its top view will be an edge view. It has no VT and edge view will represent its HT.

    (i)   Draw a regular hexagon *a'b'c'd'e'f'* of 25 mm side, keeping its corner *d'* on x-y.
    (ii)   Keep a distance of 15 mm below x-y.
    (iii) Project the points *a, b, c, d, e,* and *f* in the top view.
    (iv) Complete the projections as shown in Fig. 10.17.

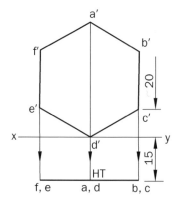

Fig. 10.17 Solution to problem 10.11

Problem 10.12 A square lamina ABCD of 20 mm side has its HT parallel to and 12 mm below x-y line. It has no VT. Draw its projections when all the sides are equally inclined to the HP.

(PTU, Jalandhar December 2004, 2007)

Solution:

(i)   Draw a square $a'b'c'd'$ of 20 mm side in the front view, keeping all its sides equally inclined to the HP.

(ii)   Keep a distance of 12 mm below x-y.

(iii)   Project the points a, b, c, and d in the top view.

(iv)   Complete the projections as shown in Fig. 10.18.

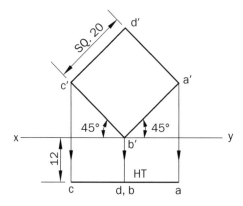

Fig. 10.18 Solution to problem 10.12

10.6 Projections of Plane Perpendicular to Both HP and VP

When a plane is perpendicular to both HP and VP, its projection will start from side view, where it gives the true shape and size. Then front and top views are to be projected from it.

Problem 10.13 A square lamina ABCD of 25 mm side is perpendicular to both HP and VP. Draw its projections in first quadrant only and locate its traces too.

Solution: As square lamina *ABCD* is perpendicular to both HP and VP, so it will give true shape and size in side view. It will have an edge view or line, both in front view and top view, which will represent its VT and HT too.

    (i)   Draw a square lamina *a"b"c"d"* of 25 mm side in side view.
    (ii)   Project the front and top views from it.
    (iii) Complete the projections as shown in Fig. 10.19.

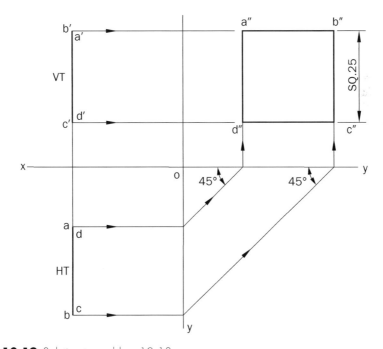

Fig. 10.19 Solution to problem 10.13

Problem 10.14 A regular hexagonal lamina ABCDEF 25 mm side is normal to both HP and VP. It is lying with one of its edges (say, *AB*) parallel to HP and perpendicular to VP. Draw its projections and locate its traces too.

Solution: As the hexagonal lamina *ABCDEF* is normal to both HP and VP, so it will provide true shape and size in side view. It will have an edge view, both in front view and top view, which will represent its VT and HT too.

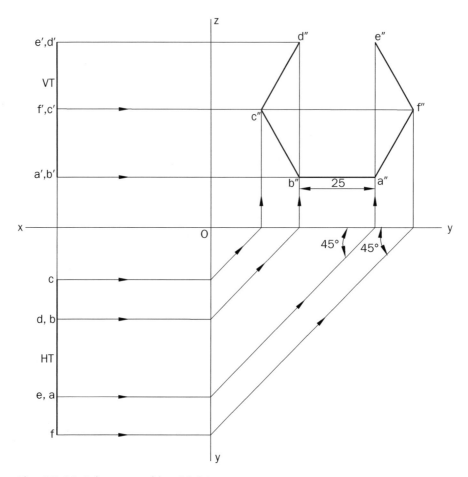

Fig. 10.20 Solution to problem 10.14

(i)   Draw a hexagonal lamina $a''b''c''d''e''f'$ of 25 mm side in side view with one of its edges (say $a''b''$) parallel to x-y.

(ii)   Project front view and top view from it.

(iii)   Complete the projections as shown in Fig. 10.20.

10.7 Projections of Plane Inclined to One of the Principal Planes and Perpendicular to the Other Plane

When a plane is inclined to one of the principal planes, its projections may be obtained in two stages. In the initial stage, the plane is assumed to be parallel to that principal plane to which it has to be made inclined. It is then titled to the required inclination in the second stage and its projections are drawn. There may be two different cases of this type of plane.

(a) **Plane inclined to the HP and perpendicular to the VP.** When a plane is inclined to the HP and perpendicular to the VP, its projections are drawn into two stages. In the initial stage, it is assumed to be parallel to the HP. Its top view will show the true shape and size in this position. The front view will be a line parallel to x-y. In the second stage, tilt the front view so that it makes the given inclination to the x-y. The new front view will be inclined to x-y at the true inclination. In the top view the corners will move along their respective paths (parallel to x-y).

Problem 10.15 A rectangular lamina *ABCD* of 50 mm × 30 mm is inclined to HP at 45° and perpendicular to VP. It rests on one of it sides say AB in HP. Draw its projections in first quadrant only and locate its traces too.

Solution:

(i) Draw a rectangular lamina abcd of 50 mm × 30 mm in the top view, assuming the lamina to be lying in HP and keep the side *ab* perpendicular to x-y.
(ii) Project the front view *a'b'c'd'* from top view, which will lie on x-y.
(iii) Tilt the front view about the *a'b'*, so that it makes 45° angle with x-y and name the points on it by adding suffix 1 to them.
(iv) Project the corresponding top view a_1 b_1 c_1 d_1 by drawing vertical projectors through the points in second stage front view and horizontal projectors through the first stage top view as shown in Fig. 10.21.

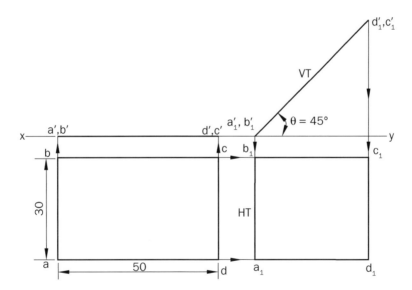

Fig. 10.21 Solution to problem 10.15

Problem 10.16 A regular pentagonal lamina ABCDE of 25 mm side has one side on the HP. Its surface is inclined at 45° to the HP and perpendicular to the VP. Draw its projections and show its traces.

Solution:

(i) Draw a pentagon *abcde* of 25 mm side in the top view with one side *ae* perpendicular to x-y, assuming the lamina to be lying in the HP.

(ii) Project the front view *a'b'c'd'e'* from the top view to x-y.

(iii) Tilt the front view about the side *a'e'*, so that it makes 45° angle with x-y and name the points on it by adding suffix 1 to them.

(iv) Project the corresponding top view a_1 b_1 c_1 d_1 e_1 and traces as shown in Fig. 10.22.

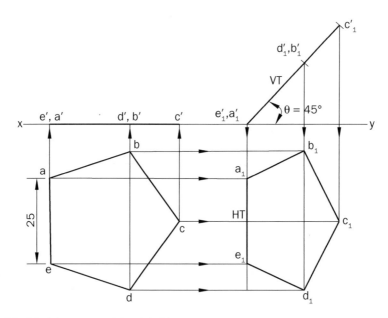

Fig. 10.22 Solution to problem 10.16

Problem 10.17 A regular pentagonal lamina ABCDE of 25 mm side, lies on one of its corner in HP, such that the surface is inclined at 45° to the HP and perpendicular to the VP. Draw its projections and locate its traces.

Solution:

(i) Draw a pentagon *abcde* of 25 mm side in the top view, assuming to be lying in the HP.

(ii) Project front view *a'b'c'd'e'* from top view to the x-y.

(iii) Tilt the front view about the corner *a'*, so that it makes an angle of 45° to x-y and name all the points by adding suffix 1 to them.

(iv) Project the corresponding top view and traces as shown in Fig. 10.23.

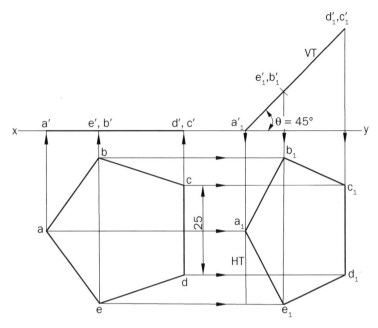

Fig. 10.23 Solution to problem 10.17

Problem 10.18 A square lamina of 70 mm sides has a circular hole of 40 mm diameter centrally placed. The lamina is resting on one of its sides on HP, this side being perpendicular to VP. The surface of the lamina is inclined at 30° to HP. Draw the projections of the lamina. Also draw its side view.

(PTU, Jalandhar May 2001)

Solution:

(i)   Draw a square *abcd* of 70 mm side in the top view, keeping side *cd* perpendicular to x-y.

(ii)   Draw a hole of φ 40 mm in the centre of the square. Divide the circular hole into sixteen equal parts.

(iii) Project the corresponding front view for the square lamina along with the circular hole.

(iv) Tilt the front view at angle of 30° to x-y.

(v)   Project the corresponding front and side views as shown in Fig. 10.24.

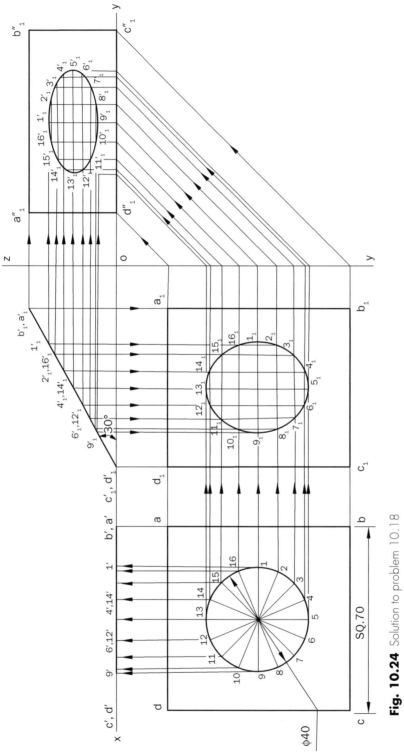

Fig. 10.24 Solution to problem 10.18

Problem 10.19 A circular plate of φ 50 mm and negligible thickness rests on HP on its rim and makes an angle of 45° to HP. Draw its projections.

Solution: A circle has no corners to project from one view to another. However a number of points, say, sixteen (equal distances apart), may be marked on its circumference.

(i) Assume the circular plate to be lying in HP, draw its projections. The top view will be a circle of diameter 50 mm and front view will be an edge view on x-y.
(ii) Divide the circumference of the circle into sixteen equal parts. Project these points in the front view.
(iii) Tilt the front view about the point 9' and makes an angle of 45° to x-y. Name all the points on it by adding suffix 1 to them.
(iv) Draw its corresponding top view. See Fig. 10.25.

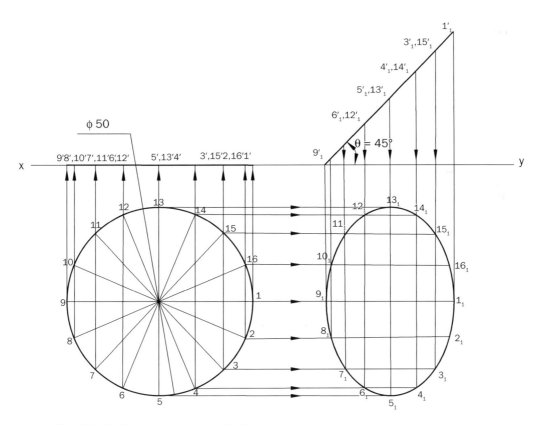

Fig. 10.25 Solution to problem 10.19

Problem 10.20 A regular hexagonal lamina ABCDEF of 25 mm side, rests on one of its sides on HP. Its surface is inclined at 45° to the HP and perpendicular to the VP. Draw its projections and locate its traces.

Solution: All the construction lines are retained to make the solution self-explanatory. See Fig. 10.26.

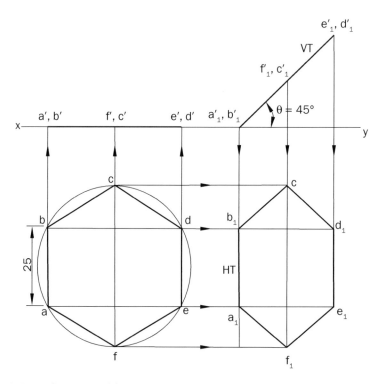

Fig. 10.26 Solution to problem 10.20

Problem 10.21 A regular pentagonal lamina ABCDEF, of 25 mm side, lies on one of its corner in HP such that the surface is inclined at 45° to the HP and perpendicular to the VP. Draw its projections and locate its traces.

Solution:

(i)   Draw a hexagon *abcdef* of 25 mm side in the top view, assuming to be lying in the HP and keep the side *bc* or *ef* to be parallel to the x-y line.

(ii)   Project the front view *a'b'c'd'e'f'* from top view.

(iii)   Tilt the front view about the corner *a'* so that it makes an angle of 45° to x-y and name all the points by adding suffix 1 to them.

(iv)   Project the corresponding top view and traces as shown is Fig. 10.27.

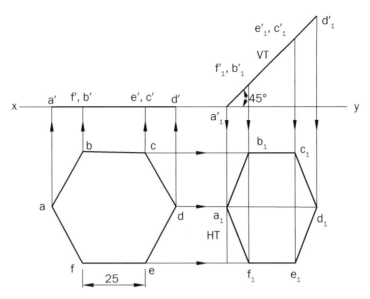

Fig. 10.27 Solution to problem 10.21

Problem 10.22 A regular pentagonal lamina ABCDE of 25 mm side has its surface inclined at 30° to the HP and perpendicular to the VP. One of its side is parallel to VP and 10 mm above HP and 15 mm in front of VP. Draw its projections and locate its traces too.

Solution: All the construction lines are retained to make the solution self-explanatory. See Fig. 10.28.

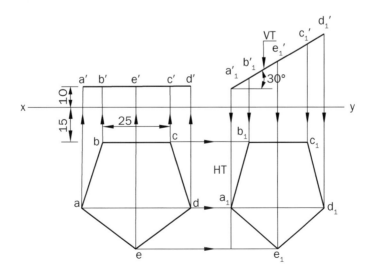

Fig. 10.28 Solution to problem 10.22

(b) **Plane, inclined to the VP and perpendicular to the HP.** When a plane is inclined to the VP and perpendicular to the HP, its projections are drawn into two stages. In the initial stage, it is assumed to be parallel to the VP and its front view will show the true shape and size in this position. The top view will be a line parallel to x-y. In the second stage, tilt the top view so that it makes the given inclination to the x-y. The new front view will be projected from it.

Problem 10.23 A rectangular lamina ABCD of 50 mm × 30 mm is inclined to the VP at 45° and perpendicular to HP. Its one of the sides say AD lies in VP. Draw its projections and locate its traces too.

Solution:

(i)   Draw a rectangular lamina $a'b'c'd'$ of 50 mm × 30 mm in the front view.
(ii) Project the top view $abcd$ from front view, which will lie on x-y.
(iii) Tilt the top view about the side da, so that it makes an angle of 45° to x-y and name the points on it by adding suffix 1 to them.
(iv) Project the corresponding front view $a_1' b_1' c_1' d_1'$ and traces as shown in Fig. 10.29.

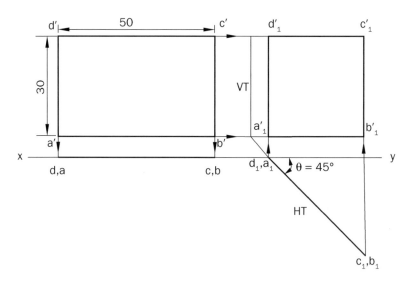

Fig. 10.29 Solution to problem 10.23

Problem 10.24 A regular hexagonal lamina ABCDEF, of 20 mm side, rests on HP on one of its sides say BC such that it is perpendicular to the HP and inclined to VP at 30°. Draw its projections and locate its traces too.

Solution:

(i) Draw a hexagon $a'b'c'd'e'f'$ in the front view with one of its sides say $(b'c')$ on the x-y.
(ii) Project top view $abcdef$ from front view to the x-y line and parallel to it.

(iii) Tilt the top view so that it makes an angle of 30° to x-y and name the points on it by adding suffix 1 to them.

(iv) Project the corresponding front view and traces as shown in Fig. 10.30.

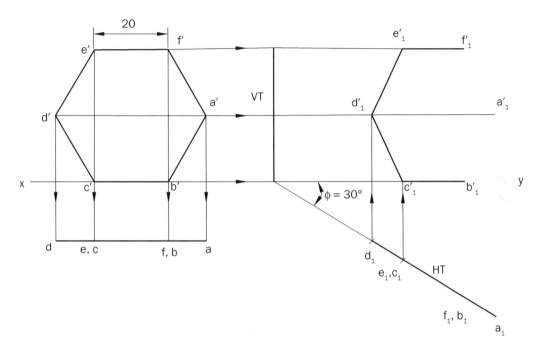

Fig. 10.30 Solution to problem 10.24

Problem 10.25 A thin circular plate of 50 mm diameter is held such that its plane is inclined at 30° to VP and perpendicular to HP. A point on its circumference is 20 mm in front of VP and 30 mm above HP. Draw its projections and locate its traces too.

Solution: A circle has no corners to project from one view to another. However a number of points, say, sixteen (equal distances apart), may be marked on its circumference.

(i) Assuming the circular plate to be parallel to the VP, draw its projections. The front view will be a circle, having its centre 30 mm above x-y and the top view will be a line, parallel to and 20 mm below x-y.

(ii) Divide the circumference into sixteen equal parts and mark the points as shown in the Fig. 10.31. Project these points in the top view.

(iii) Draw the top view in the new position, so that it makes an angle of 30° to x-y and the circumference still 20 mm away from it. Name the points on it by adding suffix 1 to them.

(iv) Draw its corresponding front view and traces too. Join a freehand curve through the sixteen points $1_1 2_1 3_1$ ..... This will be an ellipse.

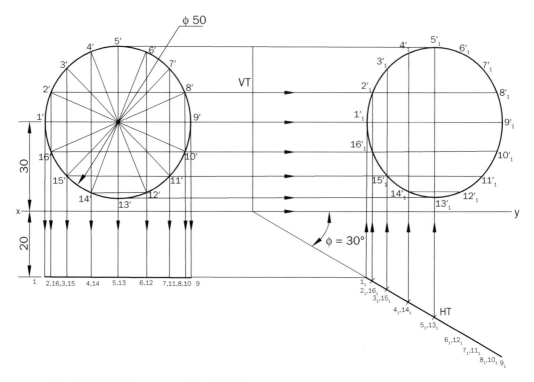

Fig. 10.31 Solution to problem 10.25

Problem 10.26 A regular hexagonal thin plate of 40 mm side has a circular hole of φ 40 mm diameter in its centre. It is resting on one of its corners in HP. Draw its projections when the plate surface is vertical, inclined at 30° to the VP and locate its traces too.

Solution:

    (i)   Draw a hexagon $a'b'c'd'e'f$ of 40 mm side in the front view, keeping one of its corners, say c', in HP.

    (ii)   Draw a hole of φ 40 mm in the centre of the hexagon. Divide the circular hole into sixteen equal parts.

   (iii) Project the corresponding top view for the hexagonal plate along with the circular hole.

   (iv) Tilt the top view at angle of 30° to x-y.

    (v) Project the corresponding front view and traces as shown in Fig. 10.32.

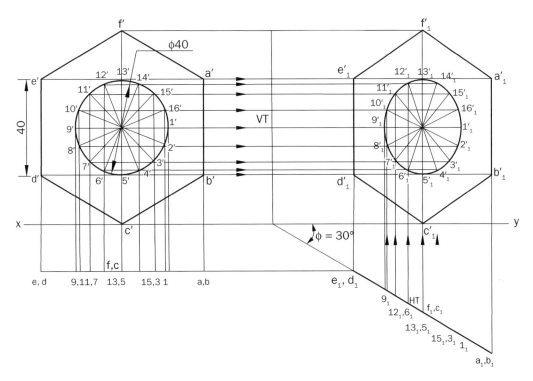

Fig. 10.32 Solution to problem 10.26

Problem 10.27 Draw the projections of a regular pentagonal lamina *ABCDE* of 25 mm side, having one of its sides AB in the VP and with its surface inclined at 60° to the VP. Locate its traces too.

Solution: Method of drawing projections have already been explained in previous problems. See Fig. 10.33.

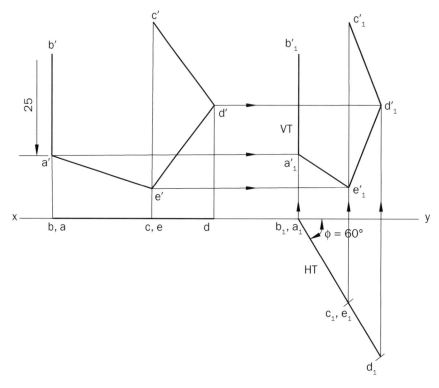

Fig. 10.33 Solution to problem 10.27

10.8 Projections of Plane Inclined to Both the Principal Planes

When a plane has its surface inclined to one plane and an edge or a diameter or a diagonal is parallel to the plane and inclined to the other plane, then its projections are drawn in three stages.

    (a) If the surface of a plane is inclined to the HP and an edge or a diameter or a diagonal is parallel to the HP and inclined to the VP.

        (i)   In the initial stage or position, the plane is assumed to be parallel to the HP and an edge perpendicular to the VP.

        (ii) It is then tilted so as to make the required angle with the HP. Its front view in this position will be a line, while its top view will be smaller in size.

       (iii) In the final stage or position, when the plane is turned to the required inclination with the VP, only the position of the top view will change. Its shape and size will not be affected. In the final front view, the corresponding distance of all the corners from x-y will remain the same as in the second stage front view.

(b) Similarly, if the surface of a plane is inclined to the VP and an edge or a diameter or a diagonal is parallel to the VP and inclined to the HP.

    (i)   In the initial position, the plane is assumed to be parallel to the VP and an edge perpendicular to the HP.

    (ii)   It is then tilted so as to make the required angle with the VP. Its top view in this position will be a line, while its front view will be smaller in size.

    (iii)   In the final position or stage, when the plane is turned to the required inclination with the HP, only the position of the front view will change. Its shape and size will not be affected. In the final top view, the corresponding distances of all the corners from x-y will remain as in the second stage top view.

Problem 10.28 A rectangular lamina ABCD, of 60 mm × 30 mm, has its side AB in HP and inclined at 45° to VP and the plane of the lamina is inclined at 60° to the HP. Draw its projections.

Solution:

    (i)   Draw a rectangular lamina *abcd* of 60 mm × 30 mm in the top view, assuming the lamina to belying in the HP and keep the side *ab* perpendicular to x-y.

    (ii)   Project front view *a'b'c'd'* from top view, which will be a line on x-y.

    (iii)   Tilt the front view about the *a'b'* so that it makes 60° angle with x-y and name the points on it by adding suffix 1 to them.

    (iv)   Project the corresponding top view $a_1 b_1 c_1 d_1$.

    (v)   Reproduce the top view $a_1 b_1 c_1 d_1$ of second stage as $a_2 b_2 c_2 d_2$ such that the side $a_2 b_2$ makes an angle of 45° to x-y.

    (vi)   Project the final front view upwards from this top view and horizontally from the second stage front view, as shown in Fig. 10.34.

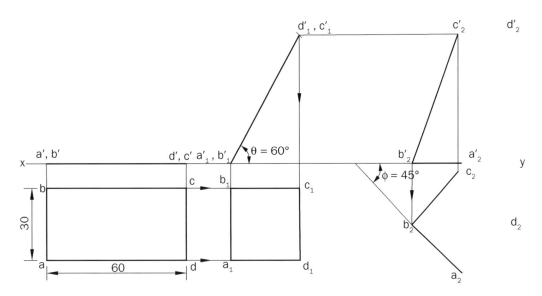

Fig. 10.34 Solution to problem 10.28

Problem 10.29 A rectangular lamina ABCD of 60 mm × 30 mm has its side AB on ground and inclined at 60° to VP and plane of the lamina is inclined at 45° to the HP. Draw its projections in third-angle.

Solution:

(i) Draw two lines x-y and gl, a suitable distance apart. Draw a rectangular lamina *abcd* in the top view, assuming the lamina to be lying in the ground and keep the side *ab* perpendicular to x-y.

(ii) Project front view *a'b'c'd'* from top view, which will be an edge view on gl.

(iii) Tilt the front view about *a'b'* so that it makes an angle of 45° with gl and name the points on it by adding suffix 1 to them.

(iv) Project the corresponding top view.

(v) Reproduce the top view $a_1 b_1 c_1 d_1$ of second stage as $a_2 b_2 c_2 d_2$ such that the side $a_2 b_2$ makes an angle of 60° to x-y.

(vi) Project the final front view as shown in Fig. 10:35.

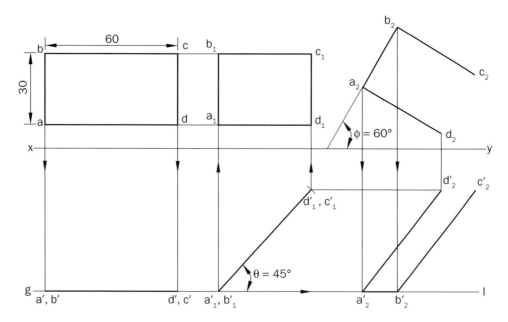

Fig. 10.35 Solution to problem 10.29 (third-angle)

Problem 10.30 A regular pentagonal lamina ABCDE, of 30 mm side, rests on HP on one of its sides such that it is inclined to the HP at 45° and the side on which it rests is inclined at 30° to the VP. Draw its projections.

Solution:

(i)   Draw a pentagon *abcde* of 30 mm side in the top view, assuming the lamina to be lying in the HP and keep the side *ab* perpendicular to x-y.

(ii)   Project the front view *a'b'c'd'e'* from top view, which will be an edge view on x-y.

(iii)   Tilt the front view about *a'b'* so that it makes 45° angle with x-y and name the points on it by adding suffix 1 to them.

(iv)   Project the corresponding top veiw $a_1 b_1 c_1 d_1 e_1$.

(v)   Reproduce the top view $a_1 b_1 c_1 d_1 e_1$ of second stage as $a_2 b_2 c_2 d_2 e_2$ such that the side $a_2 b_2$ makes an angle of 30° to x-y.

(vi)   Project the final front view as shown in Fig. 10.36.

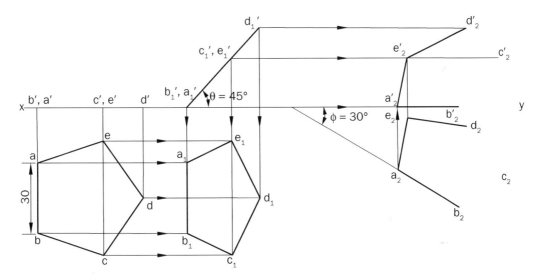

Fig. 10.36 Solution to problem 10.30

Problem 10.31 A regular pentagonal lamina ABCDE, of 30 mm side, rests on ground on one of its sides such that it is inclined to the HP at 45° and the side on which it rests is inclined at 30° to the VP. Draw its projections in third-angle.

Solution:

(i)   Draw two lines x-y and *gl*, a suitable distance apart. Draw a pentagon *abcde* of 30 mm side in the top view, assuming the lamina to be lying in the ground and keep the side as perpendicular to x-y.

(ii)   Project the front view *a'b'c'd'e'* from top view, which will be an edge view on *gl*.

(iii)   Tilt the front view about *a'b'* so that it makes an angle of 45° with *gl* and name the points on it by adding suffix 1 to them.

(iv)   Project the corresponding top view.

(v)   Reproduce the top view $a_1 b_1 c_1 d_1 e_1$ of second stage as $a_2 b_2 c_2 d_2 e_2$ such that the side $a_2 b_2$ makes an angle of 30° to x-y.

(vi)   Project the final front view as shown in Fig. 10.37.

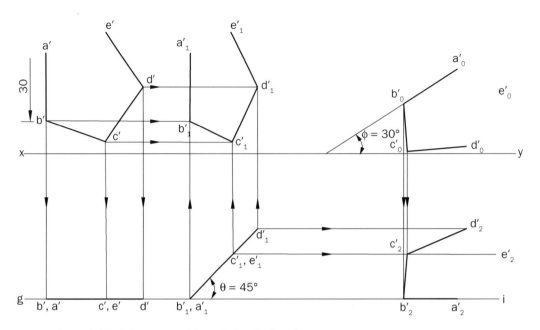

Fig. 10.37 Solution to problem 10.31 (third-angle)

Problem 10.32 A thin circular plate of φ 50 mm and negligible thickness rests on HP on its rim and makes an angle of 45° to it. One of its diameters is inclined to VP at 30°. Draw its projections keeping distance of the centre of the circular plate 35 mm in front of the VP.

Solution: A circle has no corners to project from one view to another. However a number of points, say, sixteen (at equal distances apart), may be marked on its circumference.

(i)   Assuming the circular plate to be parallel to the HP, draw its projections. The top view will be a circle, having its centre 35 mm below x-y and front view will be an edge view on x-y.

(ii) Divide the circumference into sixteen equal parts. Project these points in the front view.

(iii) Draw the front view in the new position, so that it makes an angle of 45° to x-y and name the points on it by adding suffix 1 to them.

(iv) Draw its corresponding top view.

(v)   Reproduce the top view of second stage by adding suffix 2 to them such that its diameter $13_2\,5_2$ makes an angle 30° to x-y.

(vi) Project the final front view as shown in Fig. 10.38.

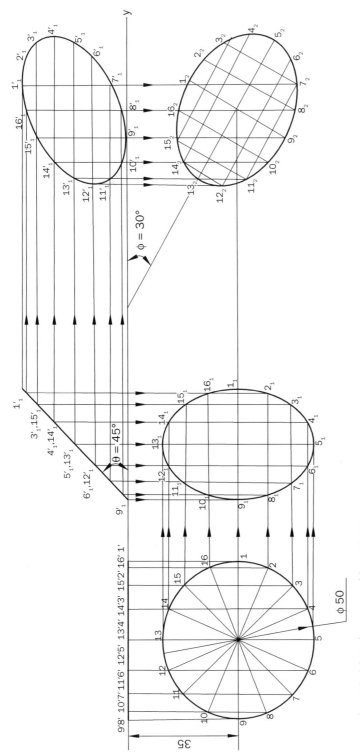

Fig. 10.38 Solution to problem 10.32

Problem 10.33 A regular hexagonal lamina ABCDEF of 25 mm side, has its plane inclined at 45° to the VP and its diagonal FC parallel to VP and inclined to HP at 45°. Draw its projections when its side DE is nearest to the VP and 10 mm in front of it.

Solution:

    (i)   Draw a hexagon $a'b'c'd'e'f'$ of 25 mm side in the front view and keep side $d'e'$ perpendicular to x-y.

    (ii)   Project the top view $abcdef$ from the front view, which will be an edge view 10 mm below x-y.

    (iii) Tilt the top view about de so that it makes an angle of 45° with x-y and name the points on it by adding suffix 1 to them.

    (iv)   Projecting the corresponding front view $a_1' b_1' c_1' d_1' e_1' f_1'$.

    (v)   Reproduce the front view $a_1' b_1' c_1' d_1' e_1' f_1'$ of second stage as $a_2' b_2' c_2' d_2' e_2' f_2'$ such that the diagonal

    (vi) Project the final top view as shown in Fig. 10.39.

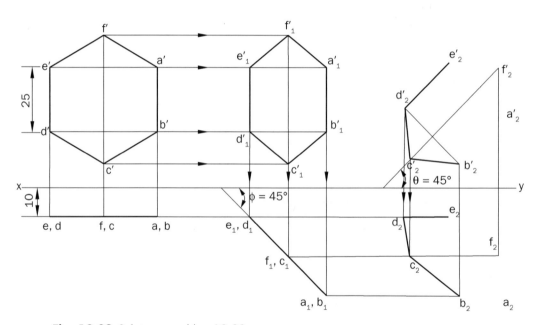

Fig. 10.39 Solution to problem 10.33

Problem 10.34 A square lamina ABCD of 25 mm side, rests on its corner C in HP. Its plane is inclined at 45° to the HP and diagonal DB inclined at 30° to the VP. Draw its projections.

(PTU, Jalandhar May 2006, December 2007)

Solution:

(i) Draw a square lamina *abcd* of 25 mm side in top view, assuming the lamina to be lying in the HP.

(ii) Project the front view *a'b'c'd'* from the top view, which will be a line on x-y.

(iii) Tilt the front view about *c'* so that it makes 45° angle with x-y and name the points on it by adding suffix 1 to them.

(iv) Project the corresponding top view $a_1 b_1 c_1 d_1$.

(v) Reproduce the top view $a_1 b_1 c_1 d_1$ as $a_2 b_2 c_2 d_2$ such that the diagonal $d_2 b_2$ makes an angle of 30° to x-y.

(vi) Project the final front view upwards from this top view and horizontally from the second stage front view as shown in Fig. 10.40.

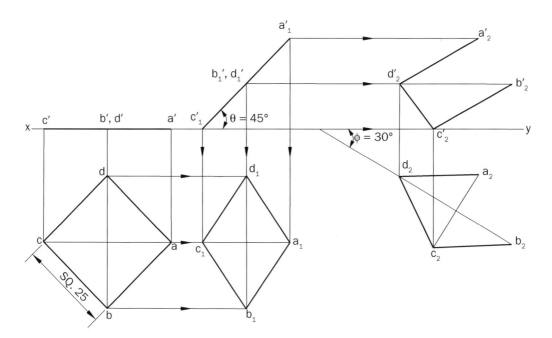

Fig. 10.40 Solution to problem 10.34

214 ENGINEERING DRAWING

Problem 10.35 A square lamina ABCD, of 25 mm side, rests on its corner C on ground plane. Its plane is inclined at 45° to the HP and diagonal DB inclined at 30° to the VP. Draw its projections in third-angle.

Solution:

   (i)   Draw two lines x-y and *gl*, at a suitable distance apart.
   (ii)   Draw a square lamina *abcd* of 25 mm side in the top view, assuming the lamina to be lying on ground plane.
  (iii) Project the front view *a'b'c'd'* from top view, which will be an edge view on *gl* line.
  (iv) Tilt the front view about *c'*, so that it makes angle of 45° with *gl* line and name the points on it by adding suffix 1 to them.
  (v)   Project the corresponding top view $a_1 b_1 c_1 d_1$.
  (vi) Reproduce the top view $a_1 b_1 c_1 d_1$ of second stage as $a_2 b_2 c_2 d_2$ such that the diagonal $d_2 b_2$ is inclined at 30° to x-y.
 (vii) Project the final front view as shown in Fig. 10.41.

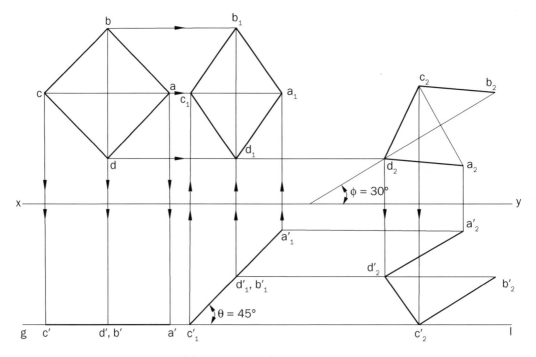

Fig. 10.41 Solution to problem 10.35 (third-angle)

Problem 10.36 Draw the projections of a rhombus having diagonals 100 mm and 50 mm long. The bigger diagonal is inclined at 30° to the HP with one of the end point in HP and the smaller diagonal is parallel to both the planes.

Solution:

(i) Draw the top view *abcd* of the rhombus, such that its bigger diagonal is parallel to the x-y line.

(ii) Project the corresponding front view *a'b'c'd'* in x-y.

(iii) Tilt the front view in such a way that the end point *c'* in the HP and the bigger diagonal *c'a'* makes an angle of 30° to the HP and name the points on it by adding suffix 1 to them.

(iv) Project the corresponding top view $a_1 b_1 c_1 d_1$.

(v) Reproduce the top view such that $b_1 d_1$ is parallel to both the planes (i.e., x-y line) and name the points on it by adding suffix 2 to them.

(vi) Project the final front view upwards from this top view and horizontally from the second stage front view, as shown in Fig. 10.42.

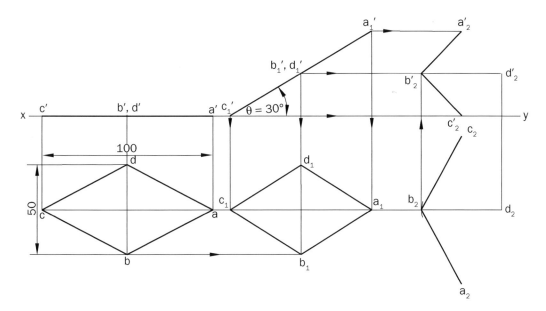

Fig. 10.42 Solution to problem 10.36

Problem 10.37 An ellipse having major axis 80 mm appears as a circle of 55 mm diameter in the front view. The top view of the major axis is perpendicular to the VP. Draw its projections and state the inclination of its surface with HP.

Solution:

(i) Draw the front view of a circle of 55 mm diameter and divide it into sixteen equal parts.

(ii) Draw the side view by taking any point 13" and cut 13"–5" equal to 80 mm and measure the inclination of the line 13"–5" with x-y line.

(iii) Complete the top view from front view and side view as shown in Fig. 10.43.

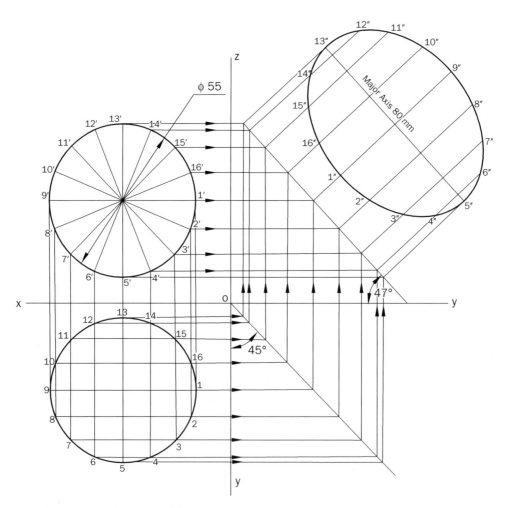

Fig. 10.43 Solution to problem 10.37

Problem 10.38 A rectangular lamina ABCD of 60 mm × 30 mm rests on its shorter side in the VP and the surface makes 30° angle with the VP. The longer side of the plane is inclined at 45° to the HP. Draw its projections.

Solution:

(i) Draw a rectangular lamina $a'\,b'\,c'\,d'$ of 60 mm × 30 mm in the front view, assuming the lamina to be lying in the *VP* and keep the side $a'\,d'$ perpendicular to x-y.

(ii) Project the top view *abcd* from front view, which will be a line on x-y.

(iii) Tilt the top view about *ad* so that it makes 30° angle with x-y and name all the points on it by adding suffix 1 to them.

(iv) Project the corresponding front view.

(v) The front view of longer side of lamina does not give the true length. So first find the apparent angle at which the longer side to be inclined. Draw a line $d_2'\,c_2'$ equal to true length (60 mm) and inclined at 45° to x-y. Draw an *arc* with centre d_2' and radius equal to longer side of the second stage ($d_1'\,c_1'$), to interrect the horizontal line at c_2'. This is the apparent angle of inclination and is greater than 45°.

(vi) Reproduce the front view along the longer edge $d_2'\,c_2'$.

(vii) Project the final top view as shown in Fig. 10.44.

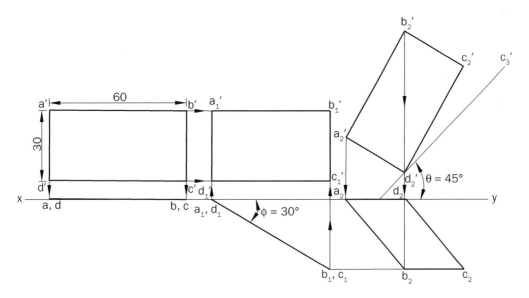

Fig. 10.44 Solution to problem 10.38

Problem 10.39 A regular pentagonal lamina of side 25 mm has one of its corner in the VP and the plane of lamina in inclined at 45° to the VP. The side of the lamina opposite to that comer is parallel to the VP and inclined at 45° to the HP. Draw its projections.

Solution: All the construction lines are retained to make the solution self-explanatory. See Fig. 10.45.

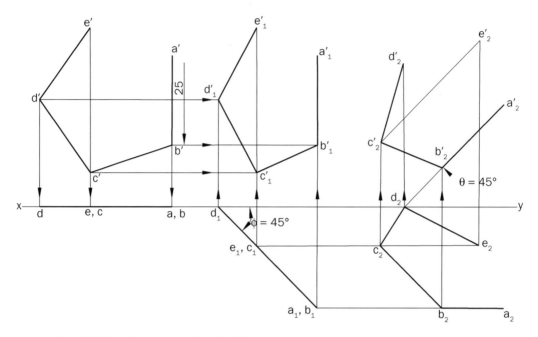

Fig. 10.45 Solution to problem 10.37

Problem 10.40 An equilateral triangular thin plate ABC of 25 mm side lies on one of its side in VP. Draw the projections of the plate when the plate surface is vertical and inclined at 45° to the VP. One of the sides of triangular plate is inclined at 30° to the HP.

Solution: All the construction lines are retained to make the solution self-explanatory. See Fig. 10.46.

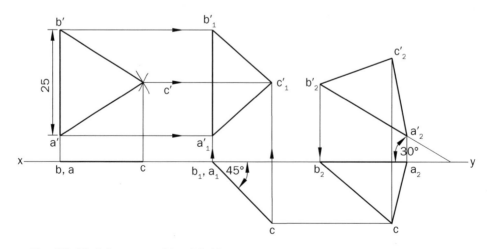

Fig. 10.46 Solution to problem 10.40

Problem 10.41 A regular hexagonal lamina ABCDEF 25 mm side has one of its corners (say D) on the HP, with the surface inclined at 45° to the HP and the top view of the diagonal through that corner is perpendicular to the VP. Draw its projections.

Solution: All the construction lines are retained to make the solution self–explanatory. See Fig. 10.47.

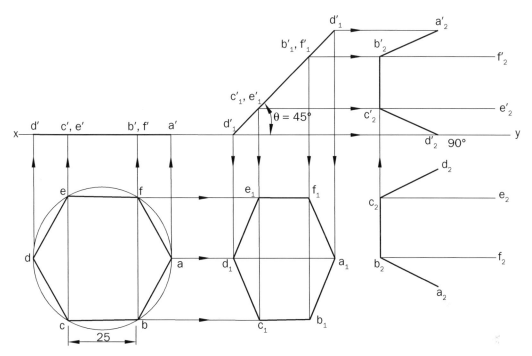

Fig. 10.47 Solution to problem 10.41

Exercises

Traces of Planes

10.1 Show by means of their traces, each of the following planes in (*i*) first quadrant and (*ii*) third quadrant only
   (a) Perpendicular to both HP and VP
   (b) Parallel to and 30 mm from VP
   (c) Parallel to and 30 mm from HP
   (d) Perpendicular to HP and inclined at 30° to VP
   (e) Perpendicular to VP and inclined at 30° to HP

Lamina Parallel to One of the Planes

10.2 An equilateral triangle *ABC* of 40 mm side is parallel to HP with one of its sides inclined at 30° to VP and the end of the side nearer to VP is 20 mm in front of the VP and 25 mm above the HP. Draw its projections and locate its traces too.

10.3   A regular pentagonal lamina *ABCDE* of 25 mm side, has its corner *A* in HP and the side *CD* parallel to the HP. Draw its projections when its plane is parallel to VP. Also locate its traces.

10.4   A regular hexagonal lamina *ABCDEF*, 20 mm side, has its corner A in HP and the side EF perpendicular to HP. Draw its projections when its plane is parallel to and 20 mm from VP. Locate it traces too.

10.5   A square lamina *ABCD* of 25 mm side is parallel to HP and is 10 mm from it. Draw its projections in first quadrant only and locate its traces.

Lamina Inclined to One of the Planes

10.6   A regular hexagonal lamina *ABCDEF* of 25 mm side, has its corner *A* in HP. Its plane is inclined at 45° to the HP and perpendicular to the VP. Draw its projections and locate its traces.

10.7   Draw the projections and traces of a thin circular sheet of 50 mm diameter and of negligible thickness, when its plane is inclined at 45° to HP and is perpendicular to VP. A point on its rim, nearest to the VP, is 20 mm above the HP.

10.8   A regular pentagonal lamina *ABCDE* of 25 mm side, has its side *BC* in HP. Its plane is perpendicular to HP and inclined at 45° to the VP. Draw its projections and locate its traces too.

10.9   A regular hexagonal thin plate of 50 mm side has a circular hole of 50 mm diameter in its centre. It is resting on one of its corners in HP. Draw its projections when the plate surface is vertical and inclined at 30° to the VP.

10.10   A regular hexagonal lamina *ABCDEF* of 25 mm side, rests on one of its sides on HP such that it is perpendicular to VP and inclined to the HP at 45°. Draw its projections and locate its traces too.

10.11   A regular pentagonal lamina *ABCDE* of 25 mm side, has its side *BC* in HP. Its plane is perpendicular to the HP and inclined at 30° to the VP. Draw its projections and locate its traces.

10.12   A thin circular plate of φ 50 mm has a square hole of 25 mm side, cut centrally through it. Draw its projections when the plate is resting on HP with its surface inclined at 30° to the HP and an edge of square hole is perpendicular to VP.

Lamina Inclined to Both HP and VP

10.13   A square lamina *ABCD* of 30 mm side, rests on one of its corners on ground. Its plane is inclined at an angle of 45° to the HP and diagonal *BD* inclined at 60° to the VP and parallel to the HP. Draw its projections.

10.14   A regular pentagonal lamina *ABCDE* of 25 mm side has its corner *A* in HP. Its side *CD* is parallel to the HP and inclined at 45° with the VP. The plane of the pentagon makes an angle of 30° with the HP. Draw its projections.

10.15   An equilateral triangular thin plate *ABC* of 65 mm side has a circle inscribed in it. Draw the projections, when its plane is vertical and inclined at 30° to the VP and one of the sides of the triangle is inclined at 45° to the HP.

10.16   A thin circular plate φ 60 mm appears as an ellipse in the front view, having its major axis 60 mm long and minor axis 40 mm long. Draw its top view when the major axis of the ellipse is horizontal.

10.17 A thin triangular sheet *ABC* has its sides *AB* = 50 mm, *BC* = 45 mm and *CA* = 35 mm. Draw its projections when its side *AB* in VP and inclined at 30° to HP, while its surfaces makes an angle of 45° with the VP.

10.18 A square lamina ABCD of sides 40 mm has one corner A and the VP and another corner D on the HP. The edge AB makes 30 to the HP and 60 to the VP. Draw the projections when the edge DC makes an angle of 30 to the HP. Draw the projections of the square lamina.

10.19 A regular hexagonal lamina of 30 mm sides is standing on an edge on the ground which makes 30 to the VP and the plane itself is at 60° to the ground. Draw the projections of the lamina.

10.20 A circular lamina of 20 mm radius appears as an ellipse of 40 mm major axis and 25 mm minor axis in the view from above. Draw the projections of the lamina.

Objective Questions

10.1 Distinguish between plane and lamina.

10.2 What is the trace of a plane?

10.3 When a plane is perpendicular to both the principal planes, its traces are ....

10.4 What is an oblique plane?

10.5 When a plane is perpendicular to the principal plane, its projection on the plane is ....

10.6 The traces of planes are ....

10.7 What is an edge view of a plane?

10.8 The lines in which the plane meet the principal planes are called .... of the plane.

10.9 Define and classify planes.

10.10 What is the difference between a quadirateral and a polygon?

Answers

10.3 Perpendicular to x-y   10.6 Straight lines

10.5 Line   10.8 Traces

Chapter 11

Auxiliary Projections

11.1 Introduction

Three views of an object, *viz.* the front view, top view and side view, are sometimes not sufficient to provide complete information regarding true shape and size of an object. Additional views are therefore projected on other planes (auxiliary planes) and are known as auxiliary views or auxiliary projections.

11.2 Types of Auxiliary Planes and Views

Auxiliary planes are of two types:

    (a)   Auxiliary Vertical Plane (AVP)
    (b)   Auxiliary Inclined Plane (AIP)

Auxiliary vertical plane (AVP): Auxiliary vertical plane (AVP) is perpendicular to the HP and inclined to the VP. The projection on an AVP is called an auxiliary front view. See Fig. 11.1.

Auxiliary inclined plane (AIP): Auxiliary inclined plane (AIP) is perpendicular to the VP and inclined to the HP. The projection on an AIP is called an auxiliary top view. See Fig. 11.2.
For showing the orthographic projections of an object, the auxiliary plane should always be rotated about the plane to which it is perpendicular.

11.3 Projections of Points

    (a) **Projection of a point on an auxiliary vertical plane (AVP):** A point A is situated above HP and in front of the VP. AVP is a plane perpendicular to the HP and inclined to the VP. The HT of this plane is inclined to x-y and VT perpendicular to x-y line. The HP and the AVP meet at right angles in the line $x_1 y_1$. From the Figs. 11.3 (a) and 11.3 (b), the following points may be observed:

(i) The distance of the auxiliary front view from x_1y_1 is equal to the distance of the front view from x-y, which in turn is the distance of the point *A* from the HP.

(ii) The line $x_1 y_1$ is inclined to x-y at an angle φ, which is the angle of inclination of the AVP with the VP.

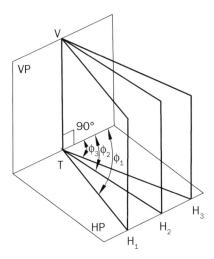

Fig. 11.1 Auxiliary vertical plane (AVP)

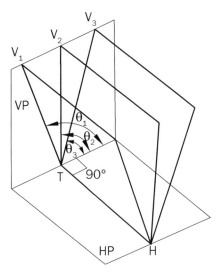

Fig. 11.2 Auxiliary inclined plane (AIP)

To draw the orthographic projections, see Fig. 11.3 (*c*).

(i)   Draw the reference line x-y and mark the front view *a'* and the top view *a*.

(ii)   Draw a new reference line $x_1 y_1$, making an angle φ with x-y.

(iii)   Through the top view *a*, draw a projector aa_1', perpendicular to $x_1 y_1$, intersecting it at a_{01}. Find out the auxiliary front view a_1', such that $a_{01} a_1' = a_0 a'$.

It may be noted that there are four possible positions for the line $x_1 y_1$ relative to the line x-y.

(b)   **Projection of a point on auxiliary inclined plane (AIP):** A point B is situated above HP and in front of the VP. AIP is a plane perpendicular to the VP and inclined to the HP. The VT of this plane is inclined to x-y and the HT perpendicular to x-y line. The VP and the AIP meet at right angles in the line $x_1 y_1$. From the Figs. 11.4 (*a*) and 11.4 (*b*), the following points may be noted:

(i)   The distance of the auxiliary top view from $x_1 y_1$ is equal to the distance of top view from x-y, which in turn is the distance of the point *B* from the VP.

(ii)   The line $x_1 y_1$ is inclined to x-y at an angle θ, which is the angle of inclination of the AIP with the HP.

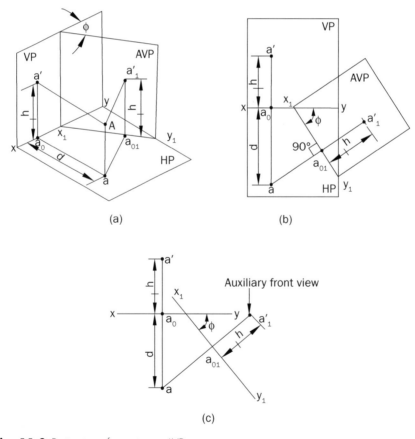

(a)   (b)

(c)

Fig. 11.3 Projection of a point on AVP

To draw the orthographic projections. See Fig. 11.4 (*c*).

(i)   Draw the reference line x-y and mark the front view b' and the top view b.
(ii) Draw a new reference line $x_1 y_1$, making an angle θ with x-y.
(iii) Through the front view b', draw a projector $b'b_1$, perpendicular to $x_1 y_1$, intersecting it at b_{01}. Find out the auxiliary top view b_1, such that $b_{01} b_1 = b_0 b$.

It may be noted that there are four possible positions for the line $x_1 y_1$ relative to the line x-y.

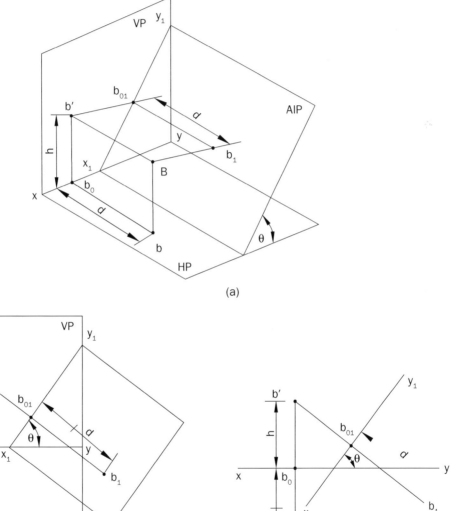

(a)

(b)   (c)

Fig. 11.4 Projection of a point on AIP

Problem 11.1 A point *A* is 30 mm above the HP and 40 mm in front of the VP. Draw an auxiliary front view on an AVP inclined at 60° to the VP.

Solution:

    (i)   Draw the reference line x-y and mark the front view a' and the top view a.

    (ii)   Draw a new reference line $x_1 y_1$, making an angle 60° with x-y. See Fig. 11.5.

    (iii) Through the top view a, draw a projector aa_1' perpendicular to $x_1 y_1$, intersecting it at a_{01}
Find out the auxiliary front view a_1' such that $30 = a_{01} a_1' = a_0 a'$.

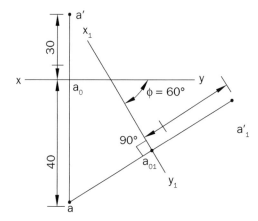

Fig. 11.5 Solution to problem 11.1

Problem 11.2 A point *B* is 30 mm above the HP and 40 mm in front of the VP. Draw an auxiliary top view on an AIP inclined at 45° to the HP.

Solution:

    (i)   Draw the reference line x-y and mark the front view b' and the top view b.

    (ii)   Draw a new reference line x-y, making an angle 45° with x-y. See Fig. 11.6.

    (iii) Through the front view b', draw a projector $b'b$ perpendicular to $x_1 y_1$, intersecting it at b_{01}.

Find out the auxiliary top view b_1 such that $40 = b_1 b_{01} = b_1 b_0$.

11.4 Projections of Straight Lines

It is similar to the projections of points on auxiliary planes. A straight line is the shortest distance between two points. When these points are projected and then these projections are joined, we obtain the projections of the straight lines.

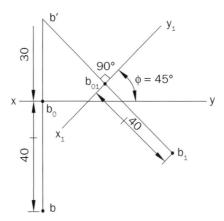

Fig. 11.6 Solution to problem 11.2

Problem 11.3 The end A of a line AB is 12 mm above HP and 30 mm in front of VP End B is 40 mm above HP and 15 mm in front of VP. Distance between the end projectors is 45 mm. Draw its front and top views and determine the true inclination with HP and VP by auxiliary plane method.

Solution:

(i)   Draw the given views of the line.

(ii) Draw a new reference line $x_1\, y_1$ to represent on AIP parallel to $a'b'$.

(iii) Project the auxiliary top view $a_1\, b_1$, which is the true length of the given line and its inclination with $x_1\, y_1$, is the true inclination of the line with VP.

(iv) Draw the another reference line $x_2\, y_2$ to represent an AVP parallel to ab.

(v)   Project the auxiliary front view $a_1'\, b_1'$ which is the true length of given line and its inclination with $x_2\, y_2$ is the true inclination of the line with HP. See Fig. 11.7.

Problem 11.4 The end A of a line AB is a 10 mm in front of VP and 40 mm above HP. The end B is 30 mm in front of VP and 10 mm above the HP. The distance between the end projectors is 45 mm. Draw the projections of the line and project an auxiliary front view on an auxiliary vertical plane (AVP) inclined to the VP at an angle of 45°.

Solution:

(i)   Draw the given views of the line.

(ii) Draw a new reference line $x_1\, y_1$ to represent an AVP inclined at 45° to the x-y line.

(iii) Draw the perpendicular projectors through a and b to new reference line $x_1\, y_1$.

(iv) Project the auxiliary front view. See Fig. 11.8.

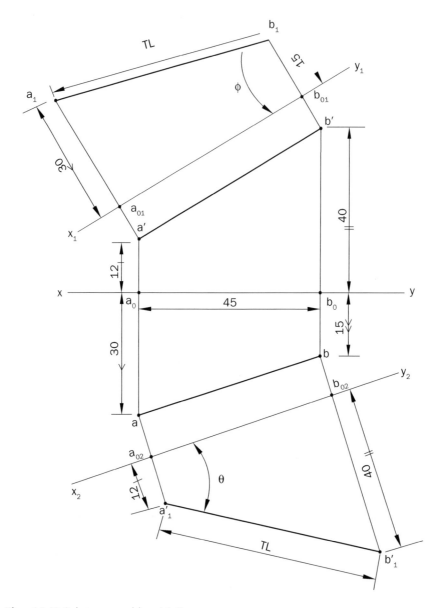

Fig. 11.7 Solution to problem 11.3

Problem 11.5 The end A of a line AB is 10 mm in front of VP and 40 mm above HP. The end B is 35 mm in front of VP and 10 mm above the HP. The distance between the end projectors is 40 mm. Draw the projections of the line and project an auxiliary top view on as auxiliary inclined plane (AIP) inclined to the HP at an angle of 60°.

Solution:

   (i)   Draw the given views of the line.

  (ii)   Draw a new reference line $x_1 y_1$ inclined at 60° to the x-y line, to represent an AIP.

 (iii)   Draw the perpendicular projectors through a' and b' to new reference line $x_1 y_1$.

 (iv)   Project the auxiliary top view. See Fig. 11.9.

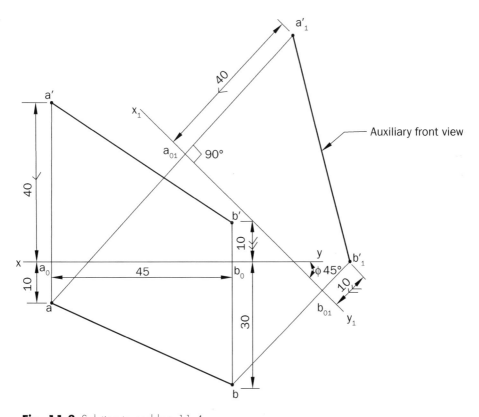

Fig. 11.8 Solution to problem 11.4

Problem 11.6 A straight line AB 70 mm long makes an angle of 45° to the HP and 30° to the VP. The end A is 15 mm in front of VP and 20 mm above HP. Draw the plan and elevation of the line AB.

(PTU, Jalandhar May 2005)

Solution:

   (i)   Draw the front view $a'b'$ and top view ab, assuming the line AB to be parallel to both HP and VP.

  (ii)   Draw a new reference line $x_1 y_1$ inclined at 45° to the x-y line, to repesent an AIP. Project a new top view $a_1 b_1$, which is still parallel to $x_1 y_1$.

(iii) The projections $a_1 b_1$ and a'b' are the primary auxiliary top view and front view of a line respectively, when it is inclined at an angle 45° to the HP and parallel to the VP.

(iv) For a line inclined to both the principal planes, neither of the projections of the line project as true length (TL). Thus none of the projections will make true angle of inclinations (i.e., θ or φ) with their reference line. Instead, the projections will make apparent inclinations (i.e., θ' or φ') with their reference line.

(v) Draw a new reference line $x_2 y_2$ inclined at φ' to the $x_1 y_1$ line. To find φ', draw a line inclined at an angle of 30° to $a_1 b_1$ at a_1 and cut $a_1 b_2 = 70$ mm (TL). With a_1 as centre and $a_1 b_1$ as radius draw an arc that intersects the locus of B at b_3. Join a_1 with b_3 and measure its inclination φ' with $a_1 b_1$.

(vi) Draw another reference line $x_2 y_2$ parallel to $a_1 b_3$ line. Project the required front view $a_1' b_1'$ on $x_2 y_2$, such that $a_1' a_{02} = a_{01}' a_1$, $b_1' b_{02} = b' b_{01}$. See Fig. 11.10.

11.5 Projections of Planes

Auxiliary projections of planes are used to obtain

(i) the edge view of a plane and
(ii) the true shape and size of an object.

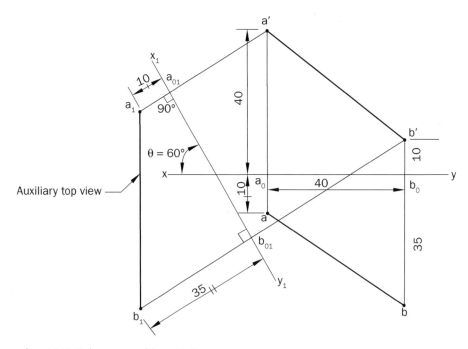

Fig. 11.9 Solution to problem 11.5

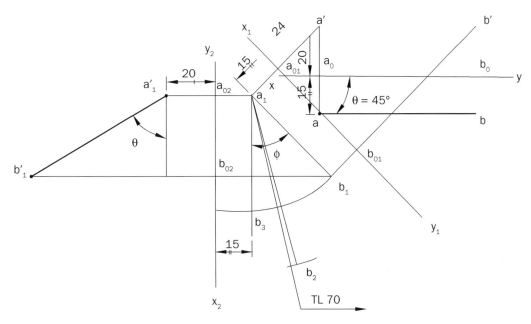

Fig. 11.10 Solution to problem 11.6

When a plane is inclined to both the principal planes, none of its projections gives the edge view or the true shape. The true shape of the plane is obtained after getting its edge view. The edge view of a plane is a straight line, when it is projected on a plane, perpendicular to the true length of any one of its elements. The projection of the line view or edge view of a plane on an auxiliary plane parallel to it shows the true shape and size of a plane figure.

Problem 11.7 A regular pentagonal lamina ABCDE of 25 mm side rests on HP on its corner D such that the plane of lamina is inclined to HP at 45°. Draw its projections by auxiliary plane method.

Solution:

    (i)   Assuming the pentagon to be in the HP, draw the top view in the initial position by keeping the side *ab* perpendicular to the x-y line. Complete the front view too.

    (ii)   Then project the primary auxiliary top view, by taking a new reference line $x_1 \, y_1$ inclined at 45° to the front view. See Fig. 11.11.

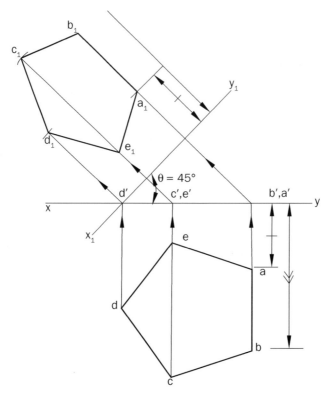

Fig. 11.11 Solution to problem 11.7

Problem 11.8 A regular hexagonal lamina ABCDEF of 20 mm side, rests on one of its sides on HP such that the plane of lamina is inclined to HP at 45°. Draw its projections by auxiliary plane method.

Solution: The solution to this problem is self-explanatory. See Fig. 11.12.

Problem 11.9 The front view and top view of a triangular lamina are represented each by an equilateral triangle with projections of side *AB* perpendicular to *XY* line in both elevation and plan. Draw the true shape of the lamina if the side in the projected view is 50 mm.

(*PTU, Jalandhar December 2002*)

Solution:

    (i)   Draw the given views of the triangular lamina.

    (ii)   Through a corner c', draw a line parallel to x-y and meeting *a'b'* at *d'*.

    (iii)   Project *d'* to *d* on the line *ab* in the top view. *cd* is the true length of the element *CD*.

(iv) Draw a reference line $x_1 y_1$ perpendicular to cd. Project a new front view from the top view. It is an edge view $a_1' b_1'$.

(v) Draw another reference line $x_2 y_2$ parallel to an edge view $a_1' b_1'$ and project on it a new top view $a_1 b_1 c_1$, which shows the true shape and size of the triangular lamina. See Fig. 11.13.

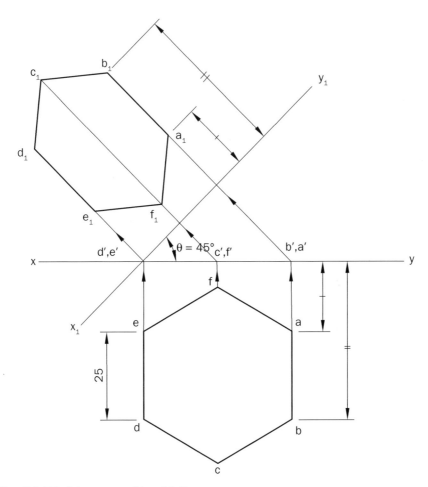

Fig. 11.12 Solution to problem 11.8

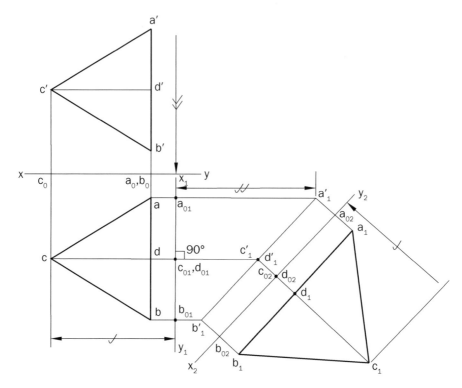

Fig. 11.13 Solution to problem 11.9

11.6 Shortest Distance between Two Skew Lines

Skew lines are lines which are non-parallel and non-intersecting. The shortest distance between two skew lines is a straight line which is perpendicular to both the lines. Therefore, to find the shortest distance between two skew lines, auxiliary views are drawn such that at least one of the lines appears as a point. Then the perpendicular distance from the point view of the line to the corresponding view of another line gives the required shortest distance.

Problem 11.10 *AB* and CD are the two skew lines with the following details:
Line *AB*: The end *B* is 20 mm above the HP and 30 mm in front of the VP. The end *A* is 10 mm above the HP and 15 mm in front of the VP. The distance between end projectors is 50 mm.

Line CD: The end C is 5 mm above the HP and 50 mm in front of the VP. The end D is 35 mm above the HP and 10 mm in front of the VP. The distance between end projectors is 25 mm. The end D is 10 mm away from the end B in the front view measured parallel to x-y line. Draw its projections and find the shortest distance between the lines.

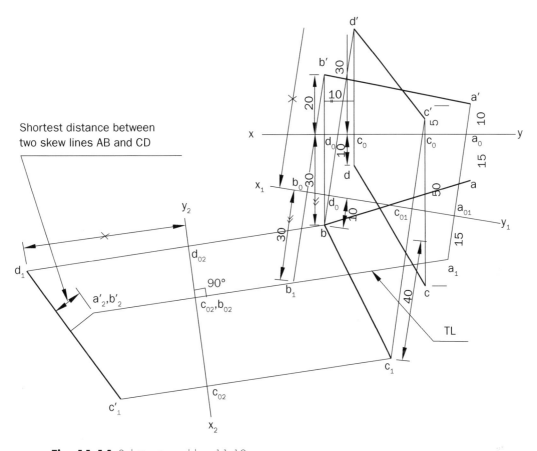

Fig. 11.14 Solution to problem 11.10

Solution:

(i)   Draw $a'b'$, $c'd'$ the front views and ab, cd the top views of the given lines AB and CD.

(ii) Draw a new reference line $x_1 y_1$ parallel to any one of the lines in the front or top views. Here $x_1 y_1$ is drawn parallel to the $a' b'$ at a suitable distance.

(iii) Draw the primary auxiliary top views $a_1 b_1$ and $c_1 d_1$ of the given lines.

(iv) Draw another reference lines $x_2 y_2$ perpendicular to the auxiliary top view $a_1 b_1$ at any convenient distance and project the secondary auxiliary front views $a_1'b_1'$ and $c_1'd_1'$ of the lines, as described in the previous problems.

(v)   Draw the perpendicular from a_1' or b_1' to $c_1' d_1'$, which gives the shortest distance between the two skew lines. See Fig. 11.14.

Additional Problems

Problem 11.11 A regular hexagonal lamina *ABCDEF* of 25 mm side rests on its side in VP. The plane of lamina is inclined at 45° to the VP and the side on which it rests in the VP is inclined at 60° to the HP. Draw its projections by auxiliary plane method.

Solution:

(i)   Draw a regular hexagon $a'b'c'd'e'f'$ in the front view such that its side $a'b'$ (or $e'd'$) perpendicular to x-y. Complete the top view *abcdef*.

(ii) Draw a new reference line $x_1 y_1$ inclined at 45° to the top view. Project the primary auxiliary front view.

(iii) Draw another new reference line $x_2 y_2$ inclined at 60° to $d_1'e_1'$.

(iv) Project all the points fron the primary auxiliary front view, perpendicular to the $x_2 y_2$ line.

(v)   Complete the secondary auxiliary top view as shown in Fig. 11.15.

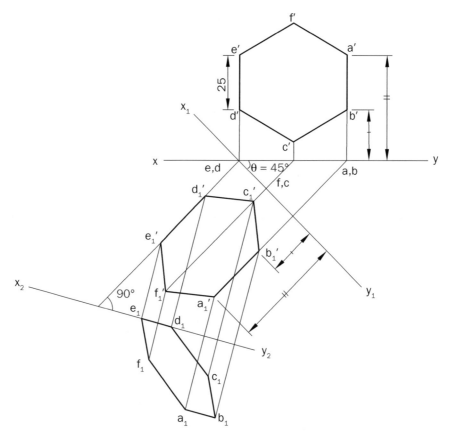

Fig. 11.15 Solution to problem 11.11

Problem 11.12 A triangular lamina *PQR* has its vertices *P*, *Q* and *R* are 60 mm, 20 mm and 45 mm above the HP and 25 mm, 60 mm and 5 mm in front of the VP, respectively. The projectors of *P* and *Q* are 40 mm apart and those of *Q* and *R* are 30 mm apart. Draw the true shape of the lamina and true inclinations with the reference planes.

Solution:

(i)   Draw triangles *p' q' r'*, *pqr* in the front view and top view respectively.

(ii)   Through a corner *r'*, draw a line parallel to x-y and meeting *p' q'* at *s'*.

(iii) Project *s'* to *s* on the line *pq* in the top view. Join *rs* which represents the true length of the element *RS*.

(iv) Draw a new reference line $x_1 y_1$ perpendcular to *rs*. Project a new front view from the top view which is a line view (or edge view) *p' q'*. Measure angle between $x_1 y_1$ and the line view (or edge view) $p_1' q_1'$ which is known as inclination of the plane with *HP*.

(v)   Draw another reference line $x_2 y_2$ parallel to an edge view *p' q'* and project on it a new top view $p_1 q_1 r_1$ which shows the true shape and size of the triangular lamina. See Fig. 11.16.

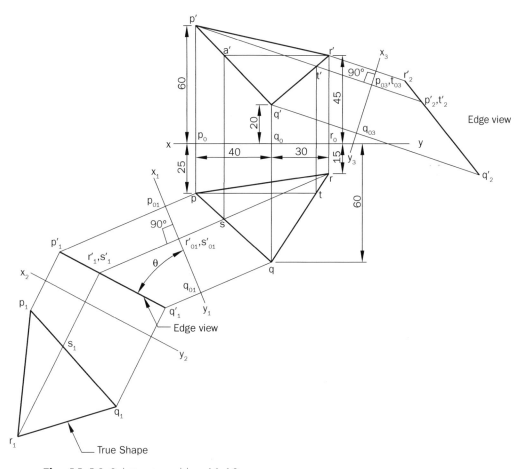

Fig. 11.16 Solution to problem 11.12

(vi) Through a corner p, draw a line parallel to x-y and meeting qr at t. Project t to t' on the line $q'r'$ in the front view. Join p' t' which represents the true length of the element PT.

(vii) Draw a new reference line x_3 y_3 perpendicular to p' t'. Project a new top view from the front view, which is a line or edge view q_3' r_3'. Measure angle between x_3 y_3 and the line view (or edge view) q_2' r_2', which is termed as angle of inclination of the lamina with the VP.

Problem 11.13 The projections a' b' c' d' and $abcd$ of a thin square metal sheet are shown in Fig 11.17. Reproduce the given views and determine the true shape of the plate by using auxiliary plane method.

Solution:

(i)   Draw the given views of the square sheet.
(ii) Through a corner b', draw a line parallel to x-y and meeting d' a' at e'.
(iii) Project e' to e on the line da in the top view. be is the true length of the element BE.
(iv) Draw a new reference line x_1 y_1 perpendicular to be. Project a new front view from the top view.
(v)   Draw another reference line x_2 y_2 parallel to an edge view $a_1'c_1'$ and project on it a new top view a_1 b_1 c_1 d_1 which shows the true shape and size of the square metal sheet. See Fig. 11.17.

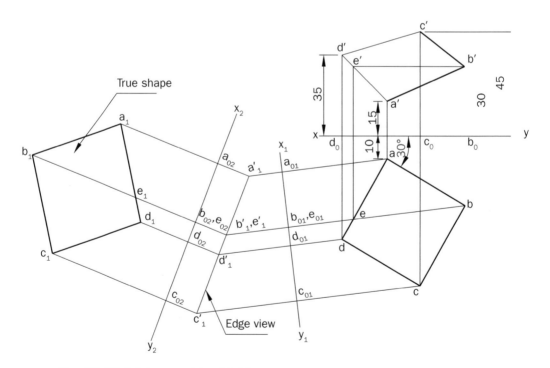

Fig. 11.17 Solution to problem 11.13

Exercises

11.1   A point P is 25 mm above the HP and 35 mm in front of the VP. Draw an auxiliary front view on an AVP inclined at 45° to the VP.

11.2   A point Q is 30 mm above the HP and 45 mm in front of the VP. Draw an auxiliary top view on an AIP inclined at 30° to the HP.

11.3   The end A of a line AB is 10 mm above HP and 25 mm in front of VP. The end B is 30 mm above HP and 15 mm in front of VP. The distance between end projectors is 40 mm. Draw its front view and top view. Also determine its TL and inclination to (*i*) the HP and (*ii*) the VP, using auxiliary plane method.

11.4   The end E of a line EF is 10 mm above HP and 30 mm in front of VP. The end F is 25 mm above HP and 12 mm in front of VP. The distance between end projectors is 40 mm. Draw the projections of the line and project an auxiliary front view on an auxiliary vertical plane inclined to the VP at an angle of 60°.

11.5   The end E of a line EF is 10 mm above HP and 30 mm in front of VP. The end F is 35 mm above HP and 15 mm in front of VP. The distance between end projectors is 45 mm. Draw the projections of the line and project an auxiliary top view on an auxiliary inclined plane inclined to the HP at an angle of 45°.

11.6   A line CD 60 mm long makes an angle of 30° to the HP and 45° to the VP. The end C is 15 mm in front of VP and 20 mm above HP. Draw the plane and elevation of the line CD.

11.7   The end C of a line CD is 40 mm above HP and 10 mm in front of VP. The end D is 15 mm above HP and 35 mm in front of VP. The distance between end projectors is 40 mm. Draw the projections of the line and determine the shortest distance of the line from x-y, using auxiliary plane method.

11.8   A square lamina ABCD of 30 mm side, has its sides equally inclined to the VP and the plane of lamina is inclined to HP at 45°. Draw its projections by auxiliary plane method.

11.9   A regular hexagonal lamina ABCDEF of 30 mm side, rests on one of its sides in HP. Its plane is inclined to the HP at 45°. Draw its projections by auxiliary plane method.

11.10   The projections *a'b'c'* and *abc* of a triangular lamina are shown in Fig. 11.18. Reproduce the given views and determine the true shape of the lamina.

11.11   The front and top views of two skew lines AB and CD are shown in Fig. 11.19. Determine the shortest distance between the skew lines.

Objective Questions

11.1   To present the view, the auxiliary plane should be rotated about the plane to which it is ....

11.2   The plane which is .... to the VP and .... to the HP, is called auxiliary vertical plane (AVP).

11.3   The plane which is inclined to the HP and perpendicular to the VP, is called ....

11.4   There are .... possible positions at which auxiliary views may be drawn.

11.5   The distance from .... to a line is known as shortest distance.

11.6   The projection obtained on an auxiliary vertical plane is known as ....

11.7   What do you mean by auxiliary views?

11.8   When are auxiliary views preferred?

11.9   What is the edge view of a plane ? How can it be obtained?

11.10 Distinguish between primary and secondary auxiliary views.

11.11 Draw the projections of a line inclined to both HP and VP. Using auxiliary planes, convert this line into a point in one of the views.

11.12 Draw the traces of auxiliary inclined plane (AIP) and auxiliary vertical plane (AVP).

11.13 What are skew lines?

11.14 When the auxiliary planes are used?

11.15 Differentiate between AVP and AIP.

Answers

11.1   Perpendicular

11.2   Inclined, perpendicular

11.3   Auxiliary inclined plane (AIP)

11.4   Four

11.5   A point

11.6   Auxiliary front view

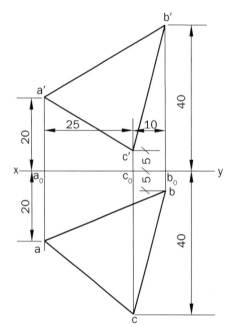

Fig. 11.18

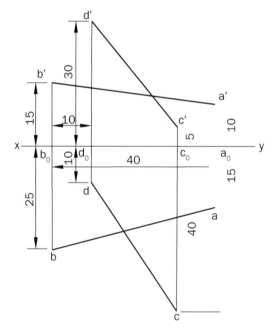

Fig. 11.19

Chapter 12

Projections of Solids

12.1 Introduction

A solid is a three-dimensional object having length, breadth and thickness. In engineering practice, one often comes across solids bounded by simple or complex geometric surfaces. To represent a solid in orthographic projections, the number and types of views necessary will depend upon the type of solid and its orientation with respect to the principal planes of projections. Sometimes, additional views projected on auxiliary planes become necessary to describe a solid completely.

12.2 Types of Solids

Solids may be divided into two main groups:

    (i)   **Polyhedra**
        (a) Regular Polyhedra
        (b) Prisms
        (c) Pyramids
   (ii) **Solids of Revolution**
        (a) Cylinders
        (b) Cones
        (c) Sphere

Polyhedra: A polyhedron is defined as a solid bounded by planes called faces.

   (a) *Regular Polyhedra*: A polyhedra is said to be regular if all its faces are similar, equal and regular. There are five regular polyhedra which may be defined as stated below:

    •   Tetrahedron: It has four equal faces, each are equilateral triangle as shown in Fig. 12.1

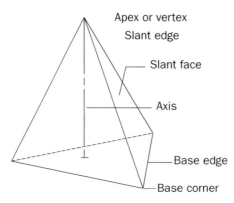

Fig. 12.1 Tetrahedron

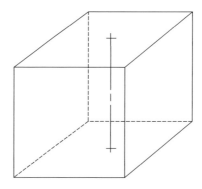

Fig. 12.2 Cube

- Cube or hexahedron: It has six equal faces, each of which is a square as shown in Fig. 12.2.
- Octahedron: It has eight equal faces, each of which is an equilateral triangle as shown in Fig. 12.3.
- Dodecahedron: It has twelve equal faces, each of which is a regular pentagon as shown in Fig. 12.4.

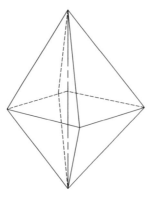

Fig. 12.3 Octahedron

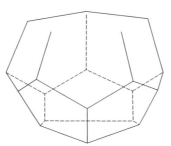

Fig. 12.4 Dodecahedron

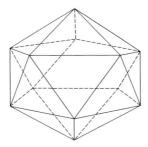

Fig. 12.5 Icosahedron

- Icosahedron: It has twenty equal faces, each of which is an equilateral triangle as shown in Fig. 12.5.
(b) *Prisms*: A prism is a polyhedron having two equal ends or bases, parallel to each other. The two bases are joined by faces which are rectangles or parallelograms. The imaginary line joining the centres of the bases is called the axis.

A right regular prism has its axis perpendicular to the base. All its faces are equal rectangles as shown in Fig. 12.6.

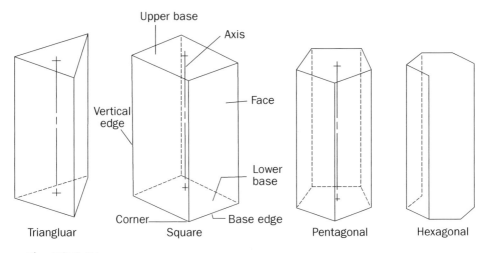

Fig. 12.6 Prisms

(c) *Pyramids*: A pyramid is a polyhedron having one base and a number of triangles as faces meeting at a point called as vertex or apex. The imaginary line joining the centre of the base with its apex or vertex is called its axis.

A right regular pyramid has its axis perpendicular to the base. All its faces are equal as shown in Fig. 12.7.

Prisms and pyramids are named according to the shape of their bases, such as triangular, square, pentagonal, hexagonal, etc., as shown in Figs. 12.6 and 12.7.

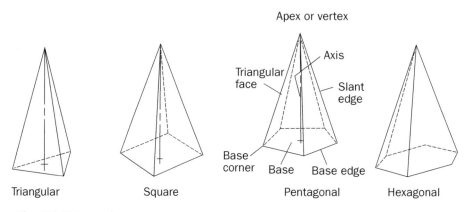

Fig. 12.7 Pyramids

Incase of oblique solids, the axis is inclined to the base. The faces of an oblique prism are parallelograms and that of pyramids are triangles, which are not similar as shown in Figs. 12.8 and 12.9. The bases of oblique prisms are parallel, equal and similar.

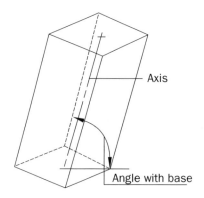

Fig. 12.8 Oblique prism

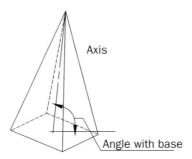

Fig. 12.9 Oblique pyramid

Solids of revolution: Solids of revolution are obtained or generated by rotating a plane figure about one of its edges.

(a) *Cylinder*: A cylinder is generated by rotating a rectangle about one of its edges, which remains fixed. It has two equal circular bases. The imaginary line joining the centres of the bases is called its axis.

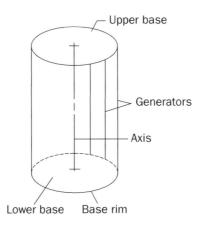

Fig. 12.10 Right cylinder

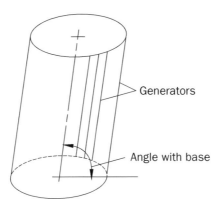

Fig. 12.11 Oblique cylinder

A cylinder is said to be right when its axis is perpendicular to the base. The lines drawn on the surface of a cylinder and parallel to the axis are known as generators. The length of the generator is equal to the height of the cylinder, as shown in Fig. 12.10.

A cylinder is said to be oblique when the axis is inclined to the base. The lateral surface is connected by two equal and parallel circular base as shown in Fig. 12.11.

    (b) *Cone*: A cone is generated by rotating a right angled triangle about one of its perpendicular sides. The lateral surface is connected by a circular base. A line drawn from the vertex to any point on the base of a cone is known as generator, where length is equal to the slant height of the cone as shown in Fig. 12.12.

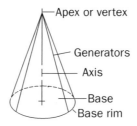

Fig. 12.12 Right cone

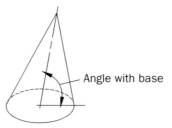

Fig. 12.13 Oblique cone

A cone is said to be oblique when the axis is inclined to the base. The base of the cone is a circle. The generators drawn on the lateral surface connecting the base and the apex are of unequal length as shown in Fig. 12.13.

    (c) *Sphere*: A sphere is a solid generated by rotating a semi-circle about its diameter as shown in Fig. 12.14. The midpoint of the diameter is the centre of the sphere. All points on the surface of a sphere are equidistant from its centre.

Fig. 12.14 Sphere

12.3 Projections of Solids in Different Positions

Various positions that a solid in space can make with reference to the principal planes of projection and the corresponding projections are illustrated in the subsequent paragraphs.

12.4 Axis Perpendicular to One of the Principal Planes and Parallel to the Other

Projections of such solids are drawn in one step only. Therefore, the projections of a solid on the plane to which its axis is perpendicular will show the true shape and size of its base.

(a) **Axis perpendicular to the HP and parallel to the VP:** When the axis is perpendicular to the HP, the top view should be drawn first and then front view is projected from it.

Problem 12.1. Draw the three views of a cube 30 mm side when it is resting on its base on HP with one of the base edges making an angle of 45° to the VP.

(PTU, Jalandhar December 2007)

Solution:

(i) Draw a square of base edge 30 mm in the top view, keeping its base edge, say, d-c or 4-3, inclined at angle of 45° to x-y line.

(ii) Project the front view and end view as shown in Fig. 12.15. Name all the corner points.

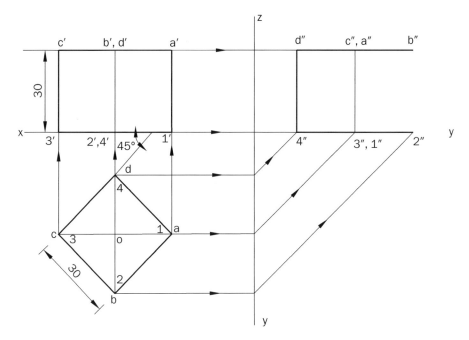

Fig. 12.15 Solution to problem 12.1

Problem 12.2 Draw the projections of the following solids, resting in HP on their bases as given, using reference line x-y.

(a) A cylinder, 40 mm base diameter and 60 mm height.
(b) A cone, 40 mm base diameter and 60 mm height.
(c) A right regular hexagonal prism, side of base 25 mm and axis 50 mm long, having one of its base edges parallel to the VP.
(d) A right regular hexagonal prism, side of base 25 mm and axis 55 mm long, having one of its base edges perpendicular to VP.
(e) A right regular pentagonal prism, side of base 25 mm and axis 55 mm long, having one of its base edges parallel to VP.
(f) A right regular pentagonal prism, side of base 25 mm and axis 55 mm long, having one of its base edges perpendicular to VP.
(g) A right regular pentagonal pyramid, edge of base 25 mm and height 60 mm long, having an edge of its base perpendicular to VP.
(h) A right regular pentagonal pyramid, edge of base 25 mm and height 60 mm long, having an edge of its base parallel to VP.
(i) A right regular hexagonal pyramid, edge of base 25 mm and height 60 mm long, having one of its base edges perpendicular to VP.
(j) A right regular pentagonal pyramid, edge of base 25 mm and axis 45 mm long, having one of its base edge inclined at an angle 30° to the VP.

(PTU, Jalandhar December 2004, May 2005)

Solution:

(a) Draw a circle of 40 mm diameter in the top view and project the front view which will be a rectangle as shown in Fig. 12.16.

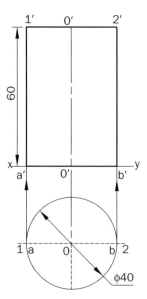

Fig. 12.16 Solution to problem 12.2 (a)

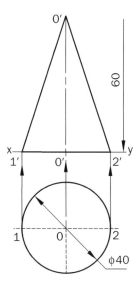

Fig. 12.17 Solution to problem 12.2 (b)

(b) Draw a circle of 40 mm diameter in the top view. Through the centre O, project the apex or vertex O', 60 mm from x-y. Complete the triangle in the front view as shown in Fig. 12.17.

(c) Draw a regular hexagon of 25 mm side in the top view, keeping one of its sides parallel to the x-y. Project the front view from top view and cut the axis 50 mm long. Name the corner points as shown in Fig. 12.18 and complete the views.

(d) Draw a regular hexagon of 25 mm side in the top view, keeping one of its sides perpendicular to the x-y. Project the front view from top view and cut the axis 55 mm long. Name the corner points as shown in Fig. 12.19 and complete the views.

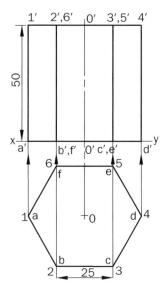

Fig. 12.18 Solution to problem 12.2 (c)

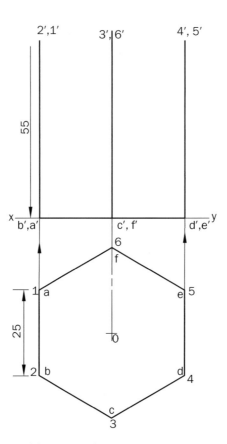

Fig. 12.19 Solution to problem 12.2 (d)

(e) Draw a regular pentagon of 25 mm side in the top view, keeping one of its base edges parallel to x-y. Project the front view from top view and cut the axis 55 mm long. Name the corner points as shown in Fig. 12.20 and complete the views.

(f) Draw a regular pentagon of 25 mm side in the top view, keeping one of its base edges perpendicular to x-y. Project the front view from top view and cut the axis 55 mm long. Name the corner points as shown in Fig. 12.21 and complete the views.

(g) Draw a regular pentagon of 25 mm side in the top view, keeping one of its base edges perpendicualr to x-y. Through the centre O, project the apex or vertex O' 60 mm from x-y. Complete the views as shown in Fig. 12.22. Name all the corner points.

(h) Draw a regular pentagon of 25 mm side in the top view, keeping one of its base edges parallel to x-y. Through the centre O, project the apex O' 60 mm from x-y. Complete the views as shown in Fig. 12.23. Name all the corner points.

(i) Draw a regular hexagon of 25 mm side in the top view, keeping one of its base edges perpendicular to x-y. Through the centre O, project the apex or vertex O', 50 mm from x-y. Complete the views as shown in Fig. 12.24.

(j) Draw a pentagonal pyramid of 25 mm base edge in the top view, with a base edge 5-1 inclined at 30° to x-y. Project the front view correspondingly as shown in Fig. 12.25.

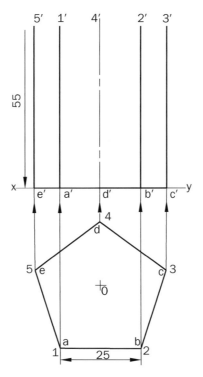

Fig. 12.20 Solution to problem 12.2 (e)

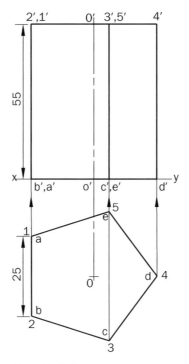

Fig. 12.21 Solution to problem 12.2 (f)

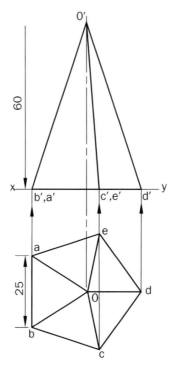

Fig. 12.22 Solution to problem 12.2 (g)

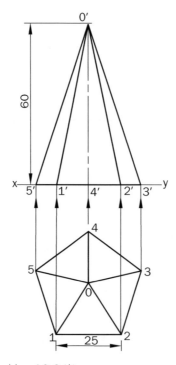

Fig. 12.23 Solution to problem 12.2 (h)

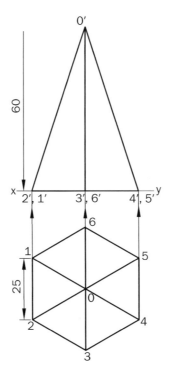

Fig. 12.24 Solution to problem 12.2 (*i*)

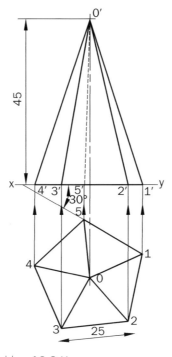

Fig. 12.25 Solution to problem 12.2 (*j*)

Problem 12.3 A right regular hexagonal prism, side of base 25 mm and axis 50 mm long, has one of its base edges parallel to theVP with its axis perpendicular to the HP. Draw its front, top and side views.

Solution: As the axis of a hexagonal prism is perpendicular to HP, hence projections need to start from top view.

    (i)   Draw a regular hexagon of 25 mm side in the top view, keeping one of its base sides parallel to x-y.

    (ii)   Project front and left side views from top view. Mark the axis to be 50 mm long. Name the corner points as shown in Fig. 12.26 and complete the views.

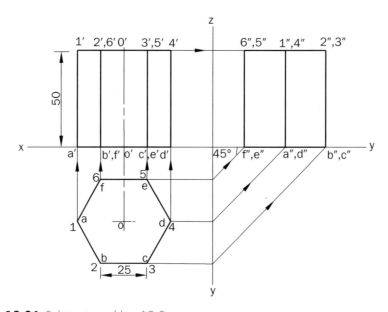

Fig. 12.26 Solution to problem 12.3

Problem 12.4 A right regular hexagonal pyramid, edge of base 25 mm and height 60 mm long, having one of its base edges perpendicular to VP, with its axis perpendicular to the HP. Draw the three views of the pyramid.

Solution: As the axis of a hexagonal pyramid is perpendicular to VP, so projections are to be started from front view.

    (i)   Draw a hexagonal pyramid of 25 mm base edge in the top view, with a base edge 2-1 or 5-4 to be perpendicular to x-y.

    (ii)   Project front and left side views from top view. Name the corner points as shown in Fig. 12.27 and complete the views.

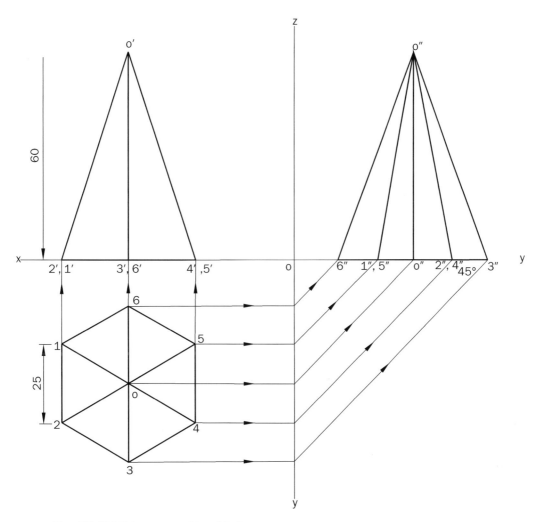

Fig. 12.27 Solution to problem 12.4

Problem 12.5 A cube of 35 mm edge is resting on the HP on one of its faces, with a vertical face inclined at 30° to the VP. Draw the three views of the cube.

(*PTU, Jalandhar May 2009*)

Solution: All the construction lines are retained to make the solution self-explanatory. See Fig. 12.28.

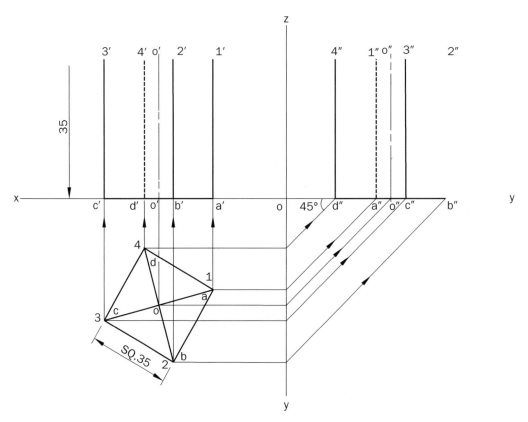

Fig. 12.28 Solution to problem 12.5

Problem 12.6 A square pyramid, edge of base 30 mm and height 45 mm, rests on its base on HP, with its base edges equally inclined to the VP. Draw the three views of the square pyramid.

Solution: The interpretation of the solution is left to the reader. See Fig. 12.29.

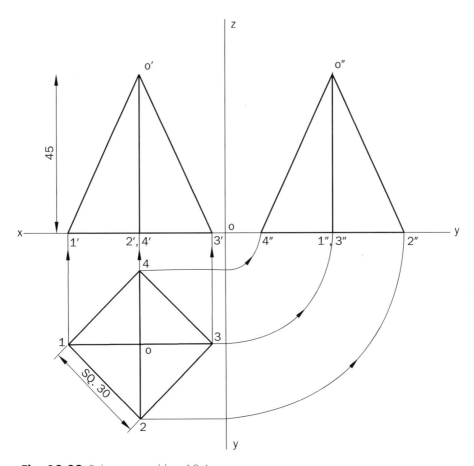

Fig. 12.29 Solution to problem 12.6

Problem 12.7 Draw the projections of a square pyramid of base edges 30 mm and axis 54 mm, resting on its base on HP with one of the base edge parallel to VP and axis perpendicular to the HP.

(*PTU, Jalandhar December 2005*)

Solution:

(i) Draw a square of base edge 30 mm in the top view, keeping one of its base edges parallel to VP.

(ii) Project front view from top view and cut the axis 54 mm long. Complete the end view too. Name all the corner points as shown in Fig. 12.30.

(b) **Axis perpendicular to the VP and parallel to the HP:** When the axis is perpendicular to the VP, the front view should be drawn first and then top view is projected from it.

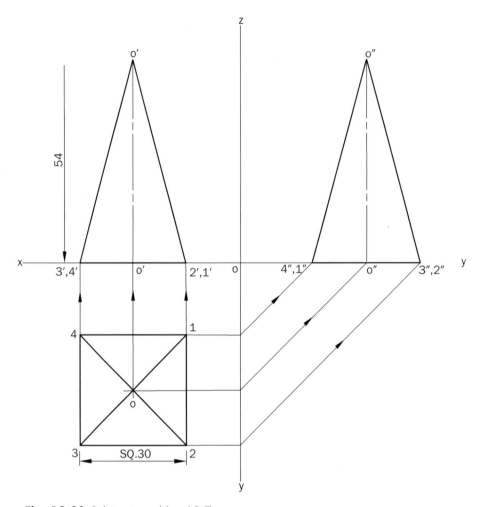

Fig. 12.30 Solution to problem 12.7

Problem 12.8 Draw the projections of the following solids

   (a) A right regular hexagonal prism, side of base 25 mm and axis 60 mm long, lies on one
        of its rectangular faces on HP with its axis perpendicular to VP.
   (b) A right regular pentagonal prism, side of base 25 mm and axis 60 mm long, lies on one
        of its rectangular faces on HP with its axis perpendicular to VP.
   (c) A right regular hexagonal pyramid, edge of base 25 mm and height 60 mm long, has its
        base parallel to VP with one of its base edges in HP.

Solution:

   (a) Draw a regular hexagon of 25 mm side in the front view, keeping one of its rectangular
        faces on x-y. Project the top view from front view and cut the axis 60 mm long. Complete
        the views as shown in Fig. 12.31. Name all the corner points.

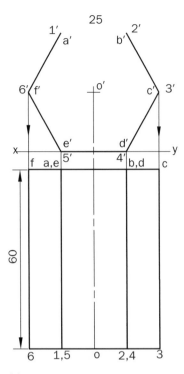

Fig. 12.31 Solution to problem 12.8 (a)

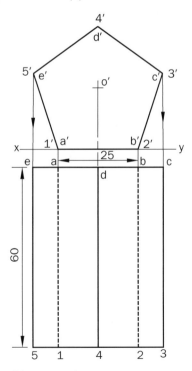

Fig. 12.32 Solution to problem 12.8 (b)

(b) Draw a regular pentagon of 25 mm side in the front view, keeping one of its rectangular faces on x-y. Project the top view from front view and cut the axis 60 mm long. Complete the views as shown in Fig. 12.32. Name all the corner points.

(c) Draw a regular hexagon of 25 mm side in the front view, keeping one of its base edges parallel to x-y. Through the centre O', project the vertex O, 60 mm long. Complete the views as shown in Fig. 12.33.

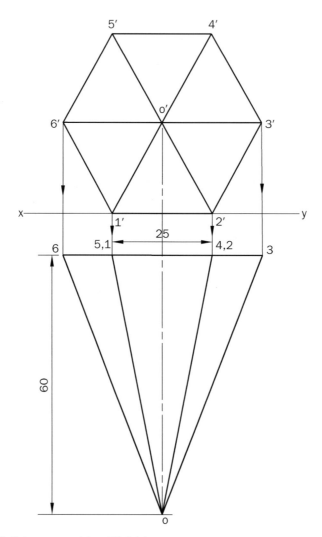

Fig. 12.33 Solution to problem 12.8 (c)

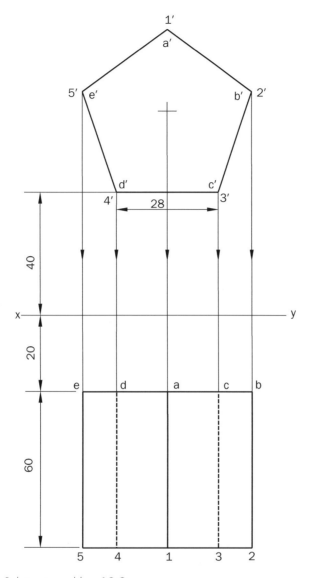

Fig. 12.34 Solution to problem 12.9

Problem 12.9 A right regular pentagonal prism side of base 28 mm and axis 60 mm long has one of its rectangular faces parallel to HP with its axis perpendicular to VP and 40 mm above HP. Draw the elevation and plan of the pentagonal prism when the nearest end is 20 mm away from the VP.

(PTU, Jalandhar December 2004)

Solution: As the axis of a pentagonal prism is perpendicular to VP, so projections are to be started from front view.

(i)   Draw a pentagonal prism with rectangular face *d'c'3'4'* parallel to HP and 40 mm above HP in the front view.

(ii) Project the corresponding top view and 20 mm away from the VP as shown in Fig. 12.34.

Problem 12.10 A right regular pentagonal prism, side of base 25 mm and axis 60 mm long, lies on one of its rectangular faces on HP with its axis perpendicular to the VP. Draw its front, top and side views.

Solution: As the axis of a pentagonal prism is perpendicular to VP, projections are to be made from front view.

(i)   Draw a regular pentagon of 25 mm side in the front, keeping one of its rectangular faces on x-y.

(ii) Project top and left side views from front view. Mark the axis to be 60 mm long. Name the corner points as shown in Fig. 12.35 and complete the views.

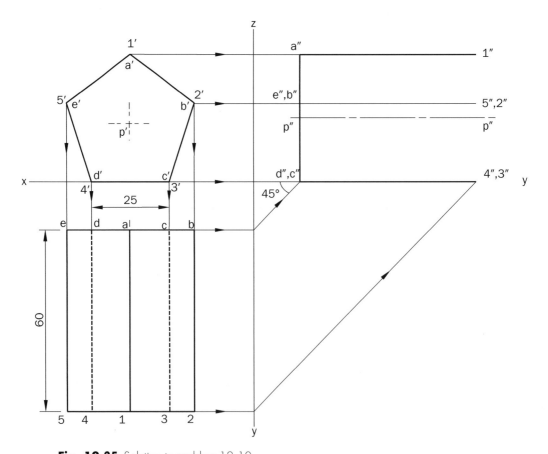

Fig. 12.35 Solution to problem 12.10

Problem 12.11 A triangular prism side of base 30 mm and axis 55 mm long lies on one of its rectangular faces in HP, with its axis perpendicular to VP. Draw its three views.

Solution: Draw an equilateral triangle of 30 mm side in the front view, keeping one of its rectangular faces on x-y. Project top view, side view from front view and mark the axis 55 mm long. Complete the projections as shown in Fig. 12.36.

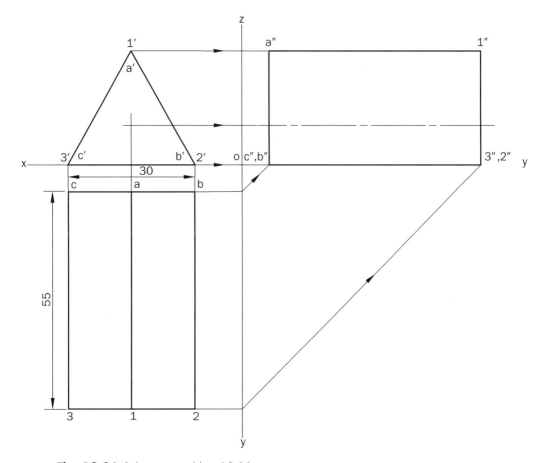

Fig. 12.36 Solution to problem 12.11

Problem 12.12 A right regular pentagonal prism, side of base 25 mm and axis 55 mm long, lies on one of its rectangular faces on HP with its axis perpendicular to VP. Draw three views of the pentagonal prism.

Solution:

    (i)   Draw a regular pentagon of 25 mm side in the front view, keeping one of its rectangular faces (*d*' c' 3' 4') on x-y.

    (ii)   Project top view from front view and cut the axis 55 mm long. Complete the end view too, as shown in Fig. 12.37. Name all the corner points.

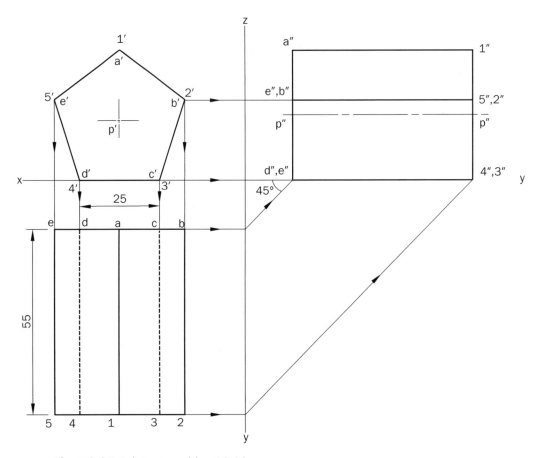

Fig. 12.37 Solution to problem 12.12

Problem 12.13 A hexagonal prism, side of base 25 mm and axis 55 mm long, has one of rectangular faces parallel to HP with its axis perpendicular to VP and 35 mm above HP. Draw the elevation, plan and side view of the hexagonal prism when the nearer end is 25 mm away from the VP.

(*PTU, Jalandhar December 2005*)

Solution: The solution to this problem has already been explained in previous problems. See Fig. 12.38.

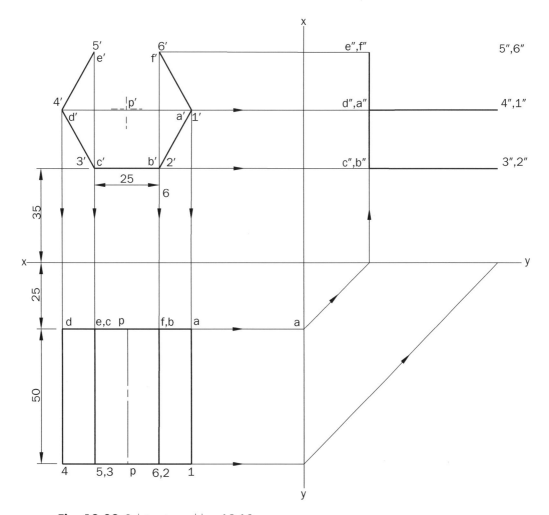

Fig. 12.38 Solution to problem 12.13

Problem 12.14 A square prism, side of base 30 mm and axis 55 mm long, is resting on HP on one of its longer edges with a face containing the longer edge inclined at 30° to the HP and its axis perpendicular to the VP. Draw its projections.

Solution: All the construction lines are retained to make the solution self-explanatory. See Fig. 12.39.

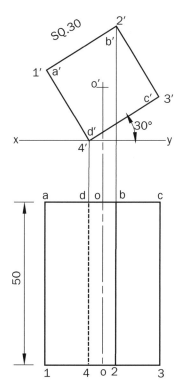

Fig. 12.39 Solution to problem 12.14

12.5 Axis Parallel to Both HP and VP

When the axis of a solid is parallel to both HP and VP, which in other words will mean the axis of a solid is perpendicular to a profile plane, the projections will be started from the profile plane, where it gives true shape and size of the base of the solid.

Problem 12.15 A right regular hexagonal prism, edge of base 30 mm and length 50 mm, lies on one of its rectangular faces on HP, such that its axis is parallel to both HP and VP. Draw its projections.

(PTU, Jalandhar December 2003, May 2004)

Solution:

(i)   As the axis is parallel to both HP and VP, begin the problem with side view, which gives the true shape and size of the solid.

(ii)   Keep one of the rectangular faces on HP, i.e., the x-y line.

(iii) Project the front and top views as shown in Fig. 12.40 and take length of the solid as 50 mm.

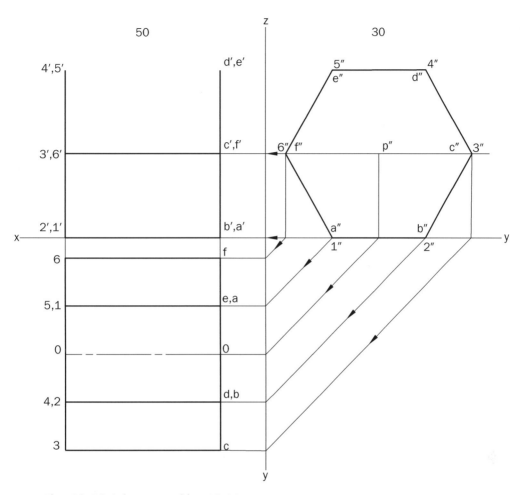

Fig. 12.40 Solution to problem 12.15

Problem 12.16 A triangular prism side of base 25 mm and axis 50 mm long lies on one of its rectangular faces (a) in HP (b) on ground plane, with its axis parallel to VP. Draw its projections.

(*PTU, Jalandhar June 2003*)

Solution:

(i) For first-angle projection draw the x-y line only. Whereas for third-angle projection draw x-y and *ground lines (gl)*, separated by a suitable distance.

(ii) As the axis is parallel to both the planes, so begin with side view.

(iii) Keep one of the rectangular faces on HP and ground plane for first-angle and third-angle projections respectively as shown in Figs. 12.41 and 12.42.

(iv) Project the front and top views. Take length of the axis is 50 mm.

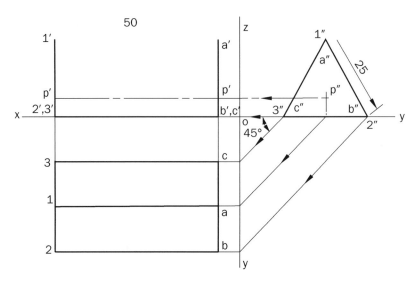

Fig. 12.41 Solution to problem 12.16 (a)

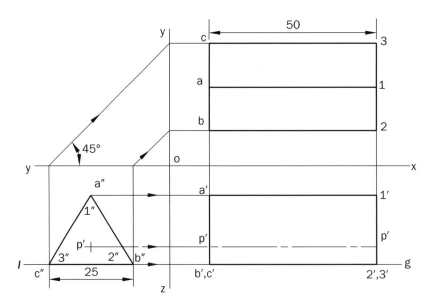

Fig. 12.42 Solution to problem 12.16 (b)

Problem 12.17 A pentagonal prism base 20 mm side and axis 75 mm long lies on one of its rectangular faces on HP with axis parallel to VP. Draw the three views of the prism.

(*PTU, Jalandhar December 2002*)

Solution: The solution to this problem is self-explanatory. See Fig. 12.43.

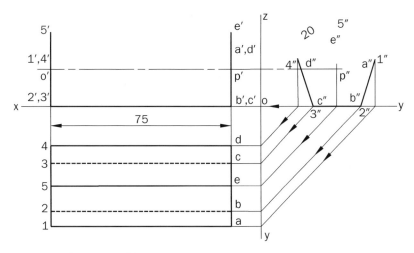

Fig. 12.43 Solution to problem 12.17

Problem 12.18. A square prism, side of base 30 mm and axis 55 mm long, is resting on HP on one of its longer edges with a face containing the longer edge inclined at 30° to the HP. Draw its projections, when the axis is parallel to both HP and VP.

Solution: All the construction lines are retained to make the solution self-explanatory. See Fig. 12.44.

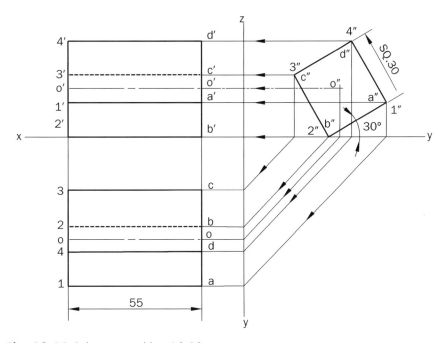

Fig. 12.44 Solution to problem 12.18

12.6 Axis Inclined to One of the Principal Planes and Parallel to the Other

When a solid has its axis inclined to one of the principal planes and parallel to the other, its projections are drawn into two stages. In the initial stage, its axis is kept perpendicular to that plane to which it is inclined. If the axis is to be inclined to the ground or HP, it is assumed to be perpendicular to the HP in the initial stage. Similarly, if the axis is to be inclined to the VP, it is assumed to be perpendicular to the VP in the initial stage. Moreover, the following two points need to be kept in mind:

(i)   If the solid has an edge of its base parallel to HP or in HP or ground, then that edge should be kept perpendicular to the VP. If the edge of the base is parallel to the VP or in VP, it should be kept perpendicular to the HP.

(ii)   If the solid has a corner of its base in the HP or on the ground, the sides of the base containing that corner should be kept equally inclined to the VP. If the corner is in the VP, the sides should be kept equally inclined to the HP.

After drawing the projections of the solid in the first stage, the final projections may be obtained by any one of the following methods:

(a) **Change of position of solids.** The position of one of the views is altered as required and the other view projected from it.

(b) **Change of reference line or auxiliary plane.** A new reference line is drawn according to the required conditions, to represent an auxiliary plane and the final view projected from it. To solve the problems on projections of solids by this method, it requires the knowledge of auxiliary projections as already described in Chapter 11.

The comparison of these methods reveals that the first method is laborious, as it take considerable time to reproduce the view, especially when the solid has curved surfaces or too many edges or corners. In such cases, it is easier and more convenient to adopt the second method. Sufficient care should be taken in transferring the distances of various points from their respective reference line.

After locating the positions of all the points representing the various corners in the final view, these are joined correctly to have the finished projection. The following sequence may be adopted for joining these points correctly:

(i)   Draw the lines for the edges of the visible base. The base which is further away from x-y in one view, will be completely visible in the other view.

(ii)   Draw the lines for the longer edges. The lines which pass through the figure of the visible base should be short dashed lines.

(iii)   Draw the lines for the edges of the other base.

It should always be remembered that when two lines representing the edges cross each other, one of them must be hidden and should therefore be drawn as short dashed line.

(a) **Axis inclined to the HP and parallel to the VP.** When a solid has its axis inclined to the HP and parallel to the VP, its projections are drawn into two stages. In the initial stage, it is assumed to be perpendicular to the HP. In such problems, the top view will be drawn first, as it will show the true shape and size of the solid and then the front view is projected from it. The front view is reproduced making the given angle with HP. Project all the points vertically from this front view and horizontally from the first top view. Join all points in the final top view, observing the rules for establishing the visibility of lines.

Problem 12.19. A right circular cone, diameter of base 50 mm and height 65 mm, is resting on HP on its base such that its axis is parallel to VP and inclined to HP at an angle of 60°. Draw its front, top and profile views.

Solution: All the construction lines are retained to make the solution self-explanatory. See Fig. 12.45.

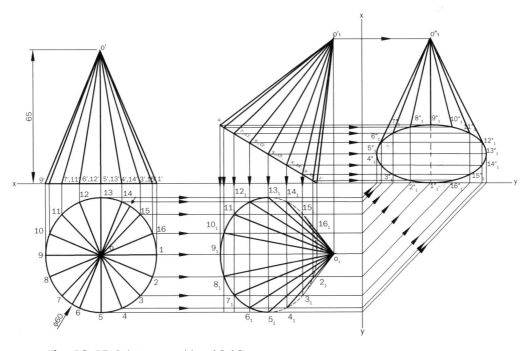

Fig. 12.45 Solution to problem 12.19

Problem 12.20 A right regular hexagonal pyramid, edge of base 30 mm and axis 60 mm long, has an edge of its base in HP, such that its axis is inclined at 30° to the HP and parallel to the VP. Draw its projections by using both the methods.

Solution: In the initial position or stage, assume the axis to be perpendicular to the HP. Draw the projections with the base in x-y and its one edge perpendicular to the VP and label it.

Method I (Change of position of solid)

(i)   Reproduce the front view so that the axis makes an angle of 30° with x-y and the edge
       2'1' remain in x-y and name the points on it by adding suffix 1 to them.

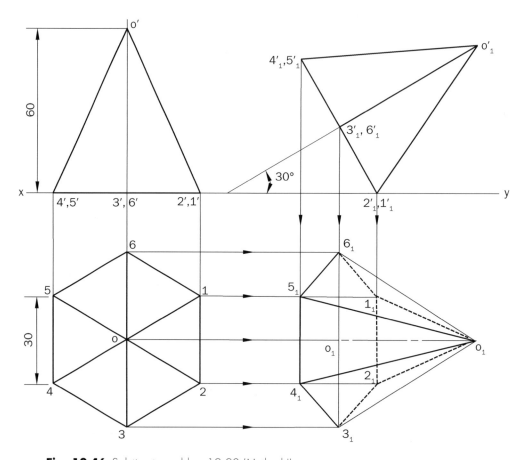

Fig. 12.46 Solution to problem 12.20 (Method I)

(ii)   Project all the points vertically from this front view and horizontally from the first top
       view. Complete the new top view as shown in Fig. 12.46. Join all the points in the final
       top view, observing the rules for establishing the visibility of lines.

Method II (Change of reference line)

(i)   Draw a new reference line $x_1 y_1$ inclined at 30° to the axis, to represent an auxiliary
       inclined plane (AIP) through edge 2'1'.

(ii)   From the front view, project the requied top view on $x_1 y_1$, keeping the distance of
       each point from $x_1 y_1$ equal to the distance of its first top view from x-y as shown in
       Fig. 12.47. Join all the points in the final top view, observing the rules establishing the
       visibility of lines.

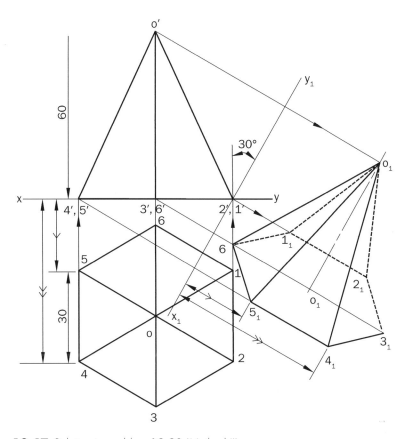

Fig. 12.47 Solution to problem 12.20 (Method II)

Problem 12.21 A right regular hexagonal prism, edge of base 30 mm and height 65 mm, rests on one of its base edges in HP such that its axis is inclined to the HP at 45° and parallel to the VP. Draw its projections by using both the methods.

Solution: In the initial stage, assume the axis to be perpendicular to the HP. Draw the projections with the base in x-y and its one edge perpendicular to the VP and label it.

Method I (Change of position of solid)

(i)   Reproduce the front view so that the axis makes an angle of 45° with x-y and the edge 2'1' remain in x-y and name the points on it by adding suffix 1 to them.

(ii) Project from it, the corresponding top view and join all the points by observing the rules establishing the visibility of lines as shown in Fig. 12.48.

Method II (Change of reference line)

(i) Draw a new reference line $x_1 y_1$ inclined at 30° to the axis, to represent an auxiliary inclined plane (AIP) through edge 2'1'.

(ii) From the front view, project the required top view on $x_1 y_1$ as shown in Fig. 12.49.

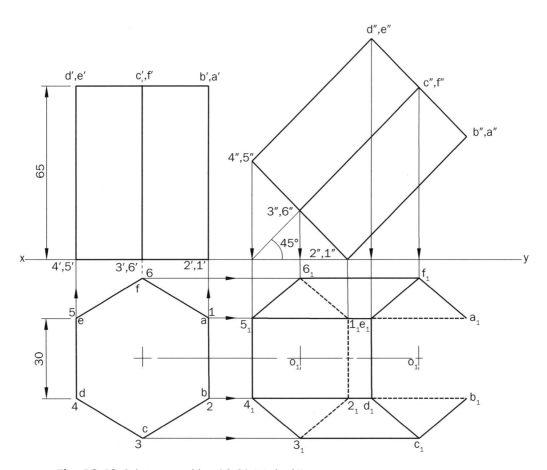

Fig. 12.48 Solution to problem 12.21 (Method I)

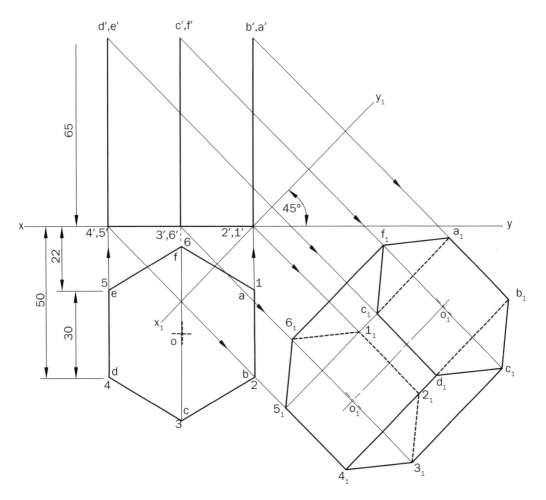

Fig. 12.49 Solution to problem 12.21 (Method II)

Problem 12.22 A right regular pentagonal prism, edge of base 25 mm and height 55 mm, rests on an edge of its base in HP such that its axis is parallel to the VP and its base makes an angle of 45° to the HP. Draw its projections.

Solution: In the initial stage, assume the axis to be perpendicular to the HP. Draw the projections with the base in x-y and perpendicular to the VP and label it.

(i) Reproduce the front view so that the axis makes an angle of 45° with x-y and the edge 3'4' remains in x-y and name the points on it by adding suffix 1 to them.

(ii) Project from it, the corresponding top view and join all the points by observing the rules establishing the visibility of faces as shown in Fig. 12.50.

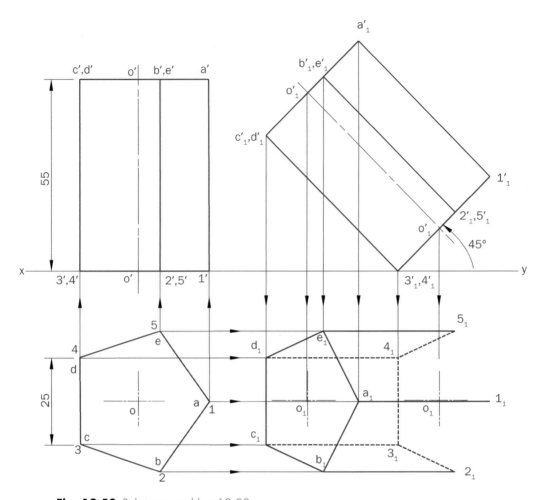

Fig. 12.50 Solution to problem 12.22

Problem 12.23 A right regular pentagonal pyramid, side of base 30 mm and height 50 mm, lies on ground plane on one of its slant edges and has its axis parallel to VP. Draw its projections in third-angle.

Solution: For third-angle projection draw x-y and gl lines, separated by a suitable distance so that the given solid may be accommodated between them.

   (i)   In the initial position, draw the top view of the pentagonal pyramid, with one of its base edge perpendicular to x-y.
   (ii)   Project the corresponding front view.
   (iii) Next change the position of the front view, so that o'1' slant edge lies on ground and add the suffix 1 to them.
   (iv)   Project all the points vertically upwards from this front view and horizontally from the first top view. Complete the new top view as shown in Fig. 12.51, observing the rules of visibility of faces.

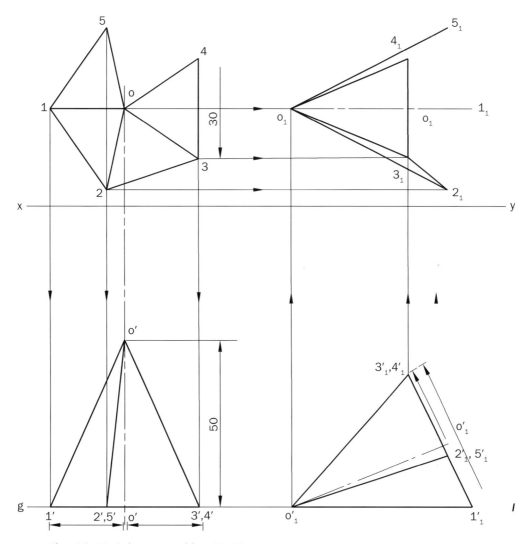

Fig. 12.51 Solution to problem 12.23

Problem 12.24 A right circular cone, diameter of base 60 mm and height 70 mm, lies on HP on one of its elements, such that its axis is parallel to VP. Draw its projections.

(*PTU, Jalandhar December 2003, May 2011*)

Solution:

(i) In the initial stage, assume the axis to be perpendicular to HP and parallel to VP. Draw the top view first. Divide the base circle into sixteen equal parts and name all the points.

(ii) Join these points to the centre of the circle o, by continuous thin lines as shown in Fig. 12.52.

  (iii) Project the corresponding front view. Here only two elements, i.e. o'1' and o'9' provides the true lengths.

  (iv) Reproduce the front view by placing o'1' element on the x-y and add suffix 1 to all of them.

  (v) Project all the points vertically from this front view and horizontally from the first top view. Complete the new top view by observing the rules of visibility of faces.

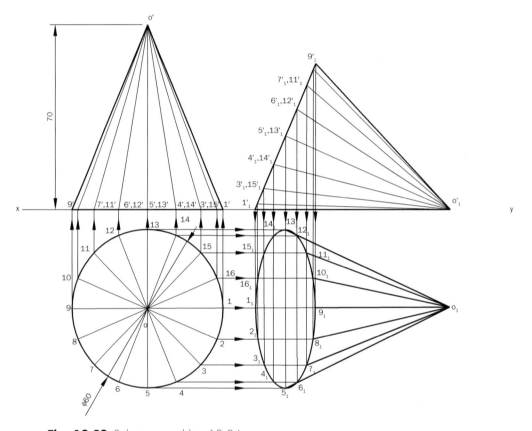

Fig. 12.52 Solution to problem 12.24

Note: Students are advised to divide the circle into sixteen equal parts to get a good curve. However, they can also divide the circle into 12 equal parts.

Problem 12.25 A right regular pentagonal pyramid, edge of base 30 mm and height 55 mm, lies on HP on one of its slant edges and has its axis parallel to VP. Draw its projections by using both the methods.

Solution: In the initial stage, assume the axis to be perpendicular to the HP and parallel to the VP. Draw the projections with the base in x-y.

Method I (Change of position of solid)

(i) Reproduce the front view so that the slant edge lies on HP and name the points on it by adding suffix 1 to them.

(ii) Project from it, the corresponding top view and join all the points by observing the rules establishing the visibility of lines as shown in Fig. 12.53.

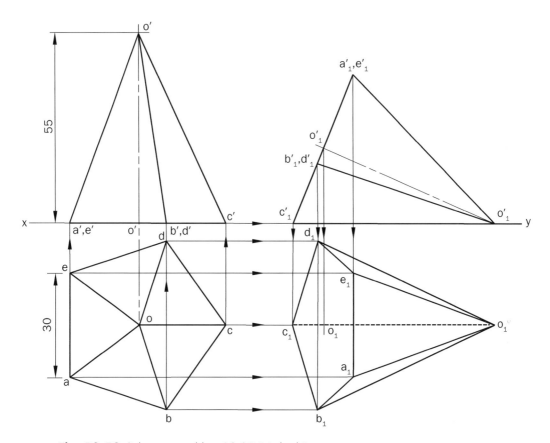

Fig. 12.53 Solution to problem 12.25 (Method I)

Method II (Change of reference line)

(i) Draw a new reference line $x_1 y_1$ to represent an auxiliary inclined plane (AIP) through slant edge $o'c'$.

(ii) From the front view, project the required top view on $x_1 y_1$ as shown in Fig. 12.54.

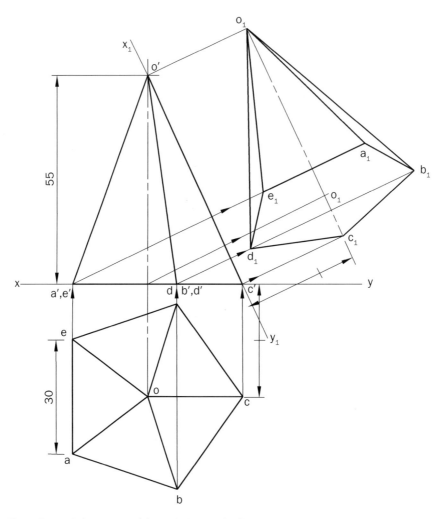

Fig. 12.54 Solution to problem 12.25 (Method II)

Problem 12.26 A right regular hexagonal prism, side of base 20 mm and height 45 mm, rests on a corner of its base on the HP with the longer edge passing through this corner making an angle of 30° to the HP. Draw the projections of the hexagonal prism.

(PTU, Jalandhar December 2002)

Solution: In the initial stage, assume the axis to be perpendicular to the HP. Draw the front and top views in this position and label it.

    (i)   Reproduce the front view such that the corner a' lies on HP and the edge $a'1'$ makes an angle 30° to the x-y and name all points by adding suffix 1 to them.

    (ii)   Project from it, the corresponding top view and join all the points by observing the rules of establishing the visibility of the faces as shown in Fig. 12.55.

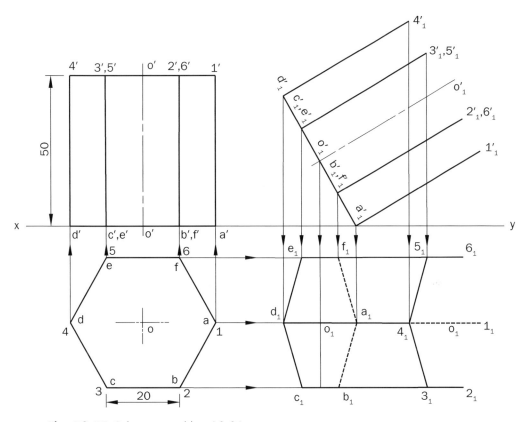

Fig. 12.55 Solution to problem 12.26

Problem 12.27 A right regular hexagonal pyramid, edge of base 30 mm and height 50 mm, is resting on one of its triangular faces on the HP with its axis parallel to the VP. Draw the projections of the hexagonal pyramid by using both the methods.

Solution: In the initial position, assume the axis to be perpendicular to the HP and parallel to the VP. Draw the projections with its base on HP and one of its base edge perpendicular to the VP. Name all the corner points on it.

Method I (Change of position of solid)

(i)   Reproduce the front view so that the triangular face o'1'2' lies on x-y line and name all
      the points on it by adding suffix 1 to them.
(ii) Project all the points vertically from this front view and horizontally from the first top
      view. Complete the new top view as shown in Fig. 12.56.

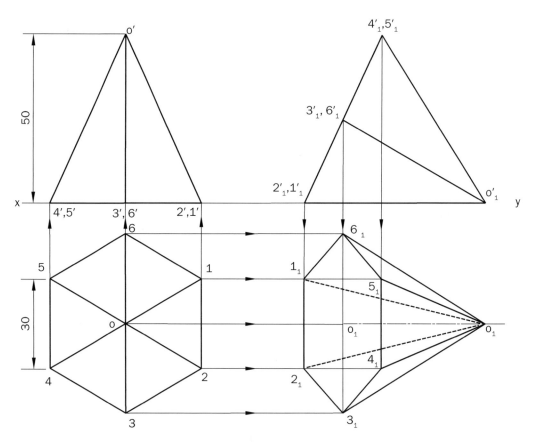

Fig. 12.56 Solution to problem 12.27 (Method I)

Method II (Change of reference line)

(i) Draw a new reference line $x_1\,y_1$ coinciding with o'1'2' in the front view. From the front view project the required top view on $x_1\,y_1$ (keeping the distance of each point from $x_1\,y_1$ equal to the distance of its first top view from x-y) as shown in Fig. 12.57.

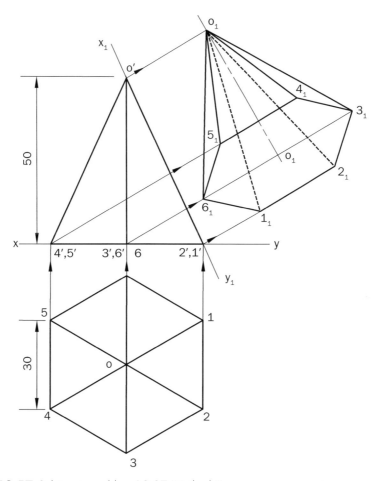

Fig. 12.57 Solution to problem 12.27 (Method II)

Problem 12.28 A right regular pentagonal prism, side of base 30 mm and height 70 mm, rests on one of its base corners on HP such that its long edge containing that corner is inclined to the HP at 45° with its axis parallel to VP. Draw its projections.

Solution:

    (i)   In the initial position, assume the axis of the prism to be perpendicular to the HP and parallel to the VP. Draw its projections keeping one of the sides of its base perpendicular to x-y and whole of the base lies on the HP. Name all the corner points on it.

    (ii)   Redraw the front view so that the longer edge $1'a'$ is inclined at 45° to x-y. Project the requried top view as shown in Fig. 12.58.

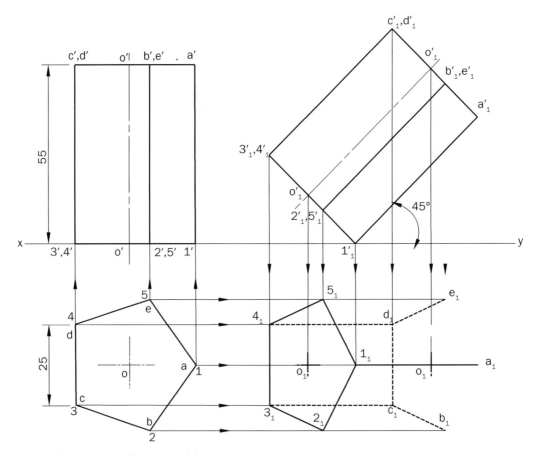

Fig. 12.58 Solution to problem 12.28

Problem 12.29 A right regular pentagonal pyramid, edge of base 25 mm and height 55 mm, is held on ground plane on one of its base corners, such that its axis is inclined at 30° to the ground plane and is parallel to the VP. Draw its projections.

(PTU, Jalandhar December 2009)

Solution:

(i) As the pyramid is given to be resting on its base corner in ground plane, therefore the projections are to be made in third-angle. Draw x-y and *gl* lines, a suitable distance apart.

(ii) Draw top and front views assuming the axis of the pyramid to be perpendicular to the ground plane.

(iii) Redraw the front view such that its axis is inclined at 30° to *gl* line. Project the required top view as shown in Fig. 12.59.

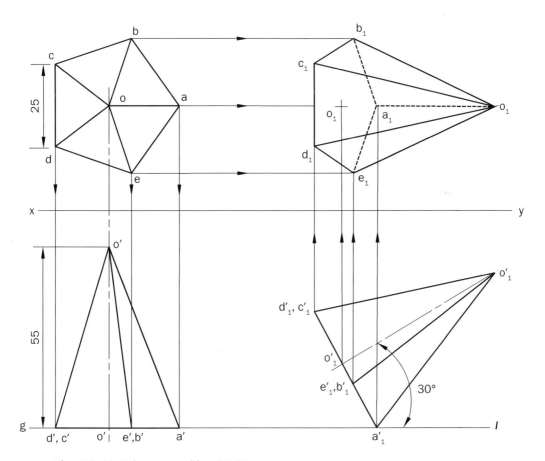

Fig. 12.59 Solution to problem 12.29

Problem 12.30 A right circular cone, diameter of base 50 mm and height 60 mm, rests on its base rim on HP such that its axis is parallel to the VP and its base inclined to the HP at 45°. Draw its projections.

Solution:

(i)   In the initial stage, assume the axis to be perpendicular to the HP and parallel to the VP. Draw the top view of the cone and project the corresponding front view. Divide the base circle in the top view into sixteen equal parts and project these points into the front view too.

(ii)   Reproduce the front view such that its base rim 1 lies on x-y and base makes an angle of 45° to the x-y. Project from it the corresponding top view as shown in Fig. 12.60, keeping in mind the principles of visibility of lines.

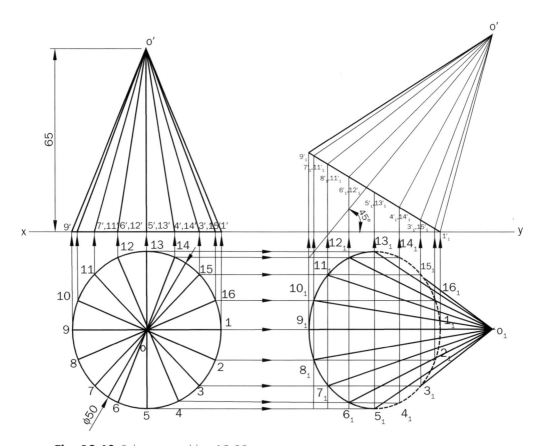

Fig. 12.60 Solution to problem 12.30

Problem 12.31 Draw the projections of a cube of 30 mm edge, when a body diagonal of the solid is kept vertical.

Solution: The interpretation of the solution is left to the reader. See Fig. 12.61.

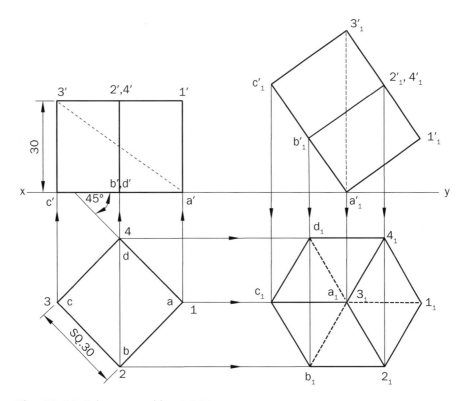

Fig. 12.61 Solution to problem 12.31

Problem 12.32 Draw the projections of a solid rectangular prism 25 mm × 50 mm × 75 mm on a plane perpendicular to one of the diagonal of the solid.

(*PTU, Jalandhar May 2004*)

Solution: All the construction lines are retained to make the solution self-explanatory. See Fig. 12.62.

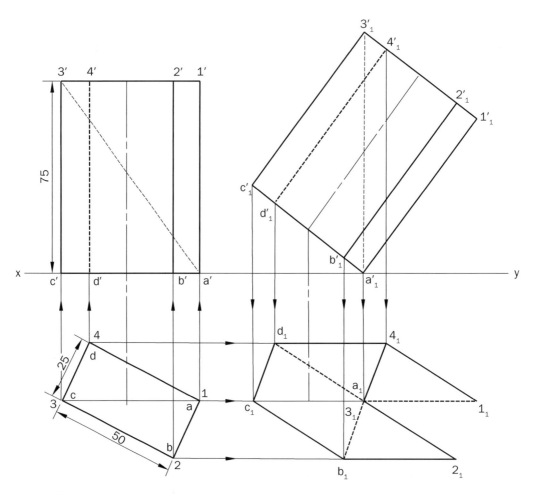

Fig. 12.62 Solution to problem 12.32

Problem 12.33 A right regular pentagonal pyramid, edge of base 30 mm and height 60 mm, rests on HP on one of its corners. Its base is inclined at 45° to HP and axis parallel to the VP. Draw its projections.

Solution: All the construction lines are retained to make the solution self-explanatory. See Fig. 12.63.

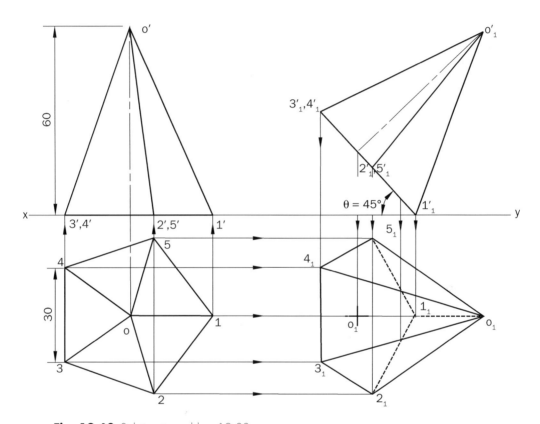

Fig. 12.63 Solution to problem 12.33

Problem 12.34 A right regular pentagonal pyramid, edge of base 30 mm and height 65 mm, lies on one of its triangular faces on HP with its axis parallel to the VP. Draw its projections by using both the methods.

Solution: In the initial stage, assume the axis to be perpendicular to the HP and parallel to the VP. Draw the projections with the base in x-y.

Method I (Change of position of solid)

(i) Reproduce the front view so that the triangular face ($o'c'd'$) lies on HP and name the points on it by adding suffix 1 to them.

(ii) Project corresponding top view from it and join all the points by observing the rules establishing the visibility of lines as shown in Fig. 12.64.

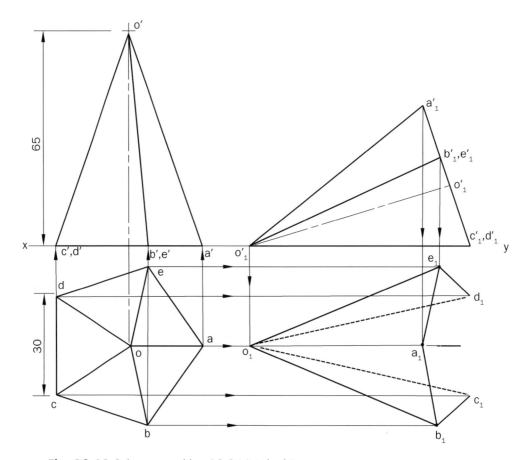

Fig. 12.64 Solution to problem 12.34 (Method I)

Method II (Change of reference line)

(i) Draw a new reference line x, y, to represent an auxiliary inclined plane (AIP) through the triangular face ($o'c'd'$).

(ii) From front view, project the required top view on $x_1 y_1$ as shown in Fig. 12.65.

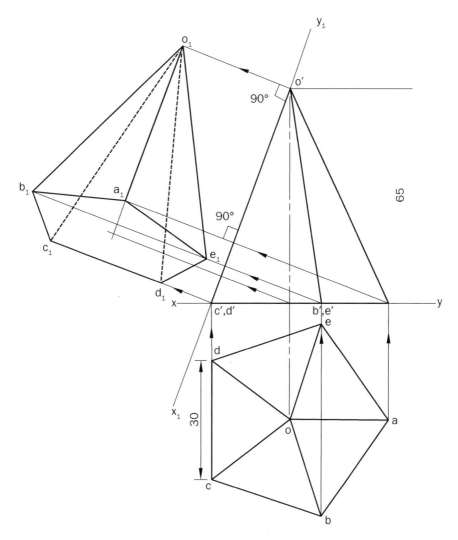

Fig. 12.65 Solution to problem 12.34 (Method II)

Problem 12.35 A right circular cylinder, diameter of base 50 mm and height 65 mm, rests on HP on its base rim such that its axis is inclined at 45° to the HP and parallel to the VP. Draw its projections.

(*PTU, Jalandhar May 2011*)

Solution:

(i)   In the initial stage, assume the axis to be perpendicular to the HP and parallel to the VP. Draw the top view of the cylinder and divide the base circle into sixteen equal parts and name all the points.

(ii) Join these points to the centre of the circle *o* by continuous thin lines as shown in Fig. 12.66.

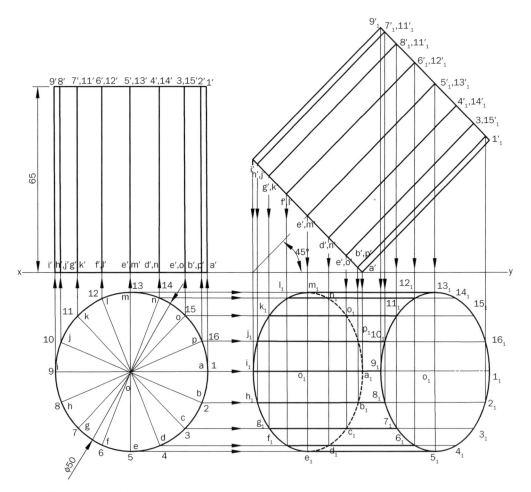

Fig. 12.66 Solution to problem 12.35

(iii) Project the corresponding front view.

(iv) Reproduce the front view by placing a' (base rim) on the x-y so that the axis makes an angle of 45° with x-y and name all the points on it by adding suffix 1 to them.

(v) Project corresponding top view from it and join all the points by observing rules establishing the visibility of elements.

Problem 12.36 A square pyramid, edge of base 45 mm and length of axis 45 mm, is resting on one its triangular faces on HP with its axis parallel to the VP. Draw its projections by using both the methods.

Solution: In the initial stage, assume the axis to be perpendicular to the HP and parallel to the VP. Draw the projections with the base in x-y.

Method I (Change of position of solid)

(i) Reproduce the front view so that triangular face (o'1'2') lies on HP and name the points on it by adding suffix 1 to them.

(ii) Project the corresponding top view, observing the rules of visibility of surfaces. See Fig. 12.67.

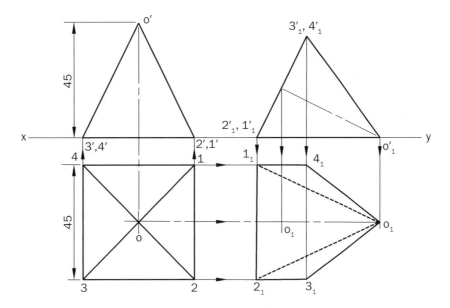

Fig. 12.67 Solution to problem 12.36 (Method I)

Method II (Change of reference line)

(i) Draw a new reference line $x_1 y_1$ to represent as auxiliary inclined plane (AIP) through the triangular face (o' 1' 2').

(ii) From the front view, project the required top view on $x_1 y_1$ as shown in Fig. 12.68.

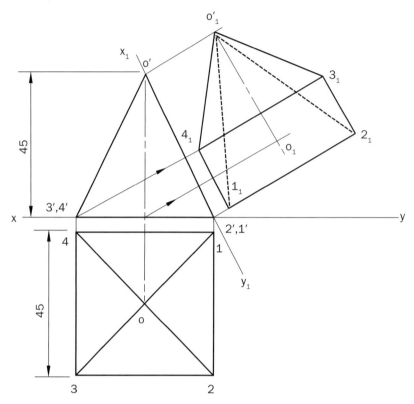

Fig. 12.68 Solution to problem 12.36 (Method II)

Problem 12.37 A right regular pentagonal prism, side of base 30 mm and 55 mm long, rests on one of its base corners on ground plane such that its long edge containing that corner is inclined to the ground plane at 45°. Draw its projections in third-angle.

Solution: The solution to this problem is self-explanatory. See Fig. 12.69.

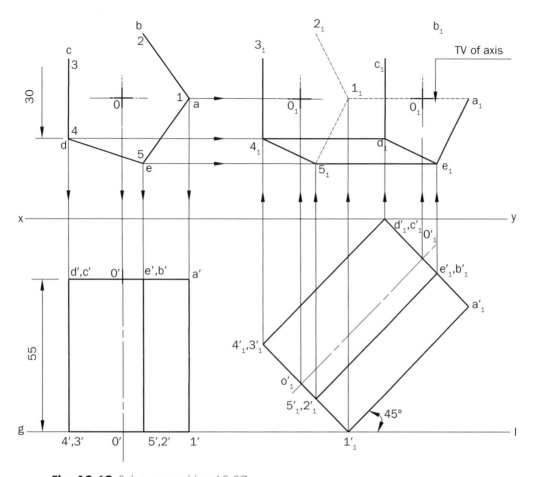

Fig. 12.69 Solution to problem 12.37

Problem 12.38 A frustum of a right regular pentagonal pyramid, edge of lower base 25 mm, edge of upper base 15 mm and axis 40 mm long, is lying on one of its slant edges on HP with its axis parallel to VP. Draw its projections.

Solution: All the construction lines are retained to make the solution self-explanatory. See Fig. 12.70.

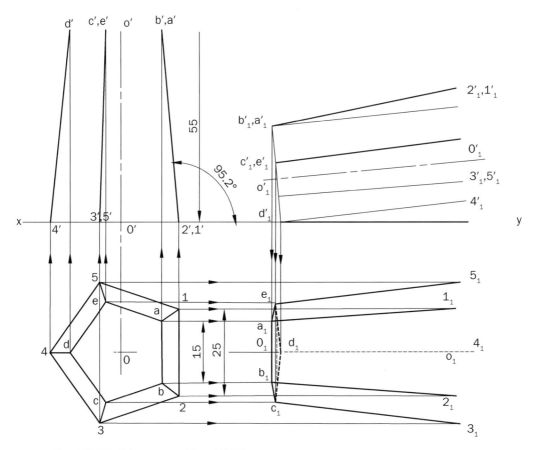

Fig. 12.70 Solution to problem 12.38

Problem 12.39 A right regular pentagonal prism, edge of base 25 mm and height 60 mm, is resting on one of its base edges in HP, such that its axis is inclined at 45° to the HP and parallel to the VP. Draw three views of the pentagonal prism.

Solution: All the construction lines are retained to make the solution self-explanatory. See Fig. 12.71.

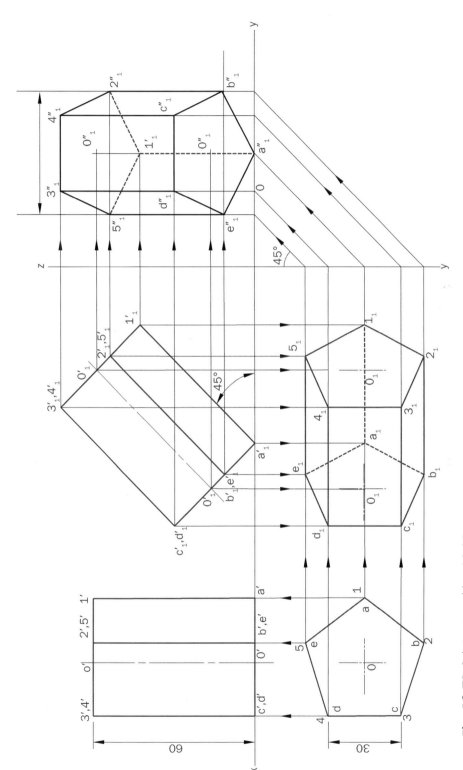

Fig. 12.71 Solution to problem 12.39

Problem 12.40 A square prism, side of base 40 mm and axis 50 mm long, has an edge of its in HP. Its axis is inclined at an angle of 45° to the HP and parallel to the VP. Draw its projections.

Solution:

(i) In the initial position, assume the axis of the prism to be perpendicular to the HP and parallel to the VP. Draw its projections keeping one of the sides of its base perpendicular to xy and whole of the base lies on the HP. Name all the corner points on it.

(ii) Redraw the front view so that the axis makes as angle of 45° to the x-y. Project the required top view as shown in Fig. 12.72.

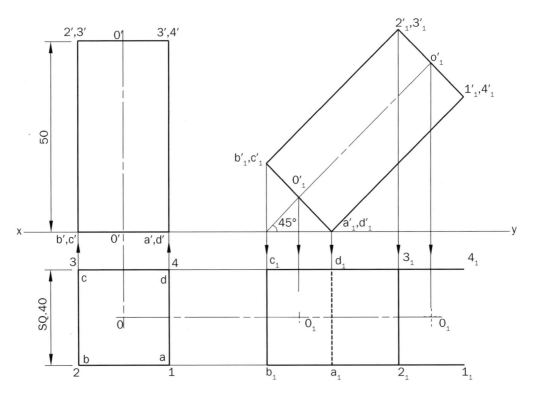

Fig. 12.72 Solution to problem 12.40

Problem 12.41 A square prism, side of base 30 mm and axis 60 mm long, has one of its base corner in HP such that its axis is inclined at 45° to the HP and parallel to the VP. Draw its projections.

Solution: All the construction lines are retained to make the solution self-explanatory. See Fig. 12.73.

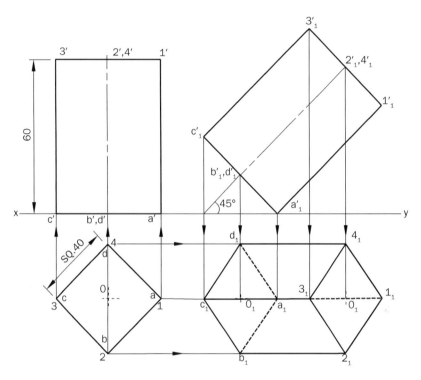

Fig. 12.73 Solution to problem 12.41

(b) **Axis inclined to the VP and parallel to the HP:** When a solid has its axis inclined to the VP and parallel to the HP, its projections are drawn into two stages. In the initial stage, it is assumed to be perpendicular to the VP. In such problems, the front view is drawn first, as it shows the true shape and size of the object and then top view is projected from it. The top view is reproduced making given angle with VP. Project all the points vertically from this top view and horizontally from the first front view. Join all points in the final front view, observing the rules for establishing the visibility of lines.

Problem 12.42 A right regular pentagonal prism, side of base 30 mm and height 70 mm, lies on one of its rectangular faces on HP with its axis inclined at 30° to VP. Draw its projections.

Solution: All the construction lines are retained to make the solution self-explanatory. See Fig. 12.74.

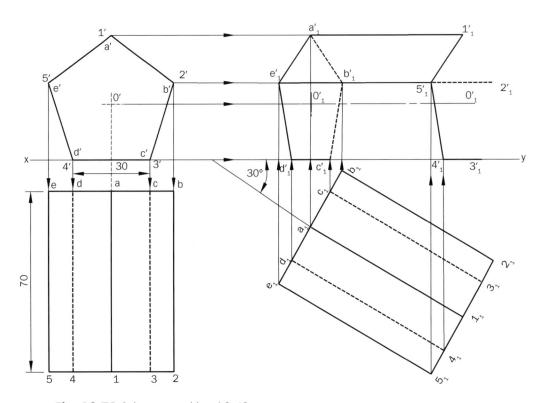

Fig. 12.74 Solution to problem 12.42

Problem 12.43 A right regular hexagonal prism, side of base 30 mm and height 70 mm, lies on one of its rectangular faces on HP and its axis inclined at 30° to the VP. Draw its projections.

Solution: In the initial stage, assume the axis to be perpendicular to VP and parallel to HP.

(i)   Draw the front view, keeping one of its faces in HP and label it.
(ii) Project the corresponding top view.
(iii) Reproduce the top view so that the axis makes an angle of 30° with x-y and name the points on it by adding suffix 1 to them.
(iv) Project from it, the corresponding front view and join all the points by observing the rules establishing the visibility of faces as shown in Fig. 12.75.

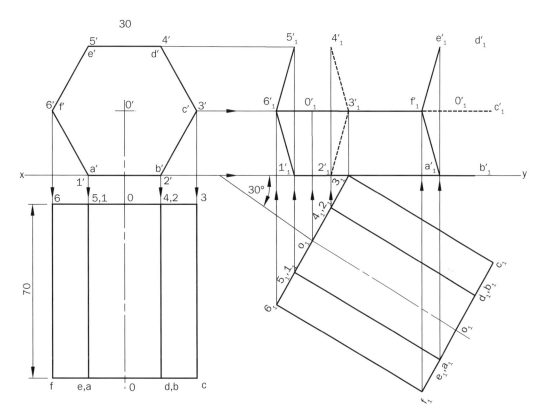

Fig. 12.75 Solution to problem 12.43

Problem 12.44 A right circular cylinder, diameter of base 60 mm and height 70 mm, lies in HP on one of its elements such that its axis is parallel to the HP and inclined to VP at 45°. Draw its projections.

(PTU, Jalandhar December 2006, December 2010)

Solution: In the initial stage, assume the axis to be perpendicular to VP and parallel to HP.

(i)   Draw the front view, keeping one of its elements in HP. Divide the circle into sixteen equal parts and name all the points.

(ii)   Join these points to the centre of the circle o' by continuous thin lines as shown in Fig. 12.76.

(iii)   Project the corresponding top view.

(iv)   Reproduce the top view so that the axis makes an angle of 45° with x-y and name all the points by adding suffix 1 to them.

(v)   Project the corresponding front view and join all the points by observing the rules of visibility of lines.

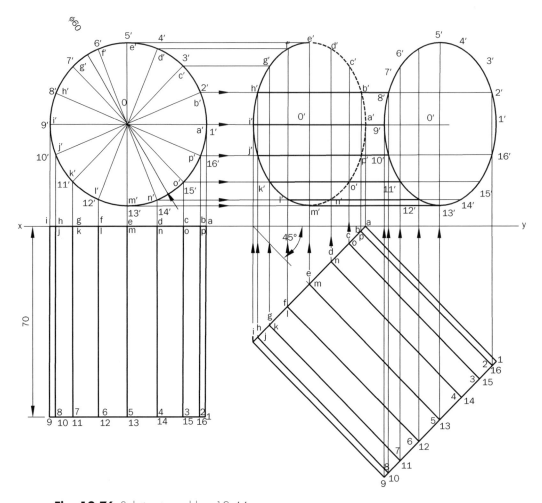

Fig. 12.76 Solution to problem 12.44

Problem 12.45 A right circular cone, diameter of base 40 mm and height 70 mm, is held on its base rim in HP such that its axis is inclined at 45° to VP and parallel to HP. Draw its projections, keeping the vertex towards the observer.

Solution:

  (i)   In the initial stage, assume the axis to be perpendicular to VP and parallel to HP. Draw the front view first, keeping its base rim on HP. Divide the base circle into sixteen equal parts and label it.

  (ii) Project the corresponding top view.

  (iii) Reproduce the top view as shown in Fig. 12.77 and add suffix 1 to them.

  (iv) Project all the points vertically from this top view and horizontally from the first front view.

Complete the new front view by observing the rules of visibility of lines.

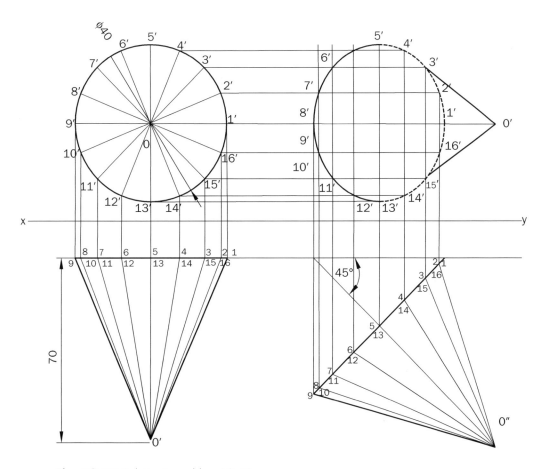

Fig. 12.77 Solution to problem 12.45

Problem 12.46 A right regular hexagonal pyramid, edge of base 25 mm and length of axis 60 mm, has one of its corners in the VP such that its axis is inclined at 45° to the VP and parallel to the HP. Draw its projections.

Solution:

(i)   In the initial stage, assume the axis to be perpendicular to VP and parallel to HP. Draw the front view first, keeping its base in VP. Name all the corner points on it.
(ii) Project the corresponding top view.
(iii) Reproduce the top view such that the corner point 4 lies on VP and the axis makes an angle of 45° to VP. Name all the points on it by adding suffix 1 to them.
(iv) From the top view, project the required front view as shown in Fig. 12.78.

25

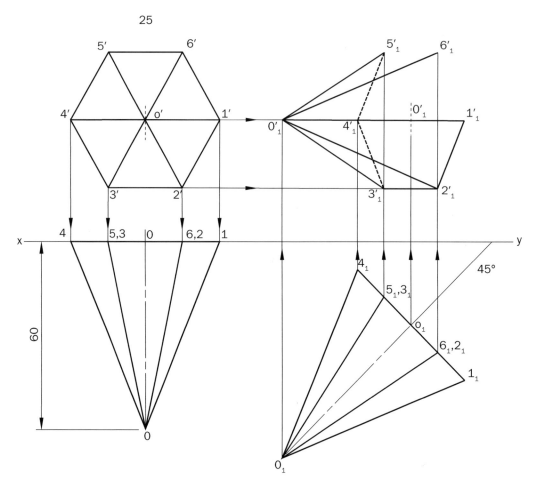

Fig. 12.78 Solution to problem 12.46

Problem 12.47 A right regular hexagonal prism, side of base 30 mm and height 60 mm, has one of its side (or face) inclined at 45° to the VP and one of its longer edges on the HP. Draw its projections.

Solution: All the construction lines are retained to make the solution self-explanatory. See Fig. 12.79.

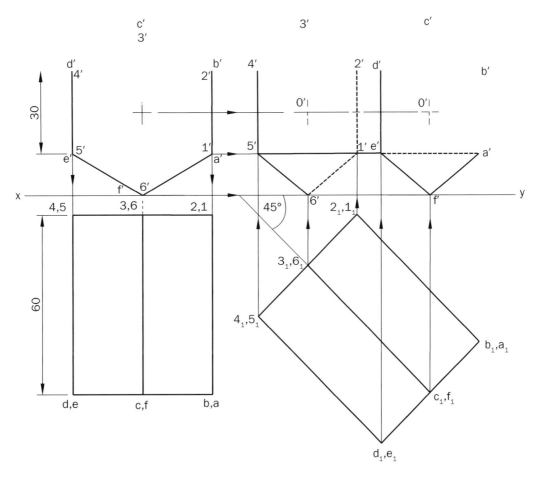

Fig. 12.79 Solution to problem 12.47

Problem 12.48 Draw the three views of a frustum of cone, 50 mm diameter at the top, 30 mm diameter at the bottom, 60 mm high, when its axis makes an angle of 30° with the VP and parallel to the HP.

Solution: All the construction lines are retained to make the solution self-explanatory. See Fig. 12.80.

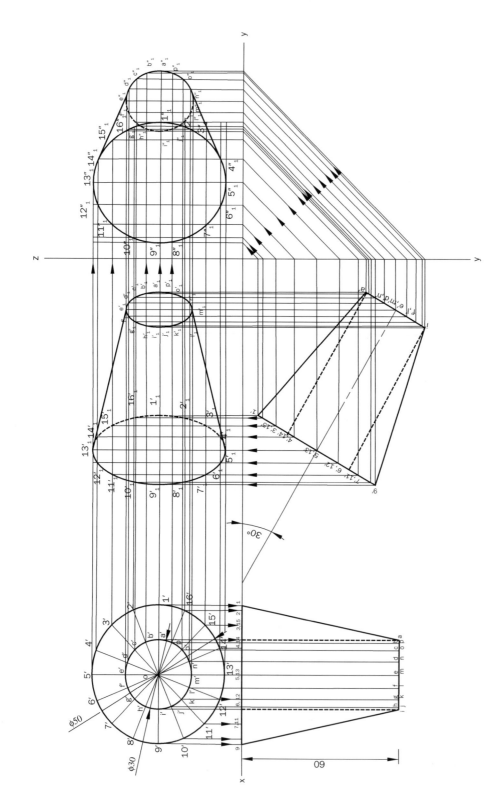

Fig. 12.80 Solution to problem 12.48

Problem 12.49 A square prism, side of base 30 mm and axis 55 mm long, is resting on HP on one of its longer edges with a face containing the larger edge inclined at 30° to the HP and its axis inclined at 30° to the VP. Draw its projections.

Solution: All the construction lines are retained to make the solution self-explanatory. See Fig. 12.81.

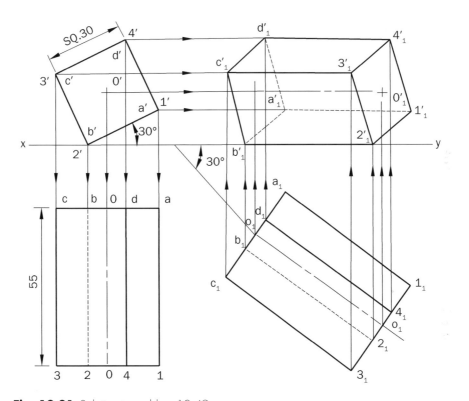

Fig. 12.81 Solution to problem 12.49

12.7 Axes Inclined to Both HP and VP

The projections of a solid with its axis inclined to both the principal planes are drawn in three stages:

Stage I: The two views are drawn keeping the solid in its simple position.
Stage II: One of the views is titled at the given angle, keeping the axis inclined to one of the principal planes and parallel to the other. Project the other corresponding view from it.
Stage III: In the final position, angle with the other plane is made by turning the later drawn view of the second stage at the given angle and the other view of the final stage is then projected.

The second and final positions may be drawn by either of the two methods already discussed.

Problem 12.50 A right regular pentagonal prism, side of base 30 mm and height 70 mm, rests on one of its base corners on HP such that its long edge containing the corner is inclined to the HP at 45° and the side of base opposite to the corner is inclined at 45° to the VP. Draw its projections.

Solution:

(i)   In the initial position, assume the prism to be resting on its base on the horizontal plane. Draw its projections, keeping one of the sides of its base perpendicular to x-y.

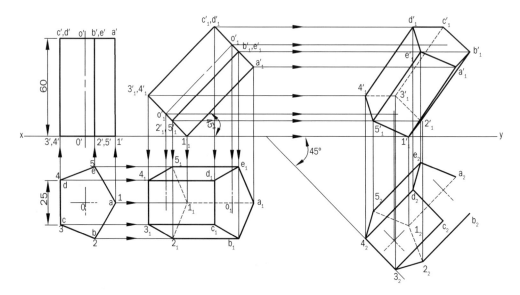

Fig. 12.82 Solution to problem 12.50

(ii)   Redraw the front view so that the longer edge 1'a' is inclined at 45° to x-y. Project the required top view and add suffix 1 to all the points. Complete the top view, keeping in mind the principles of visibility of faces.

(iii) Redraw the top view so that the side $4_1 \, 3_1$ (which is still true length) at an angle of 45° to the x-y.

(iv)   Project all the points vertically from this top view and horizontally from the second front view. Complete the new front view by observing the rules of visibility of lines as shown in Fig. 12.82.

Problem 12.51 A right regular pentagonal prism, side of base 30 mm and height 55 mm, rests on one of its base corners on ground plane such that its long edge containing that corner is inclined to the ground plane at 45° and the top view of the axis is inclined at 30° to VP. Draw its projections.

Solution:

(i)   As the prism is given to be resting on its base corner in ground plane, therefore the projections will be made in third-angle. Draw x-y and *gl* lines, a suitable distance apart.

(ii)   Draw top and front views assuming the axis of the prism to be perpendicular to the ground plane.

(iii)   Redraw the front view so that the longer edge $1_1'a_1'$ is inclined at 45° to *gl*. Project the required top view, keeping in mind the principles of visibility of lines.

(iv)   Redraw the top view by turning its axis at 30° to x-y. Project from it the final front view. See Fig. 12.83.

Problem 12.52 A right circular cone diameter of base 50 mm and height 60 mm lies on one of its elements in HP such that the element is inclined to the VP at 30°. Draw its projections.

Solution:

(i)   In the initial stage, assume the axis to be perpendicular to the HP and parallel to the VP. Draw the top and front views of the cone assuming the base on HP. Divide the base circle in top view into sixteen equal parts and project these points in the front view.

(ii)   Redraw the front view such that the element $o_1'1_1'$ (which is true length) lies on x-y. Project from it the corresponding top view, keeping in mind the principles of visibility of lines.

(iii)   Redraw the top view, as the element $o_2 1_2$ (in second stage, which is true length) is inclined at 30° to x-y.

(iv)   Project all the points vertically from this top view and horizontally from the second front view. Complete the new front view by observing the rules of visibility of lines as shown in Fig. 12.84.

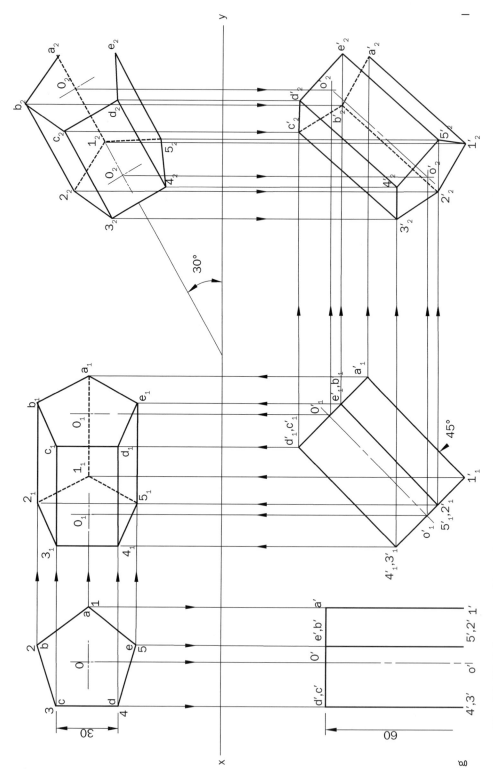

Fig. 12.83 Solution to problem 12.51

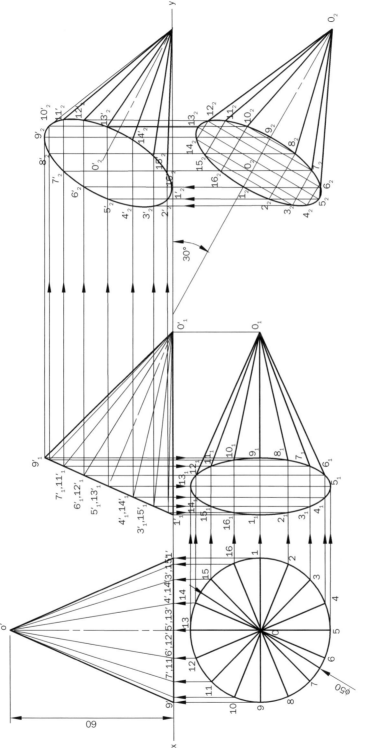

Fig. 12.84 Solution to problem 12.52

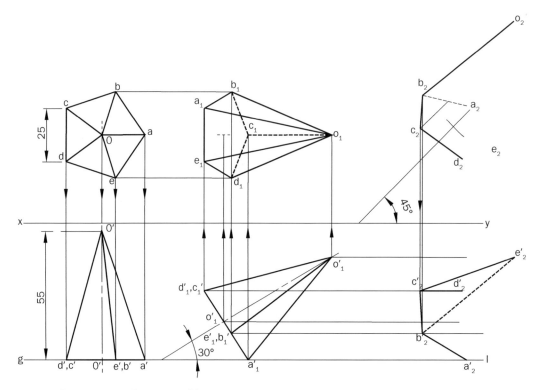

Fig. 12.85 Solution to problem 12.53

Problem 12.53 A right regular pentagonal pyramid, edge of base 25 mm and height 55 mm, is held on ground plane on one of its base corners, such that its axis is inclined at 30° to ground plane and 45° to VP. Draw its projections.

Solution:

(i)　As the pyramid is given to be resting on its base corner in ground plane, therefore the projections will be made in third-angle. Draw x-y and *gl* lines, a suitable distance apart.

(ii)　Draw top and front views assuming the axis of the pyramid to be perpendicular to the ground plane.

(iii)　Redraw the front view such that its axis is inclined at 30° to *gl*. Project the required top view, keeping in mind the principles of visibility of lines.

(iv)　Redraw the top view by turning its axis at 45° to x-y, but in the second stage top view, the projection of the axis is not true length. So first find out the apparent angle at which the top view of the axis is to be inclined so that the axis should make an angle of 45° to the x-y.

(v)　Mark any point o_2 above x-y. Draw a line $o_2 o_3$ equal to the true length of the axis, i.e., $o_1'o_1'$ and inclined at 45° to x-y. With o_2 as centre and radius equal to $o_1 o_1$ (length of the top view of the axis in second stage), draw an arc cutting the locus of o_3 at o_2. Then α is the apparent angle of inclination and is greater than 45°. Take $o_2 o_2$ as axis, reproduce the second top view and project the final front view as shown in Fig. 12.85.

Problem 12.54 A right circular cone, diameter of base 50 mm and height 60 mm, rests on its base rim on HP with its axis inclined at 45° to it such that (a) top view of the axis inclined at 30° to VP (b) axis inclined at 30° to the VP. Draw its projections.

(PTU, Jalandhar May 2005, May 2010)

Solution:

(i) In the initial stage, assume the axis to be perpendicular to the HP and parallel to the VP. Draw the top and front views of the cone assuming the base on HP. Divide the base circle in the top view into sixteen equal parts and project these points in the front view.

(ii) Reproduce the front view such that its axis makes an angle of 45° to the x-y. Project from it the corresponding top view, keeping in mind the principles of visibility of lines.

(iii) (a) Reproduce the top view of the axis such that it is inclined at 30° to the x-y. Project the final front view.

(b) The top view of the axis in the second stage does not give true length. So first the apparent angle at which the top view of the axis is to be inclined so that the axis should make an length of the axis i.e. $o_1'o_1'$ and inclined at 30° to x-y. With o_3 as centre and radius equal to $o_1 o_1$ (length of the top view of the axis in second stage), draw an arc cutting the locus of o_4 at o_3. Then is the apparent angle of inclination and is greater than 30°. Take $o_3 o_3$ as axis, reproduce the second top view and project the final front view as shown in Fig. 12.86.

Problem 12.55 A right regular pentagonal pyramid, edge of base 30 mm and height 50 mm, rests on one of its base corners on HP and the base is inclined at 25° to the HP. The side opposite the corner is inclined at 30° to the VP. Draw its projections.

Solution: The interpretation of the solution is left to the student. See Fig. 12.87.

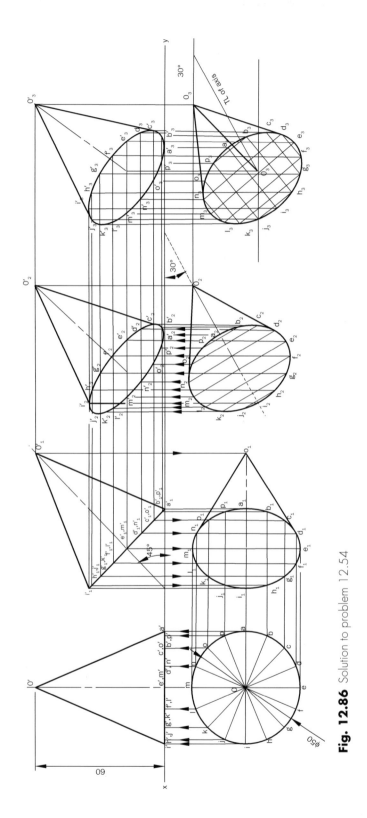

Fig. 12.86 Solution to problem 12.54

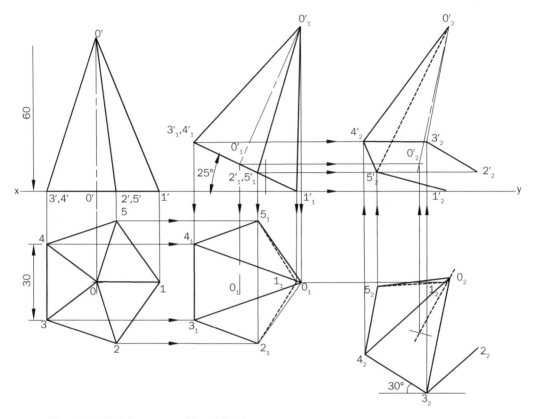

Fig. 12.87 Solution to problem 12.55

Problem 12.56 A right circular cylinder, diameter of base 50 mm and height 65 mm, rests on HP on its base rim such that its axis is inclined at 45° to the HP and the top view of the axis is inclined at 60° to the VP. Draw its projections.

Solution: The interpretation of the solution is left to the student. See Fig. 12.88.

Problem 12.57 A cube of 30 mm side rests on HP on one of its corners with a body diagonal perpendicular to the VP. Draw its projections.

Solution:

(i)   In the initial position, assume the cube to be resting on one of its faces on the HP with a body diagonal parallel to the VP. Draw a square 1234 in the top view with its sides inclined at 45° to x-y. Project the front view too.

(ii)   Tilt the front view about the corner a' so that the line $c'1'$ becomes parallel to x-y. Project the corresponding top view. The body diagonal $c'1'$ is now parallel to both HP and VP. Name the points on it by adding suffix 1 to them.

(iii)   Reproduce the second top view such that the top view of the body diagonal, i.e., $c_1 1_1$, is perpendicular to x-y and add suffix 2 to all the points.

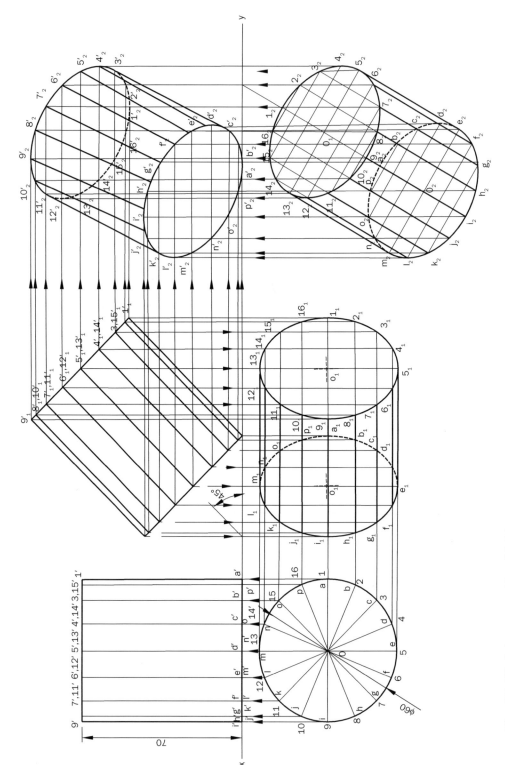

Fig. 12.88 Solution to problem 12.56

(iv) Project all the points vertically from this top view and horizontally from the second front view. Complete the new front view by observing the rules of visibility of lines or surfaces as shown in Fig. 12.89.

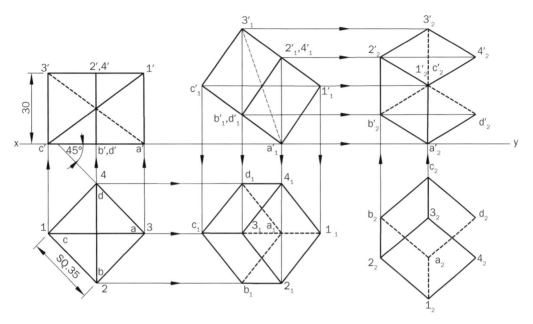

Fig. 12.89 Solution to problem 12.57

Problem 12.58 A right regular pentagonal prism 70 mm height with each side of the base 30 mm is resting on one of the base edges on the horizontal plane and inclined at 30° to VP and the face containing that edge is inclined at 45° to the HP. Draw the projections of the pentagonal prism.

(PTU, Jalandhar December 2004)

Solution:

(i) In the initial position, draw the pentagon in top view with one of its base sides (*ae*), perpendicular to x-y line. Project the front view from it.

(ii) Tilt the front view about the base side *ae* i.e. *a'e'1'5'* (face) at an angle of 45° to the HP. Project the corresponding top view observing the rules of visibility of surfaces. Name all the points on it by adding suffix 1 to them.

(iii) Reproduce the second top view aboout the side $a_1 e_1$ (true length) at an angle of 30° to the VP, i.e., x-y line, and add suffix 2 to all the points.

(iv) Project all the points vertically from this top view and horizontally from the second front view. Complete the new front view as shown in Fig. 12.90.

Problem 12.59 A right regular pentagonal pyramid of base 30 mm sides and height 60 mm rests on one of its slant edges on HP. The plan of the axis is inclined at 30° to VP. Its apex is nearer to VP. Draw the projections of the pyramid.

(PTU, Jalandhar May 2001)

Solution: The solution to the problem is left to the reader. All construction lines are retained to make it easy to understand the solution. See Fig. 12.91.

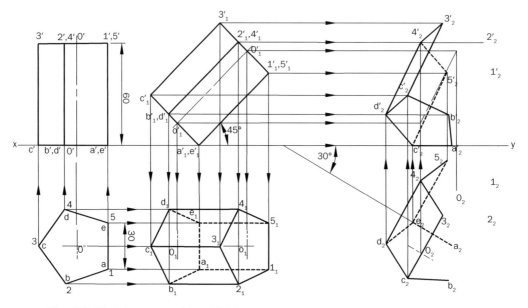

Fig. 12.90 Solution to problem 12.58

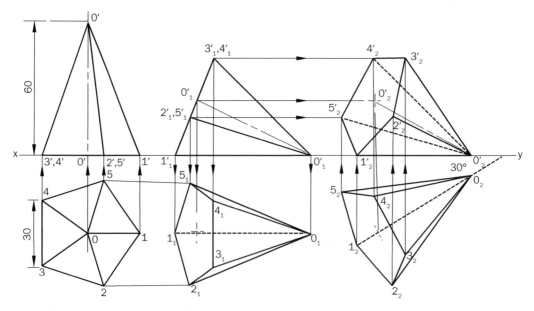

Fig. 12.91 Solution to problem 12.59

Problem 12.60 A square pyramid, edge of base 45 mm and length of axis 45 mm, is resting on one of its triangular faces on the HP having a slant edge containing that face parallel to the VP. Draw the projections of the pyramid by using both the methods.

Solution: In the initial stage, assume the axis to be perpendicular to the HP. Draw the top view and label it.

Method I (Change of position of solid)

(i)   Project the corresponding front view.
(ii) Reproduce the front view so that o'1'2' triangular face lies on HP and project the corresponding top view. Name all the points on it by adding suffix 1 to them.
(iii) Reproduce the top view such that the slant edge $o1_1$ or $o2_2$ is parallel to the x-y. Project the new front view as shown in Fig. 12.92, by observing the rules of visibility of lines.

Method II (Change of reference line)

(i)   Draw a new reference line $x_1\, y_1$ coinciding with o'2' or o'1' in the front view. Project new top view, keeping the distance of $1_1, 2_1, 3_1, 4_1, o_1$ from $x\, y$ equal to the distance of 1, 2, 3, 4, o from x-y.
(ii) Draw another reference line $x_2\, y_2$ coinciding with the slant edge $o_1\, 1_1$ or $o_1\, 2_1$. Project new front view as shown in Fig. 12.93.

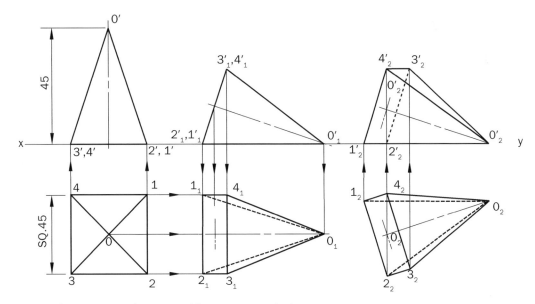

Fig. 12.92 Solution to problem 12.60 (Method I)

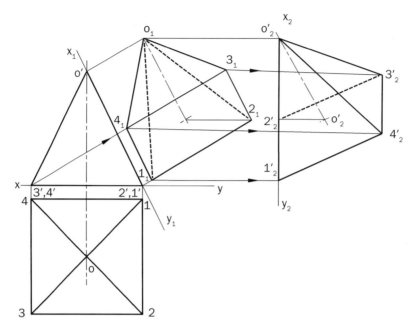

Fig. 12.93 Solution to problem 12.60 (Method II)

Problem 12.61 A hexagonal prism, base 30 mm side and axis 75 mm long, has an edge of the base parallel to the HP and inclined at 45° to the VP. Its axis makes an angle of 60° with the HP. Draw the projections.

(PTU, Jalandhar December 2002)

Solution:

(i)   In the initial position, draw the hexagon in the top view with one of its base sides (*ab*) perpendicular to x-y. Then project the corresponding front view from it.

(ii) Tilt the front view about the axis at angle of 60° to the HP. Project the corresponding top view, observing the rules of visibility of surfaces.

(iii) Reproduce the second top view about the side $a_1 b_1$ (true length) at an angle of 45° to the VP. Name all the points on it by adding suffix 2 to them.

(iv) Project the corresponding front view as shown in Fig. 12.94.

Problem 12.62 Draw the top view and front view of a square pyramid, side of base 30 mm, axis 55 mm long, that is freely suspended from one of the corners of its base with its axis in a vertical plane and making an angle of 45° with the VP.

(PTU, Jalandhar December 2004)

Solution: All construction lines are retained to make the solution self-explanatory. See Fig. 12.95.

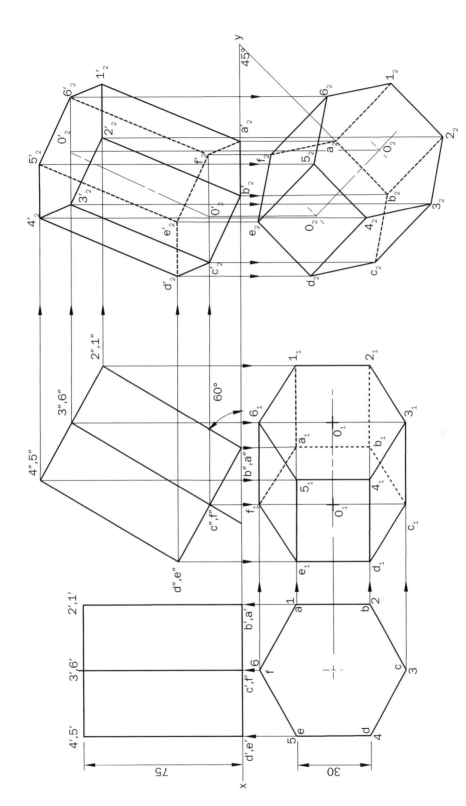

Fig. 12.94 Solution to problem 12.61

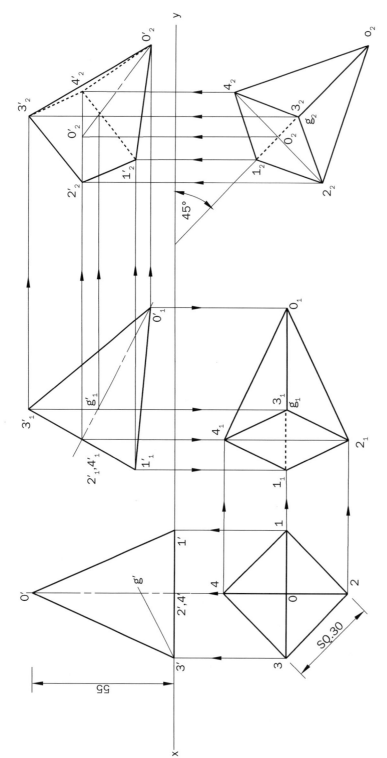

Fig. 12.95 Solution to problem 12.62

Problem 12.63 A right regular pentagonal prism, side of base 25 mm and axis 60 mm long, is resting on one of its base edges in the HP with its axis inclined at 45° to the HP and the top view of the axis is inclined at 60° to the VP. Draw its projections.

Solution:

(i) In the initial position, draw a pentagon in the top view with one of its base edges perpendicular to x-y. Then project the corresponding front view from it. Name all the corner points on it.

(ii) Tilt the front view about the axis at angle of 45° to the HP. Project the corresponding top view, observing the rules of visibility of surfaces. Name all the points on it by adding suffix 1 to them.

(iii) Reproduce the top view such that the top view of the axis is inclined at 60° to the VP. Project the corresponding front view. See Fig. 12.96.

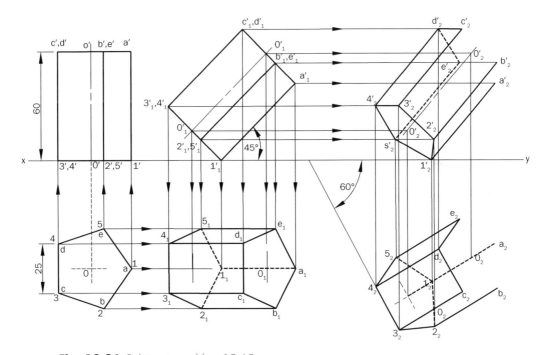

Fig. 12.96 Solution to problem 12.63

Problem 12.64 A right regular pentagonal pyramid, edge of base 25 mm and height 50 mm, lies on one of its triangular faces in HP such that (a) top view of the axis is inclined at 45° to the VP and (b) axis is inclined at 45° to the VP. Draw its projections.

Solution:

(i)   In the initial stage, assume the axis to be perpendicular to the HP and parallel to the VP. Draw elevation and plan of the pentagonal pyramid in the required position. Name all the corner points on it.

(ii)   Reproduce the front view such that one of its triangular faces say (0'3'4') lies on x-y. Project from it the corresponding top view.

(iii) (*a*) Reproduce the top view, such that the top view of the axis is inclined at 45° to the VP. Project the final front view.

(*b*) The top view of the axis in the second stage does not give true length. So, first find out the true angle of inclination with x-y. Complete the projections as shown in Fig. 12.97.

Problem 12.65 A frustum of a right regular pentagonal pyramid, edge of lower base 25 mm, edge of upper base 15 mm and axis 40 mm long, is lying on one of its slant edges on HP with its axis parallel to VP. The top view of axis is inclined at 45° to the VP. Draw its projections.

Solution:

(i)   In the initial stage, assume the axis to be perpendicular to the HP. Draw the top view and label it.

(ii)   Project the corresponding front view.

(iii) Reproduce the front view so that slant edge (d'' 4") lies on HP and project the corresponding top view. Name all the points on it by adding suffix 1 to them.

(iv)   Reproduce the top view such that axis (o_1 o_1) makes an angle of 45° to x-y. Project the final front view by observing the rules of visibility of lines. See Fig. 12.98.

Problem 12.66 A right regular hexagonal prism, side of base 25 mm and axis 60 mm long, is resting on one its base corners in the HP with its axis inclined at 30° to the HP and the top view of the axis is inclined at 45° to the VP. Draw its projections.

Solution: All the construction lines are retained to make the solution self-explanatory. See Fig. 12.99.

Problem 12.67. A square prism, side of base 30 mm and axis 70 mm long, is resting on a edge of its base in HP with its axis inclined at 45° to the HP. The top view of the axis is inclined at 30° to the VP. Draw its projections.

Solution: All the construction lines are retained to make the solution self-explanatory. See Fig. 12.100

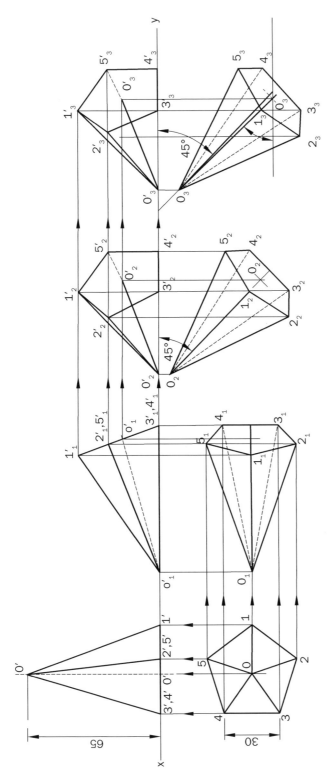

Fig. 12.97 Solution to problem 12.64

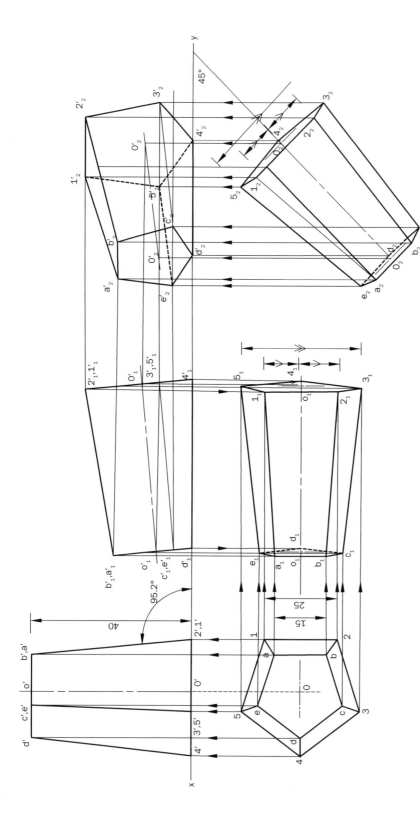

Fig. 12.98 Solution to problem 12.65

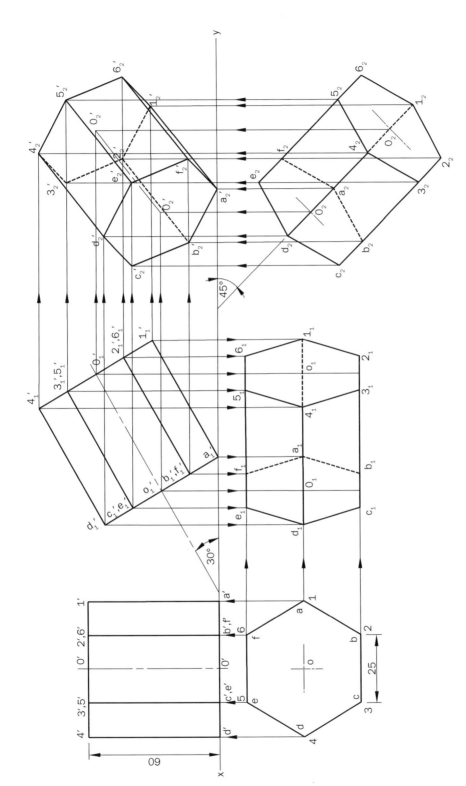

Fig. 12.99 Solution to problem 12.66

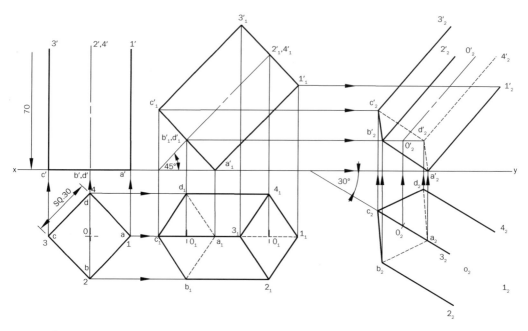

Fig. 12.100 Solution to problem 12.67

Exercises

Axis Perpendicular to One of the Planes

12.1 Draw the projections of the following solids, resting on HP on their bases, using a common reference line.

(a) A cube, side of base 30 mm and having its vertical faces equally inclined to the VP.

(b) A tetrahedron, 45 mm edge, having an edge of its base parallel to the VP.

(c) A square pyramid, side of base 30 mm and height 50 mm, resting on its base on HP with one of its edges parallel to the VP.

(d) A cone, 45 mm diameter and height 60 mm.

12.2 A triangular prism, side of base 30 mm and axis 60 mm long, lies on one of its rectangular faces in HP with its axis parallel to the VP. Draw its projections.

12.3 A right regular hexagonal prism, edge of base 30 mm and length 50 mm, lies on one of its rectangular faces on ground, such that its axis is parallel to both HP and VP. Draw its projections.

Axis Inclined to One of the Planes

12.4 A right regular pentagonal prism, edge of base 25 mm and height 50 mm, rests on an edge of its base in HP such that its axis is parallel to VP and one of its rectangular faces is perpendicular to VP and inclined to HP at 45°. Draw its projections.

12.5 Draw the projections of a cube, of 40 mm edge, when a body diagonal of the solid is kept vertical.

12.6   A right regular pentagonal pyramid, side of base 20 mm and height 40 mm, rests on a corner of its base on HP, such that its axis is inclined at 45° to the HP and is parallel to the VP. Draw its projections by using both the methods.

12.7   A right regular hexagonal pyramid, edge of base 25 mm and height 50 mm, rests on one of its base edges in HP with its axis parallel to VP. Draw its projections when the base makes an angle of 45° to the HP.

12.8   A square pyramid, edge of base 30 mm and axis 50 mm long, rests on HP on one of its base corners, such that the slant edge containing that corner is perpendicular to the HP. Draw its projections.

12.9   A right circular cone diameter of base 45 mm and height 50 mm rests on ground on its base rim such that its axis is parallel to the VP and inclined to the HP at 45°. Draw its projection.

12.10 A right circular cylinder diameter of base 45 mm and height 60 mm is resting in ground on its base rim such that its axis is parallel to VP, inclined at 30° to the HP. Draw its projections in third-angle.

12.11 A right regular pentagonal prism, side of base 25 mm and axis 50 mm long, lies on HP on one of its rectangular faces, such that its axis is parallel to HP and inclined at 30° to the VP. Draw its projections.

12.12 A right regular hexagonal pyramid, edge of base 25 mm and height 50 mm, is held on one of its base edges in HP. Its axis is parallel to the HP and inclined at 30° to the VP. Draw its projections.

12.13 A right regular hexagonal prism, side of base 30 mm and height 60 mm, has one of its rectangular faces in VP. The edge of base contained by that face is making an angle of 30° to HP. Draw its projections.

Axis Inclined to Both HP and VP

12.14 A square prism, edge of base 30 mm and axis 60 mm, rests on HP, on one of its base edges, such that its axis is inclined to HP at 45° and the edge on which the prism rests is inclined at 30° to the VP. Draw its projections.

12.15 A right regular hexagonal pyramid, side of base 25 mm and height 50 mm, is resting on one of its base edges on ground with its axis inclined at 30° to the HP and the edge of the base on which its rests is inclined at 45° to VP. Draw its projections.

12.16 A tetrahedron of 40 mm edge rests on HP on one of its edges such that the face containing that edge is inclined at 30° to HP and the edge is inclined at 45° to VP. Draw its projections.

12.17 A cube of 30 mm side rests on HP on one of its corners with a body diagonal perpendicular to the VP. Draw its projections.

12.18 A right regular hexagonal prism, edge of base 30 mm and axis 70 mm long, is held so that an edge of its base parallel to HP and inclined at 45° to VP, while its axis makes an angle of 60° to the HP. Draw its projections.

12.19 A right regular hexagonal prism, edge of base 30 mm and axis 65 mm long, rests on one of its base corners on HP with its axis inclined at 45° to HP and the top view of the axis inclined at 30° to the VP. Draw its projections.

12.20 A right regular hexagonal pyramid, edge of base 25 mm and axis 50 mm long, has one of its triangular faces in VP and the edge of base contained by that face is inclined at 30° to HP. Draw its projections.

12.21 A right regular pentagonal pyramid, edge of base 30 mm and height 60 mm, is held on ground plane on one of its base corners such that the slant edge containing the corner is inclined to HP at 45° and the base edge opposite the corner is inclined at 30° to VP. Draw its projections.

12.22 A square pyramid, whose faces are isoceles triangles of 45 mm base and 65 mm attitude, is lying on one of its triangular faces in HP. The projection of the axis of the pyramid on the HP makes an angle of 30° with the x-y line. Draw its projections.

12.23 A tetrahedron of 30 mm edge and resting on one of its base corners on the HP has one of its edges perpendicular to the HP. The edge opposite to this in the view from above makes 45 with the VP. Draw the projections.

12.24 A right regular pentagonal prism, side of base 25 mm and axis 60 mm long, is resting on one of its base edges in the HP with its axis inclined at 45° to the HP and the top view of the axis is inclined at 60° to the VP. Draw its three views.

12.25 A hexagonal prism, base 30 mm side and axis 75 mm long, has an edge of the base parallel to the HP and inclined at 60° to the VP. Its axis makes an angle of 45° with the HP. Draw the projections.

Objective Questions

12.1 To represent a solid in orthographic projections, at least .... views are required.

12.2 A solid having four equal equilateral triangular faces is called ....

12.3 When the axis of a solid is parallel to both HP and VP, ... shows the true shape of the base.

12.4 If a solid rests on its base on HP, its axis must be kept .... to HP.

12.5 Differentiate between right and oblique solids.

12.6 Differentiate between prism and pyramid.

12.7 In the projections of a solid, when two lines representing the edges cross each other, one of them must be ....

12.8 An object having .... dimensions is called a solid.

12.9 An oblique solid is one, which has its axis .... to its base.

12.10 If a solid rests on an edge of its base on VP, it must be kept .... to VP.

12.11 What are the different types of solids?

12.12 Illustrate solids of revolution with simple sketches.

12.13 A cone is resting with its apex on the HP. Draw its projections.

12.14 What is the side view of a cylinder if its axis is parallel to both HP and VP?

12.15 If a solid rests on an edge of its base on VP, it must be kept .... to HP.

12.16 A solid with two identcal ends is called ....

12.17 A cone is generated by the revolution of a .... about its altitude.

12.18 What do you mean by right regular prism? Show it by a sketch.

Answers

12.1 Two	12.7 Invisible	12.15 Perpendicular
12.2 Tetrahedron	12.8 Three	12.16 Prism
12.3 Side view	12.9 Inclined	12.17 Right angle triangle
12.4 Perpendicular	12.10 Perpendicular	

Chapter 13

Sections of Solids

13.1 Introduction

In engineering practice, it is often required to make a drawing that shows the interior details of the object. If the object is simple in its construction, the interior portion of the object can be easily interpreted by dotted lines in the orthographic projections. When the dotted lines of hidden parts are too many, the views become confusing and hard to read. In such cases, views can be drawn by cutting the object by an imaginary cutting plane so as to expose its interior or hidden details. The part of the object between the cutting plane and the observer is assumed to be removed so as to show the internal constructional features or details of the invisible surface. The exposed interior details are drawn in continuous thin lines instead of dotted lines. Such views are known as sectional views or views in section. The section surfaces are indicated by section lines, evenly spaced and inclined at 45° to the reference line.

13.2 Section Planes

These are generally perpendicular planes. These may be perpendicular or parallel to one of the principal planes and either perpendicular, parallel or inclined to the other plane. These planes are usually described by their traces.

13.3 Sections

Basically, sections are of two types:

    (i)   Apparent Section
    (ii)   True Section

(i) **Apparent Section:** The projection of the section on the principal plane to which the section plane is perpendicular is a straight line coinciding with the trace of the section plane on it, whereas its projection on the other plane to which it is inclined is called apparent section.

(ii) **True Section:** The projection of the section on a plane parallel to the section plane shows the true shape of the section. When the section plane is parallel to the HP or ground plane, the true shape of the section is seen in the sectional top view. When it is parallel to the VP, the true shape is projected in the sectional front view.

But when the section plane is inclined to one of the principal planes, the section has to be projected on an auxiliary plane parallel to the section plane to obtain its true shape.

When the section plane is perpendicular to both the principal planes, the sectional side view shows the true shape of the section.

13.4 Frustum of a Solid and a Truncated Solid

When a cone or a pyramid is cut by a plane parallel to its base, thus removing the top portion, the remaining lower portion is called its frustum, as shown in the Fig. 13.1. The section obtained is called the true section of a solid.

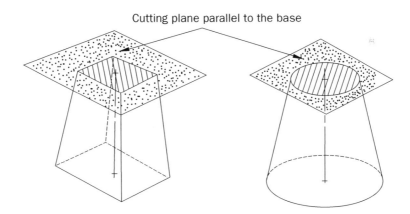

Cutting plane parallel to the base

Fig. 13.1 Frustum of a solid

When a solid is cut by a plane inclined to its base, thus removing the top portion, the remaining lower portion is called a truncated solid, as shown in the Fig. 13.2. The section obtained is called the apparent section of a solid.

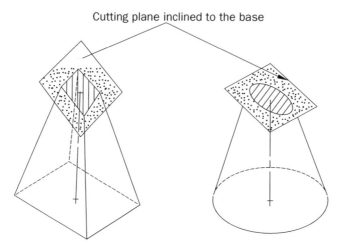

Cutting plane inclined to the base

Fig. 13.2 Truncated of a solid

13.5 Classification of Sections of Solids

Solids may be assumed to be cut by the section planes in many ways to obtain the sectional views. These are as follows:

- Section plane parallel to the HP
- Section plane parallel to the VP
- Section plane perpendicular to the VP and inclined to the HP
- Section plane perpendicular to the HP and inclined to the VP
- Section plane perpendicular to both HP and VP

13.6 Section Plane Parallel to the HP

As the section plane is perpendicular to the VP and parallel to the HP, therefore its VT will be a straight line parallel to x-y and has no HT. As the section plane is parallel to the HP, the projection of the section on the HP is its true shape and size. Its projection on the VP is a line and coincides with VT of the plane.

Problem 13.1 A right regular pentagonal pyramid, edge of base 30 mm and height 50 mm, rests on its base on HP with one of its base edges perpendicular to VP. A section plane parallel to the HP cuts the pyramid bisecting its axis. Draw its front view and sectional top view.

(PTU, Jalandhar December 2003)

Solution:

(i) Draw the projections of the pentagonal pyramid in the required position and label it.

(ii) As the section plane is parallel to the HP and perpendicular to the VP, it is represented by its VT.

(iii) The slant edges o'1', o'2', o'3', o'4' and o'5' intersect at the points a', b', c', d' and e', respectively, in the front view.

(iv) Project these points on the corresponding edges in the top view. Join these points in proper order and draw section lines in it. It will give the required sectional top view as shown in Fig. 13.3.

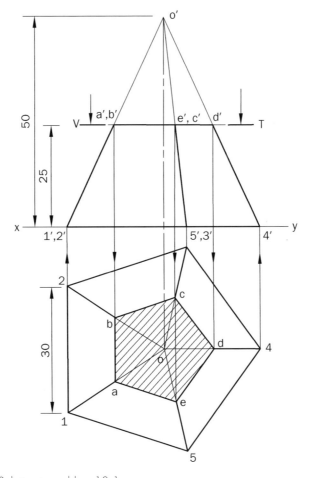

Fig. 13.3 Solution to problem 13.1

Problem 13.2 A right regular hexagonal pyramid, edge of base 30 mm and height 50 mm, rests on its base on ground plane with one of its base edges parallel to VP. A section plane parallel to the HP cuts the pyramid bisecting its axis. Draw its front view and sectional top view.

Solution:

(i)   Draw two lines x-y and *gl*, a suitable distance apart. Draw the projections of the hexagonal pyramid in the required position and label it.

(ii) Draw the VT to represent the section plane at a distance of 25 mm from *gl* and parallel to it.

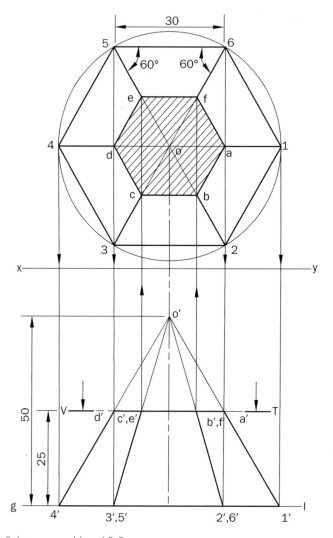

Fig. 13.4 Solution to problem 13.2

(iii) The slant edges o'1', o'2', o'3', o'4', o'5', o'6' cut the points *a'*, *b'*, *c'*, *d'*, *e'*, *f'*, respectively, in the front view.

(iv) Project these points of intersection on the corresponding edges in the top view. Join these points in proper order and draw section lines in it. It will give the required sectional top view as shown in Fig. 13.4.

Problem 13.3 A right regular pentagonal pyramid, side of base 30 mm and height 65 mm, lies on one of its triangular faces in HP such that its axis is parallel to VP. A section plane parallel to HP cuts the axis at a point 10 mm away from its base. Draw its front view and sectional top view.

(*PTU, Jalandhar May 2004*)

Solution:

- (i) Draw the projections of the given pyramid in the required position and label it.
- (ii) As the section plane is parallel to HP and perpendicular to the VP, it is represented by its VT.
- (iii) In the front view, at a distance of 10 mm from its base, find the point along its axis and draw a cutting plane line coinciding with VT, passing through the above mentioned point.
- (iv) Mark the points of intersection of the section plane with different edges of the pyramid.
- (v) Project these points on the corresponding edges in the top view. Join these points in their proper order and draw section lines in it. It will give the required sectional top view as shown in Fig. 13.5.

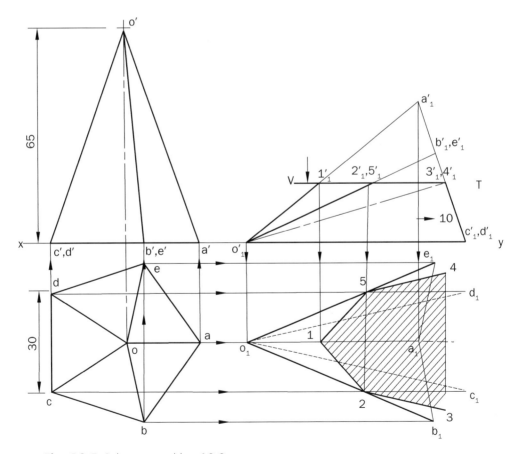

Fig. 13.5 Solution to problem 13.3

Problem 13.4 A right circular cone, diameter of base 50 mm and height 60 mm, lies on one of its elements in HP, such that its axis is parallel to VP. A section plane parallel to the HP and perpendicular to the VP cuts the cone, meeting the axis at a distance of 15 mm from the base. Draw its front view and sectional top view.

Solution:

(i)   Draw the projections of the given cone in the required position and label it.

(ii)   As discussed above, in the front view, at a distance of 15 mm from its base, find the point along its axis and draw a cutting plane line coinciding with VT, passing through the given point.

(iii) Mark the points of intersection of the section plane with different elements or base of the cone.

(iv)   Project these points on the corresponding elements or base in the top view. Join these points in their proper order and draw section lines in it. It will give the required sectional top view as shown in Fig. 13.6.

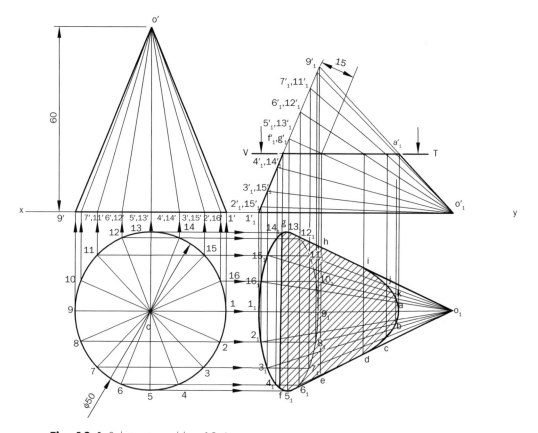

Fig. 13.6 Solution to problem 13.4

Problem 13.5 A right regular hexagonal prism, side of base 20 mm and length of axis 55 mm, is lying on one of its rectangular faces in HP. Its axis is parallel to both HP and VP. It is cut by a section plane parallel to and at a distance of 20 mm from the HP. Draw its front view and sectional top view.

Solution:

(i)   Draw the projections of the given prism in the required position and label it. Here in this case, the projections will be started from the side view and then project the front and top views from it.

(ii)   Draw a cutting plane coinciding with VT parallel to x-y and 20 mm away from it.

(iii)   Mark the points of intersection of the section plane with different elements of the prism.

(iv)   Project these points on the corresponding elements in the top view. Join these points in their proper order and draw section lines in it. It will give the required front and sectional top views as shown in Fig. 13.7.

Problem 13.6 A triangular prism, side of base 45 mm and length of axis 60 mm, is lying on one of its rectangular faces in HP. Its axis is parallel to both HP and VP. It is cut by a section plane parallel to and at a distance of 20 mm from the HP. Draw its front view and sectional top view.

(PTU, Jalandhar December 2006, December 2007)

Solution: The procedure followed to solve this problem is the same as explained in problem 13.5. The interpretation of the solution is left to the student. See Fig. 13.8.

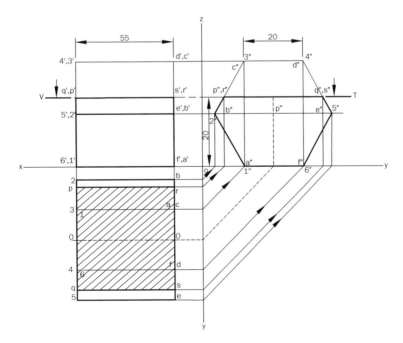

Fig. 13.7 Solution to problem 13.5

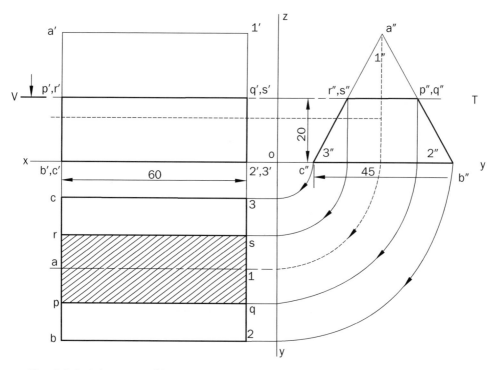

Fig. 13.8 Solution to problem 13.6

Problem 13.7 A right circular cylinder, diameter of base 50 mm and height 60 mm, rests on its base rim on ground plane such that its axis is inclined at 45° to the ground plane and is parallel to the VP. A section plane parallel to the HP cuts the cylinder bisecting its axis. Draw its front view and sectional top view.

Solution:

(i)   Draw two lines x-y and *gl* by a suitable distance apart. Draw the projections of the cylinder in the required position and label it.

(ii) Draw the VT to represent the section plane at a distance of 30 mm from base and parallel to it.

(iii) The section plane line cuts the elements $1_1{}'a_1{}'$, $2_1{}'b_1{}'$, and so on, at points I', H', and so on. Project these points on the corresponding elements in the top view. Join these points in proper order and draw section lines in it. It will give the required sectional top view as shown in Fig. 13.9.

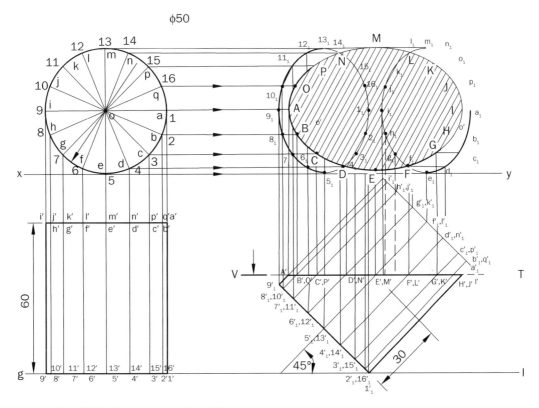

Fig. 13.9 Solution to problem 13.7

Problem 13.8 A hexagonal prism of base edge 20 mm long and height 60 mm is resting on one of its corners on HP with the base making 60° to HP. Axis is parallel to the VP. A section plane parallel to HP and perpendicular to VP cuts the object such that it is 15 mm away from the base measured along its axis. Draw the front view and sectional top view of the solid.

(PTU, Jalandhar December 2002)

Solution:

(i) Draw the projections of the prism in the given position and label it.

(ii) Draw the section plane line VT parallel to the x-y and 15 mm away from the base along its axis.

(iii) Mark the points of intersection of the section plane with different elements of the prism.

(iv) Project these points of intersection, say, e_1, g_1, i_1, h_1, c_1 in the top view to their corresponding elements of the prism. Join these points in their proper order and draw section lines in it. It will give the required sectional top view as shown in Fig. 13.10.

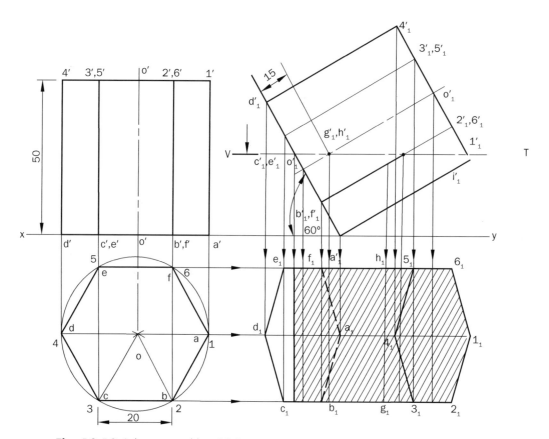

Fig. 13.10 Solution to problem 13.8

Problem 13.9 A right regular hexagonal prism, side of base 20 mm and height 45 mm, rests on a corner of its base on the HP with the longer edge passing through this corner making an angle of 30° to the HP. A section plane parallel to the HP cuts the axis at a distance of 15 mm from top base. Draw its sectional top view and front view.

Solution:

The interpretation of the solution is left to the reader. See Fig. 13.11.

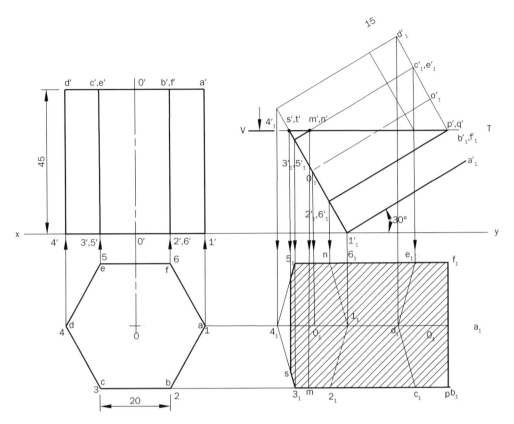

Fig. 13.11 Solution to problem 13.9

Problem 13.10 A right regular hexagonal pyramid, edge of base 30 mm and height 60 mm, rests on its base in HP with one of its base edges perpendicular to VP. A section plane parallel to the HP cuts the pyramid bisecting its axis. Draw its front view and sectional top view.

Solution:

(i)   Draw the projections of the pyramid in the given position and name the corner points on it.

(ii) Draw the cutting plane line VT parallel to x-y and bisecting its axis.

(iii) The cutting plane live VT cuts the various edges as shown in Fig. 13.12.

(iv) Project these points in the top view. The projections c, f of c', f in the top view are to lie on o3 and o6. By direct intersection these points cannot be plotted. Project c', f horizontally on a slant edge (which gives true length) or o'1' $(o_1'2_1')$ and then project point of intersection vertically into in the top view. With 0 as centre, rotate these points to lie on o3 and o6 at c and f, respectively.

(v)   Join the points in proper order and draw section lines in it.

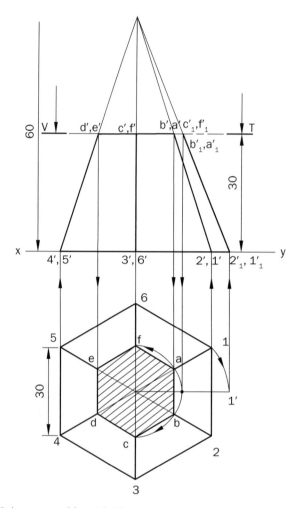

Fig. 13.12 Solution to problem 13.10

Problem 13.11 A right regular pentagonal prism, side of base 25 mm and length of axis 60 mm, is lying on one of its rectangular faces in HP. Its axis is parallel to both HP and VP. It is cut by a section plane parallel to and at a distance of 15 mm from the HP. Draw its front view and sectional top view.

Solution: The interpretation of the solution is left to the reader. See Fig. 13.13.

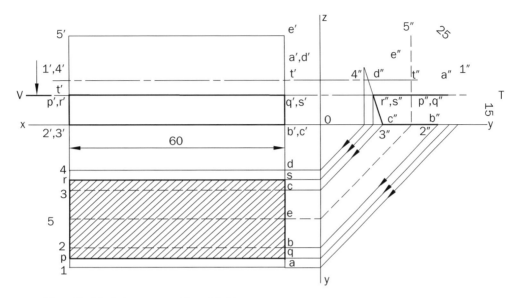

Fig. 13.13 Solution to problem 13.11

Problem 13.12 A right regular pentagonal pyramid, edge of base 30 mm and height 60 mm, rests on its base on HP with one of its base edges parallel to VP. A section plane parallel to the HP cuts the pyramid bisecting its axis. Draw its front view and sectional top view.

(*PTU, Jalandhar May 2009*)

Solution: All the construction lines are retained to make the solution self-explanatory. See Fig. 13.14.

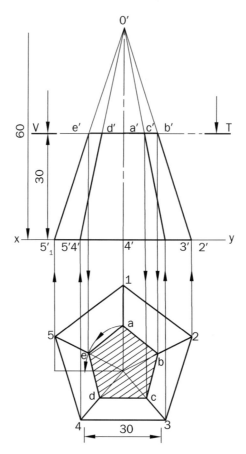

Fig. 13.14 Solution to problem 13.12

13.7 Section Plane Parallel to the VP

As the section plane is perpendicular to the HP and parallel to the VP, therefore its HT will be a straight line parallel to x-y and has no VT. As the section plane is parallel to the VP, projection of the section on the VP is its true shape and size. Its projection on the HP is a line, coinciding with the HT of the plane.

Problem 13.13 A right regular pentagonal pyramid, edge of base 30 mm and height 55 mm, rests on its base on HP, such that one of its base edges is perpendicular to the VP. A section plane parallel to the VP cuts the pyramid at a distance of 10 mm from the axis. Draw its top view and sectional front view.

(PTU, Jalandhar June 2003, May 2004)

Solution:

    (i)   Draw the projections of the given pyramid in the required position and label it.

    (ii)   As the section plane is parallel to VP and perpendicular to the HP, hence it is represented by its HT. In the top view, cut a distance of 10 mm from its axis, coinciding with HT.

   (iii)   Mark the points of intersection of the section plane with different edges of the pyramid.

   (iv)   Project these points on the corresponding edges in the front view. Join these points in proper order and draw section lines in it. It will give the required sectional front view as shown in Fig. 13.15.

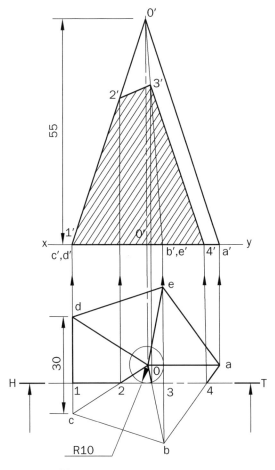

Fig. 13.15 Solution to problem 13.13

Problem 13.14 A right regular pentagonal pyramid, edge of base 30 mm and height 50 mm, rests on its base on ground plane, such that one of its base edges is perpendicular to VP. A section plane parallel to the VP cuts the pyramid at a distance of 10 mm from the axis. Draw its top view and sectional front view.

Solution: The interpretation of the solution is left to the student. See Fig. 13.16.

Problem 13.15 A right pentagonal pyramid of base side 25 mm and height 50 mm rests on the HP with one edge of the base at 45° to the VP. Draw the sectional elevation of the solid when it is cut by a plane parallel to the VP containing the apex.

(*PTU, Jalandhar December 2004*)

Solution:

   (i)   Draw the projections of the given pyramid in the required position and label it.

   (ii)   As the section plane is parallel to the VP, hence it is represented by its HT. In the top view, draw a cutting plane line passing through the apex coinciding with the HT.

  (iii) Mark the points of intersection of the section plane with different edges of the pyramid.

  (iv) Project these points on the corresponding front view. Join these points in the proper sequence and draw section lines in it as shown in Fig. 13.17.

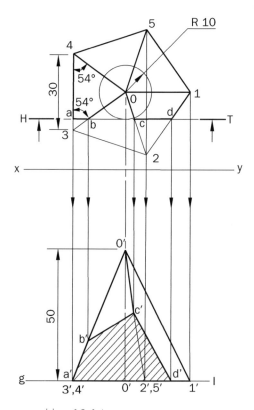

Fig. 13.16 Solution to problem 13.14

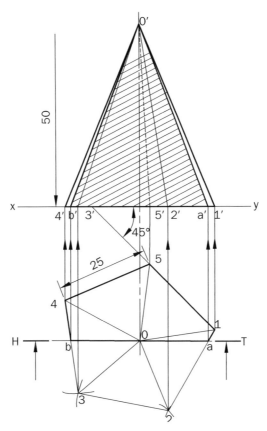

Fig. 13.17 Solution to problem 13.15

Problem 13.16 A cube of 35 mm long edge is resting on the HP on one of its faces with a vertical face inclined at 30° to the VP. It is cut by a section plane parallel to the VP and 10 mm away from the axis and further away from the VP. Draw its sectional front view and top view.

(PTU, Jalandhar December 2007, May, 2008)

Solution:

(i) Draw the projections of the cube in the required position and label it.

(ii) As the section plane is parallel to the VP and is perpendicular to the HP, hence the section plane is represented by its HT. Draw a line HT in the top view parallel to x-y and 10 mm from its axis.

(iii) Mark the points of intersection of the section plane with different edges of the cube.

(iv) Project these points on the corresponding edges in the front view and join these points in the correct sequence and draw section lines in it as shown in Fig. 13.18.

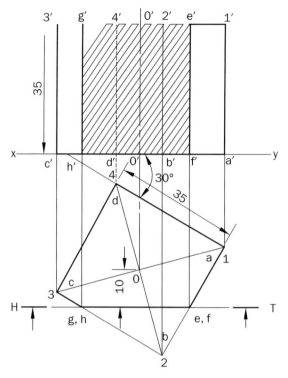

Fig. 13.18 Solution to problem 13.16

Problem 13.17 A right regular pentagonal pyramid, edge of base 30 mm and height 60 mm, rests on HP on one of its corners. Its base is inclined at 45° to HP and axis is parallel to the VP. A section plane parallel to the VP cuts the pyramid at a distance of 10 mm from its axis. Draw its top view and sectional front view.

Solution:

(i) Draw the projections of the given pyramid in the required position and label the corner points.

(ii) Draw the section plane line HT, to represent the section plane in top view, coinciding with HT and passing through a distance of 10 mm from its axis as shown in Fig. 13.19.

(iii) Project these points of intersection on the corresponding front view. Join these points in proper order and draw section lines in it. Complete the required top view and sectional front view.

Problem 13.18 A right circular cylinder, diameter of base 40 mm and height 60 mm, is lying on HP on one of its elements, such that its axis is inclined at 30° to the VP. A section plane parallel to VP cuts the cylinder at a distance of 10 mm from its end face meeting its axis. Draw its sectional front view and top view.

Solution:

(i)   Draw the projections of the given cylinder in the required position and label it.

(ii)   Draw the section plane line HT, to represent the section plane in top view, at a distance of 10 mm from its end face and parallel to the x-y.

(iii)   Mark the points of intersection of the section plane with different elements of the cylinder.

(iv)   Project these points on the corresponding elements in the front view. Join these points in proper order and draw section lines in it. The required sectional front view and top view are as shown in Fig. 13.20.

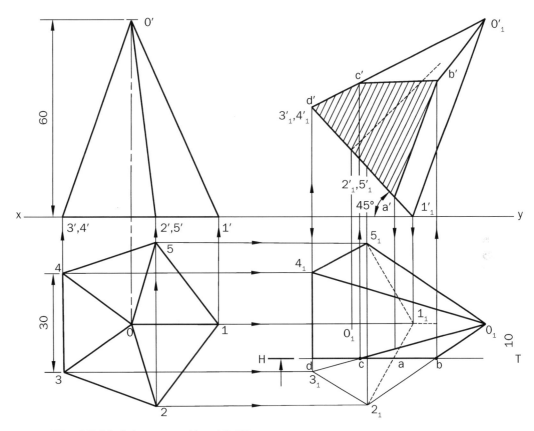

Fig. 13.19 Solution to problem 13.17

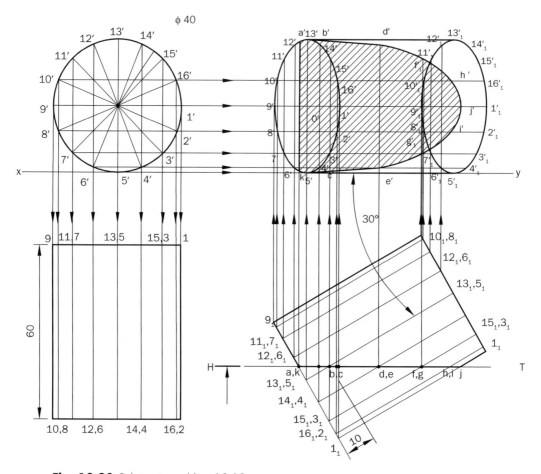

Fig. 13.20 Solution to problem 13.18

Problem 13.19 A right circular cone, diameter of base 50 mm and height 60 mm, lies on one of its elements in HP with its axis parallel to the VP. A section plane perpendicular to the HP and parallel to the VP cuts the cone and is 10 mm away from the axis. Draw its top view and sectional front view.

Solution: The interpretation of the solution is left to the students. See Fig. 13.21.

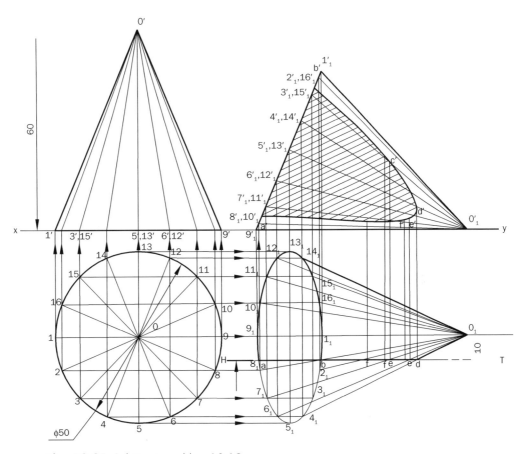

Fig. 13.21 Solution to problem 13.19

Problem 13.20 A right regular hexagonal pyramid, edge of base 25 mm and height 60 mm, rests on the HP with one edge of the base at 45° to the VP. A section plane parallel to the VP cuts the pyramid at a distance of 12 mm from the axis. Draw its top view and sectional front view.

Solution:

(i)   Draw the projections of the given pyramid in the required position and label it.

(ii) In the top view, cut a distance of 12 mm from its axis, coinciding with HT.

(iii) Mark points of intersection of the section plane with different edges of the pyramid.

(iv) Project these points on the corresponding edges in the front view. Join these points in proper order and draw section lines in it. It will give the required sectional front view as shown in Fig. 13.22.

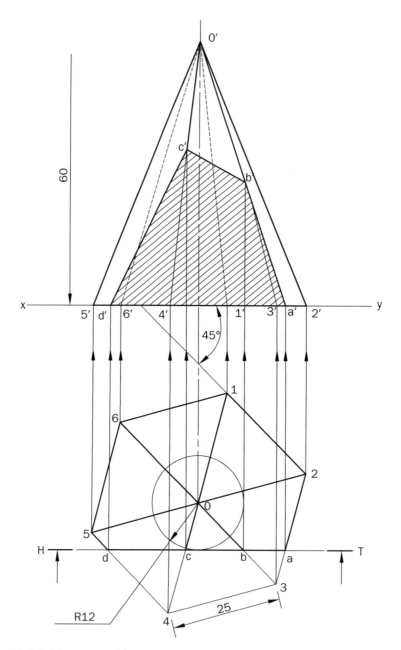

Fig. 13.22 Solution to problem 13.20

Problem 13.21 A right regular hexagonal prism, side of base 25 mm and height 60 mm, lies on one of its rectangular faces on HP and its axis inclined at 45° to the VP. A section plane parallel to the VP cuts the prism bisecting its axis. Draw its top view and sectional front view.

Solution: All the construction lines are retained to make the solution self-explanatory. See Fig. 13.22.

13.8 Section Plane Perpendicular to the VP and Inclined to the HP

When a section plane passing through a solid is perpendicular to the VP and inclined to the HP, its VT is inclined to the x-y line. Whereas its HT, which is perpendicular to the x-y line, serves no purpose in drawing the section views, so it is omitted. The projection of such a section in front view is a line, coincident with the cutting plane line VT. As the section plane is inclined to HP, its projection on the HP does not show its true shape and size and is called apparent section.

The true shape of section may be obtained on an auxiliary inclined plane (AIP) parallel to the given section plane.

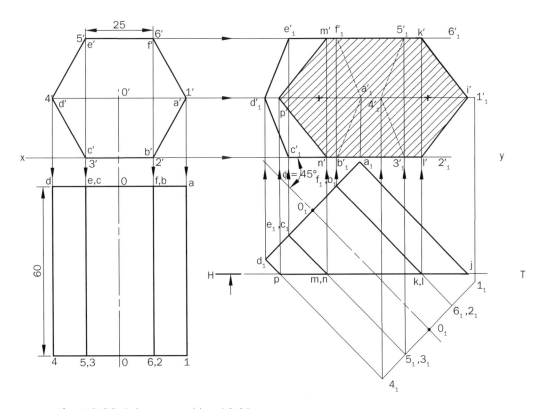

Fig. 13.23 Solution to problem 13.21

Problem 13.22 A cylinder of 45 mm diameter and 60 mm length is resting on one of its bases on HP. It is cut by a section plane inclined at 60° with HP and perpendicular to VP passing through a point on the axis 15 mm from it top end. Draw its front view, sectional top view and true shape of the section.

(*PTU, Jalandhar December 2009*)

Solution: The interpretation of the solution is left to the reader. See Fig. 13.24.

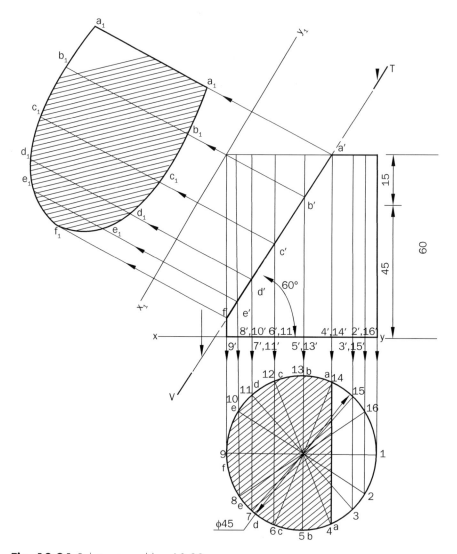

Fig. 13.24 Solution to problem 13.22

Problem 13.23 A right circular cone, diameter of base 50 mm, height 60 mm, rests on its base on HP. A section plane perpendicular to VP and inclined to HP at 30° cuts the cone bisecting its axis. Draw its front view, sectional top view and true shape of the section.

(*PTU, Jalandhar May 2006*)

Solution:

   (i)   Draw the projections of the cone in the given position and name the points on it.

   (ii) Draw the cutting plane line VT inclined at 30° to x-y and bisecting its axis.

   (iii) The cutting plane line VT cuts the various elements as shown in Fig. 13.25.

(iv) Project all the points on the corresponding elements in the top view. Join these points by a smooth curve and draw section lines in it.

(v) To draw the true shape of the section, draw a new reference line $x_1 y_1$ at a convenient distance and parallel to the cutting plane line.

(vi) Through the points on the section in the front view, draw perpendicular projectors to $x_1 y_1$. On the perpendicular projector through 1', from $x_1 y_1$ cut the distance of point 1 in the top view from x-y.

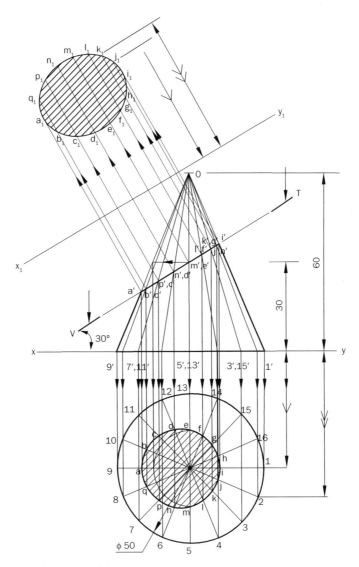

Fig. 13.25 Solution to problem 13.23

Similarly plot the other points. Join these points by a smooth curve and draw section lines in it. The required figure is called the true shape of the section.

Problem 13.24 A right regular hexagonal pyramid, edge of base 25 mm, height 50 mm, rests on its base on HP, with one of its base edges parallel to VP. A section plane perpendicular to VP and inclined to HP at 30° cuts the pyramid, bisecting its axis. Draw its front view, sectional top view and true shape of the section.

Solution:

    (i)   Draw the projections of the pyramid in the given position and name all the corner points on it.

    (ii)   Draw the cutting plane line VT inclined at 30° the x-y and bisecting its axis.

    (iii)   The cutting plane line VT cuts the various edges as shown in Fig. 13.26.

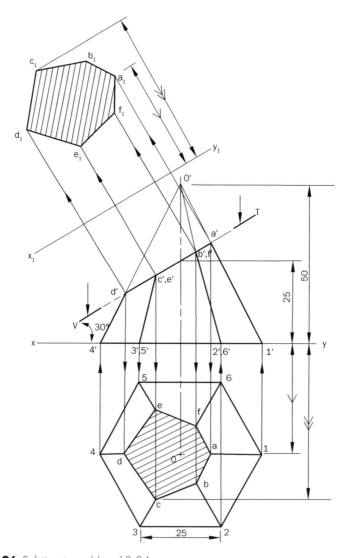

Fig. 13.26 Solution to problem 13.24

(iv) Project all these points on the corresponding edges in the top view. Join these points in proper order and draw section lines in it.

(v)   To draw the true shape and size of the section, draw a new reference line $x_1 y_1$, at a convenient position and parallel to the cutting plane line.

(vi) Through the points on the section in front view, draw perpendicular projectors to x_1 y_1. On the perpendicular projectors cut the distances of the points from x-y in the top view. Join these points in proper order and draw section lines in it. This is the required true shape of the section.

Problem 13.25 A square pyramid, edge of base 35 mm, height 50 mm, rests on its base on HP with its base edges equally inclined to VP. A section plane perpendicular to the VP and inclined to the HP at 30° cuts the pyramid bisecting its axis. Draw its front view, sectional top view and true shape of the section.

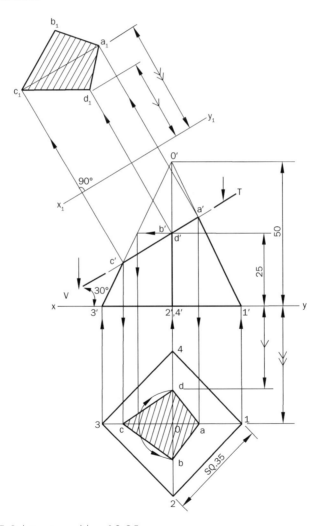

Fig. 13.27 Solution to problem 13.25

Solution:

(i)   Draw the projections of the pyramid in the given position and name the corners points on it.

(ii)   Draw the cutting plane line VT inclined at 30° to x-y and bisecting its axis.

(iii)   The cutting plane line VT cuts the various edges of the pyramid as shown in Fig. 13.27.

(iv)   Project these points in the top view. The projections b, d of b', d' in the top view are to lie on 02 and 04. By direct intersection these points cannot be plotted. Project b', d' horizontally on a slant edge (which gives true length) $0'1'$ or $0'3'$ and then project the point of intersection vertically into 01 or 03 in the top view. With o as centre, rotate these points to lie on 02 and 04 at b and d, respectively.

(v)   Join these points in proper order and draw section lines in it.

(vi)   To draw the true shape of the section, draw a new reference line $x_1 y_1$ at a convenient position and parallel to the cutting plane line.

(vii)   Project all the points as discussed in the earlier problems. Repeat the same procedure here to get the true shape of the section.

Problem 13.26 A square pyramid, edge of base 30 mm, height 45 mm, rests on its base on HP with its base edges equally inclined to the VP. A section plane perpendicular to the VP and inclined to the HP at 30° cuts the pyramid bisecting its axis. Draw its front view, sectional top view, sectional left side view and true shape of the section.

(PTU, Jalandhar May 2006)

Solution: The interpretation of the solution is left to the student. See Fig. 13.28.

Problem 13.27 A right regular pentagonal pyramid, edge of base 25 mm and height 50 mm, rests on its base on HP such that one of its base edges to be perpendicular to VP. A section plane perpendicular to the VP and inclined to the HP at 30° cuts the pyramid bisecting its axis. Draw its front view, sectional top view and true shape of the section.

(PTU, Jalandhar May 2010)

Solution:

(i)   Draw the projections of the pyramid in the given position and label it.

(ii)   Draw the cutting plane line VT inclined at 30° to x-y and bisecting its axis.

(iii)   The cutting plane line VT cuts the various edges of the pyramid as shown in Fig. 13.29.

(iv)   Project all the points of intersection in the top view. Join these points in proper order and draw section lines in it.

(v)   To draw the true shape of the section, draw a new reference line $x_1 y_1$ at a convenient distance and parallel to the section plane line.

(vi)   Project all the points as discussed in earlier problems. Repeat the same procedure to get the true shape of the section.

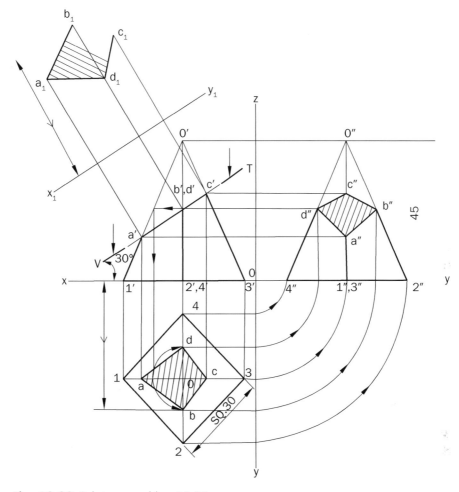

Fig. 13.28 Solution to problem 13.26

Problem 13.28 A cylinder of 45 mm diameter and 60 mm length is resting on one of its bases on HP. It is cut by a section plane inclined at 60° with HP and perpendicular to VP passing through a point on the axis 15 mm from its top end. Draw its sectional top view, front view and sectional end view.

(*PTU, Jalandhar May 2001*)

Solution:

(i) Draw the projections of the cylinder in the given position and label it.

(ii) Draw the cutting plane line VT inclined at 60° to x-y and passing through a point on the axis 15 mm from the top end of the cylinder.

(iii) The cutting plane line VT cuts the various edges as shown in Fig. 13.30.

(iv) Project the points of intersection in the top view and end view to their corresponding edges. Join these points in proper order and draw section lines in it.

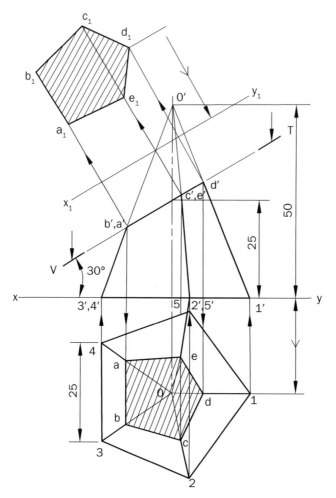

Fig. 13.29 Solution to problem 13.27

Problem 13.29 A cube of 65 mm long edges has its vertical faces equally inclined to the VP. It is cut by a section plane perpendicular to the VP, so that the true shape of the section is a regular hexagon. Determine the inclination of the cutting plane with the HP and draw the sectional top view and true shape of the section.

(PTU, Jalandhar June 2003)

Solution:

(i) The true shape of the section is a regular hexagon, as all the edges of the cube are equal, therefore the section plane should pass through the mid-points of all the six edges, which are cut by a section plane.

(ii) The cube must be kept in such a way that its vertical faces make equal angles with the VP. Therefore draw the top and front views of the cube in this position.

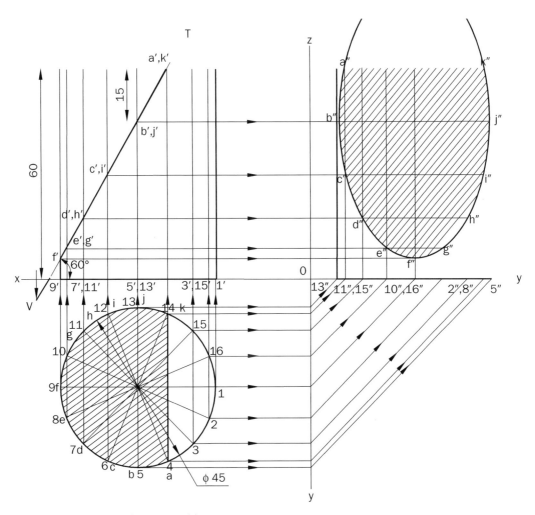

Fig. 13.30 Solution to problem 13.28

(iii) The cutting plane now passes through the points *f, g, e* and *h* which are the mid-points of the edges in the top view. Project these points in the front view to get the section plane through *f'g'* and *e'h'*.

(iv) Join the points *e, b, f, g, j, h* in the top view by straight lines to obtain the sectional top view and draw the section lines in it.

(v) Then draw $x_1 y_1$ line parallel to the section plane line at any convenient distance; the true shape of the section can be drawn on an auxiliary top view as shown in Fig. 13.31.

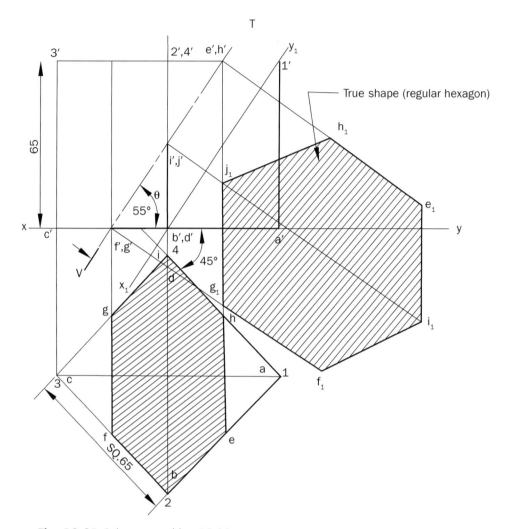

Fig. 13.31 Solution to problem 13.29

Problem 13.30 A right regular hexagonal pyramid, base 30 mm side and axis 65 mm long, is resting on its base on the HP with one of its base edges parallel to the VP. It is cut by a section plane perpendicular to the VP and inclined to the HP at 60° and intersecting the axis at a point 25 mm above the base. Draw the front view, sectional top view and sectional left side view.

Solution: The interpretation of the solution is left to the reader. See Fig. 13.32.

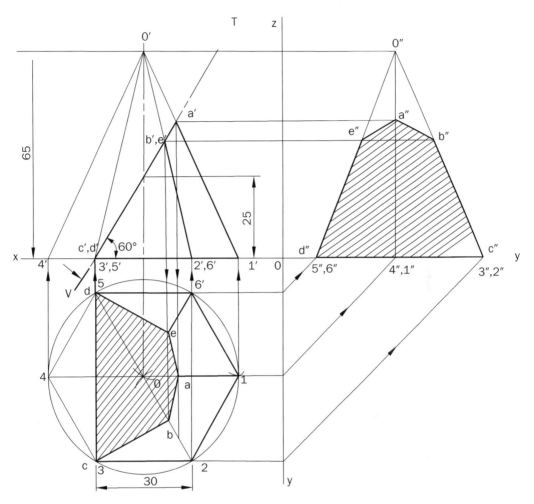

Fig. 13.32 Solution to problem 13.30

Problem 13.31 A square pyramid, edge of base 40 mm and height 60 mm, is resting on its base on HP with one of its base edges perpendicular to VP. A section plane perpendicular to the VP and inclined to the HP cuts the pyramid in such a way that the true shape of the section is a trapezium where parallel sides measure 30 mm and 15 mm. Draw the front view, sectional top view and true shape of the section. Also determine the inclination of the section plane with HP.

Solution: The interpretation of the solution is left to the student. See Fig. 13.32.

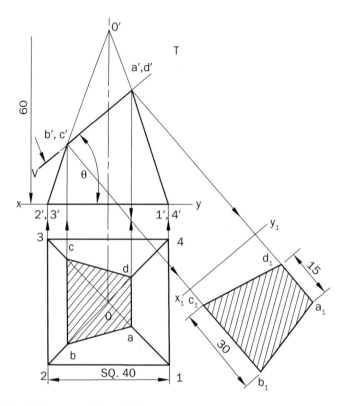

Fig. 13.33 Solution to problem 13.31

Problem 13.32 A square prism edge of base 25 mm and 70 mm long axis is resting on its base on HP. The edges of the base are equally inclined to the VP. It is cut by a plane inclined to the HP and perpendicular to the VP passing through the mid-point of the axis in such a way that the true shape of the section is rhombus having diagonals of 70 mm and 35 mm. Draw the projections and determine the inclination with HP.

Solution:

    (i)   Draw the projections of the square prism in the given position and name it.

    (ii)   Mark the mid-point of the axis in the front view. With the mid-point of the axis as centre and radius equal to 35 mm, draw an arc cutting the two opposite sides of the prism as shown in Fig. 13.33.

    (iii) Project these points in the top view and complete the true shape of the section. Draw the section lines in it.

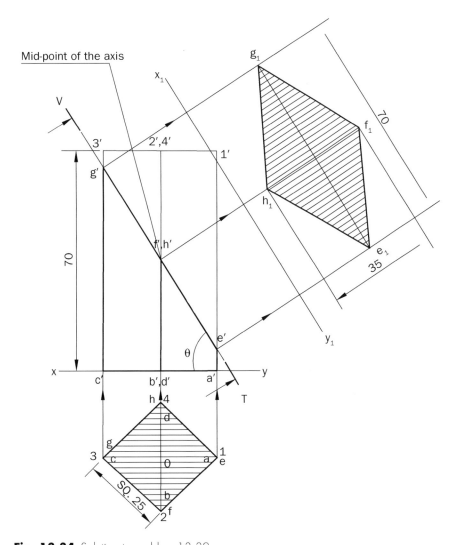

Fig. 13.34 Solution to problem 13.32

Problem 13.33 A right circular cylinder, diameter of base 60 mm and axis 60 mm long, rests on its base in HP. It is cut by a section plane which is perpendicular to the VP and inclined to the HP at 30° cuts the cylinder bisecting its axis. Draw the front view, sectional top view, sectional left side view and true shape of the section.

(PTU, Jalandhar December 2009)

Solution: The interpretation of the solution is left to the reader. See Fig. 13.35.

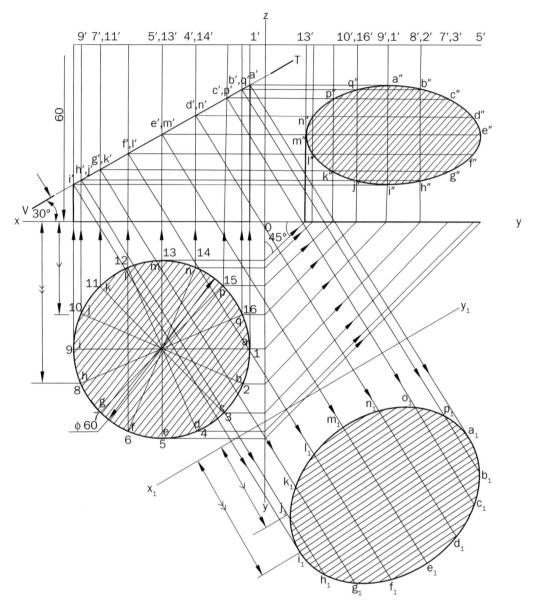

Fig. 13.35 Solution to problem 13.33

Problem 13.34 A right regular hexagonal pyramid, edge of base 20 mm and height 40 mm, rests on its base in HP, with one of its base edges perpendicular to VP. A section plane inclined to HP at 30° and perpendicular to VP cuts the pyramid, bisecting its axis. Draw its front view, sectional top view and true shape of the section.

Solution:

(i)   Draw the projections of the pyramid in the given position and name the corner points on it.

(ii) Draw the cutting plane line VT inclined at 30° to x-y and bisecting its axis.

(iii) The cutting plane line VT cuts the various edges of the pyramid as shown in Fig. 13.36.

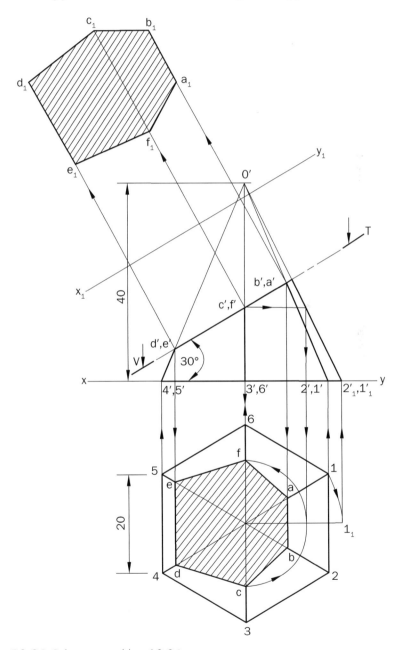

Fig. 13.36 Solution to problem 13.34

(iv) Project these points in the top view. The projections *c, f* of *c', f* are to lie on o3 and o6. By direct intersection, these points cannot be plotted. Project *c', f* horizontally on a slant edge (which gives true length) o' 1₁' (o' 2₁') and then project point of intersection vertically in the top view. With o as centre, rotate there points to lie on o3 and o6 at *c* and *f*, respectively.

(v) Join these points in proper order and draw section lines in it.

(vi) To draw the true shape of the section, draw a new reference x_1 y_1 at a convenient position and parallel to the cutting plane line.

(vii) Project all these points as discussed in the previous problems. Repeat the same procedure to get the true shape of the section.

Problem 13.35 A square pyramid, edge of base 30 mm, height 60 mm, rests on its base on HP with one of its base edge inclined at 45° to the VP. A section plane perpendicular to the VP and inclined to the HP at 30° cuts the pyramid at a distance of 25 mm from its apex on the axis. Draw its front view, sectional top view and true shape of the section.

Solution: All the construction lines are retained to make the solution self-explanatory. See Fig. 13.37

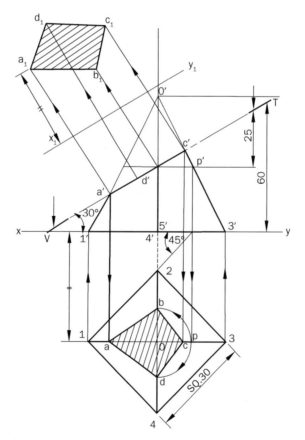

Fig. 13.37 Solution to problem 13.35

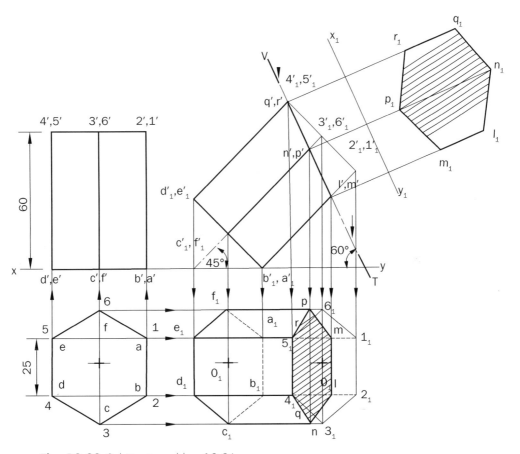

Fig. 13.38 Solution to problem 13.36

Problem 13.36 A right regular hexagonal prism of base 30 mm and axis 60 mm long is resting on HP on one of its base edge with its axis inclined at 45° to HP and parallel to the VP. A section plane inclined to the HP at 60° and perpendicular to the VP is passing through the topmost edges of the prism. Draw its front view, sectional top view and true shape of the section.

Solution: All the construction lines are retained to make the solution self-explanatory. See Fig. 13.38.

Problem 13.37 A cube of 55 mm edge rests on HP with one of its vertical faces inclined at 30° to the VP. A section plane inclined to the HP at 60° and perpendicular to the VP cuts the solid at a distance of 48 mm from the base along its axis. Draw its front view, sectional top view and true shape of the section.

Solution: All the construction lines are retained to make the solution self-explanatory. See Fig. 13.39.

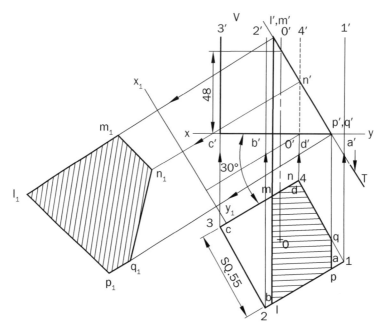

Fig. 13.39 Solution to problem 13.37

13.9 Section Plane Perpendicular to the HP and Inclined to the VP

When a section plane passing through a solid is perpendicular to the HP and inclined to the VP, its HT is inclined to the x-y line. Whereas VT, which is perpendicular to the x-y line, serves no purpose in drawing the section views, so it is omitted. The projection of such a section in top view is a line, coincident with the cutting plane line HT. As the section plane is inclined to VP, its projections on the VP do not show its true shape and size and is called apparent section.

The true shape of the section may be obtained on an auxiliary vertical plane (AVP) parallel to the given plane.

Problem 13.38. A right regular hexagonal pyramid, edge of base 25 mm, height 50 mm, lies on one of its slant edges on HP with its axis parallel to VP. A section plane perpendicular to HP and inclined to VP at 30° cuts the pyramid bisecting its axis. Draw its top view, sectional front view and true shape of the section.

Solution: All the construction lines are retained to make the solution self-explanatory. See Fig 13.40.

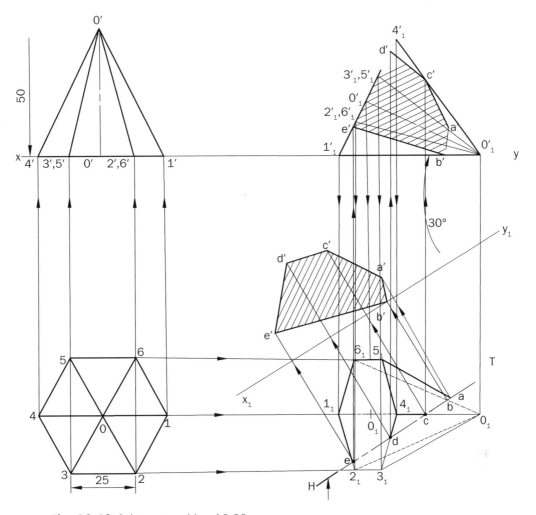

Fig. 13.40 Solution to problem 13.38

Problem 13.39 A right regular hexagonal pyramid, edge of base 25 mm, height 60 mm, rests on its base on HP with one of its base edges parallel to VP. A section plane perpendicular to HP and inclined to the VP at 30° cuts the pyramid and is 10 mm away from the axis. Draw its top view, sectional front view and true shape of the section.

Solution:

(i) Draw the projections of the pyramid in the given position and name the corner points on it.

(ii) Draw the cutting plane line HT inclined at 30° to x-y and 10 mm away from the axis. This can be done by drawing a circular arc of 10 mm radius with o as centre and draw the cutting plane line HT tangential to the arc.

(iii) The points *a, b, c, d, e* of the cutting plane line intersect with various edges of the pyramid in the top view.

(iv) Project these points in the front view to their corresponding edges. Join these points in proper order and draw section lines in it. As the section plane is inclined to the VP, projections of the section in the front view is an apparent section.

(v) To draw true shape of the section, draw an auxiliary plane parallel to the section plane, as shown in the Fig. 13.41 by the method already explained.

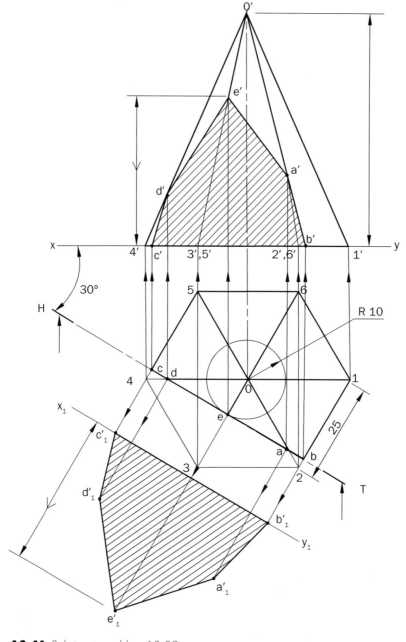

Fig. 13.41 Solution to problem 13.39

Problem 13.40 A right regular pentagonal pyramid, edge of base 25 mm and height 50 mm, rests on its base on HP such that one of its base edges is perpendicular to VP. A section plane perpendicular to the HP and inclined to the VP at 30° cuts the pyramid and is 10 mm away from the axis. Draw its top view, sectional front view and true shape of the section.

(PTU, Jalandhar December 2008)

Solution:

 (i)   Draw the projections of the pyramid in the given position and label it.
 (ii) Draw the cutting plane line HT inclined at 30° to x-y and 10 mm away from the axis. This can be done by drawing a circular arc of 10 mm radius with 0 as centre and draw the cutting plane line HT tangential to it.
 (iii) The points of intersection *a*, *b*, *c*, *d* of the cutting plane line with various elements of the pyramid in the top view are shown in Fig. 13.42.

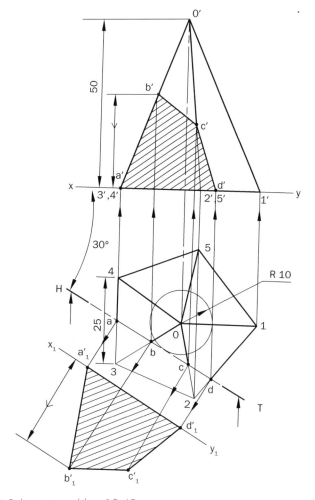

Fig. 13.42 Solution to problem 13.40

(iv) Project these points in the front view to their corresponding elements. Join these points in proper order and draw section lines in it.

(v) To get the true shape of the section, draw an auxiliary plane parallel to the section plane, and project all the points as discussed in the previous problems.

Problem 13.41 A right circular cone, diameter of base 50 mm and height 65 mm, rests on its base in HP. A section plane perpendicular to the HP and inclined to the VP at 45° cuts the cone and is 12 mm away from the axis. Draw its top view, sectional front view and true shape of the section.

(PTU, Jalandhar December 2004)

Solution:

(i) Draw the projections of the cone in the given position and label it.

(ii) Draw the section plane line HT inclined at 45° to x-y and 12 mm away from the axis. This can be done by drawing a circular arc of 12 mm radius with o as centre and draw the section plane line HT tangential to it.

(iii) The points of intersection a, b, c, d, e, f, g of the cutting plane line makes with various elements of the cone in the top view. Project these points in the front view. The projection of point b of b' in the front view are to lie on 0'5'. This point cannot be plotted by direct intersection. Rotate the point b, either on 01 or on 09 (which gives the true length 0'1' or 0'9') and then project the point of intersection vertically into 0'1' or 0'9' in the front view and b' horizontally on the 0'5' or 0'13'.

(iv) Join these points in proper order and draw section lines in it.

(v) To draw the true shape of the section, draw a new reference line $x\ y$ at a convenient distance and parallel to the section plane line. Project all the points as discussed in the earlier problems. Repeat the same procedure to get the true shape of the section as shown in Fig. 13.43.

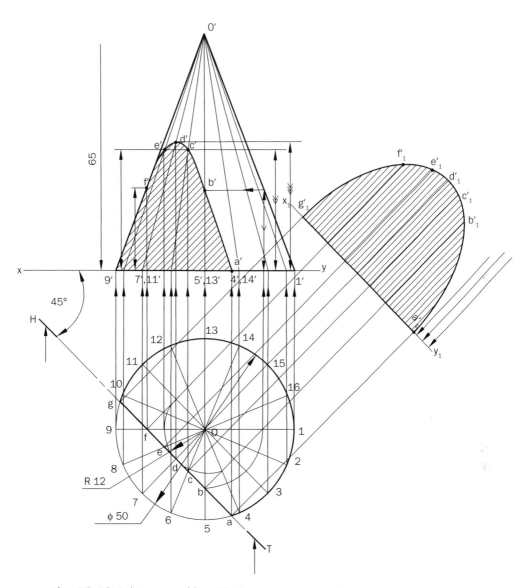

Fig. 13.43 Solution to problem 13.41

Problem 13.42 A right circular cone, 45 mm diameter axis 65 mm long, is resting on its base on
HP. It is cut by a plane, the HT of which makes an angle 45° with the VP and is passing 15 mm
from the top view axis. Draw the sectional front view and true shape of the section.

(*PTU, Jalandhar December 2003*)

Solution: The interpretation of the solution is left to the student. See Fig. 13.44.

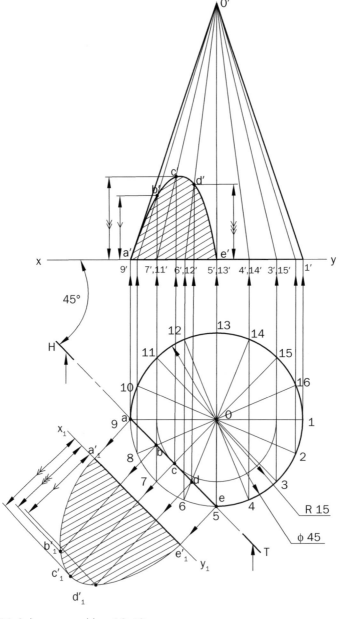

Fig. 13.44 Solution to problem 13.42

Problem 13.43 A right regular pentagonal pyramid, edge of base 25 mm, height 50 mm, lies on one of its triangular faces on HP with its axis parallel to VP. A section plane perpendicular to the HP and inclined to the VP at 30° cuts the pyramid bisecting its axis. Draw its top view, sectional front view and true shape of the section.

Solution:

(i) Draw the projections of the pyramid in the given position and label it.

(ii) Mark the mid-point of the axis in the front view and project it in the top view.

(iii) Draw the cutting plane line HT inclined at 30° the x-y and passing through the centre of the axis.

(iv) The cutting plane line HT cuts the various edges as shown in Fig. 13.45.

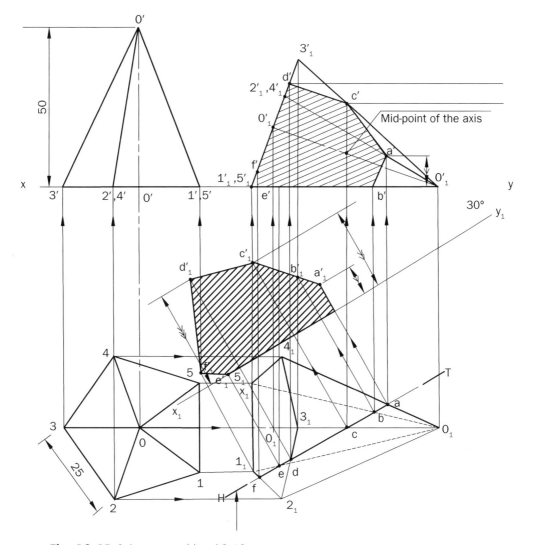

Fig. 13.45 Solution to problem 13.43

(v)   Project the points of intersection in the front view of their corresponding edges. Join these points in proper order and draw section lines in it. As the section plane is inclined to the VP, projections of the section in the front view is an apparent section.

(vi)   To draw true shape of the section, draw an auxiliary plane parallel to the section plane by the method already explained.

Problem 13.44 A right regular pentagonal prism, side of base 25 mm and height 60 mm, rests on an edge of its base on HP, such that one of its base corners lies on HP and its axis is inclined at 45° to the HP and parallel to the VP. A section plane perpendicular to the HP and inclined to the VP at 45° cuts the prism bisecting its axis. Draw its top view and sectional front view.

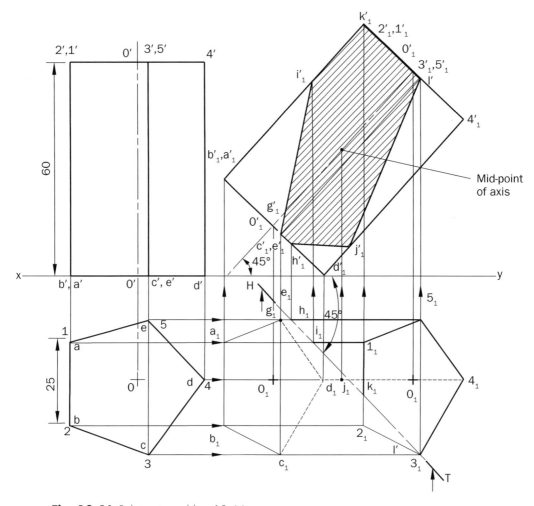

Fig. 13.46 Solution to problem 13.44

Solution:

   (i)   Draw the projections of the prism in the given position and name it.

   (ii)   Mark the mid-point of the axis in the front view and project it in the top view.

   (iii)   Draw the cutting plane line HT inclined at 45° to x-y and passing through the centre of the axis.

   (iv)   The cutting plane line HT cuts the various edges as shown in Fig. 13.46.

   (v)   Project the points of intersection in the front view to their corresponding edges. Join these points in proper order and draw section lines in it. As the section plane is inclined to the VP, the projections of the section in the front view is an apparent section.

Problem 13.45 A vertical cylinder of 40 mm diameter is resting on its base in HP. It is cut by a section plane which is perpendicular to the HP and inclined to the VP at 30° such that the true shape of the section is a rectangle of 30 mm × 60 mm. Draw the front view, sectional top view and true shape of the section.

Solution:

   (i)   Draw the projections of the cylinder in the given position and name it.

   (ii)   Draw a line $x_1 y_1$ in such a way that the chord length in the top view is 30 mm at 30° to x-y line

   (iii)   Project the points of intersection and draw the rectangle of 30 mm × 60 mm as shown in Fig. 13.47.

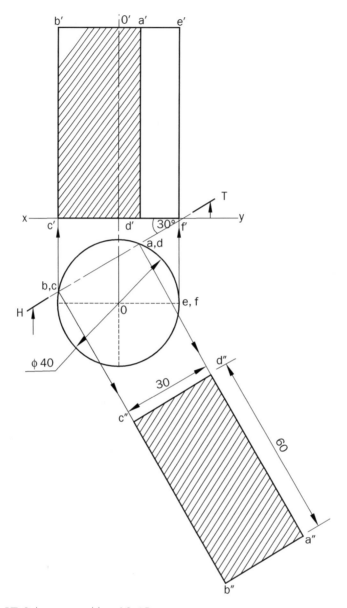

Fig. 13.47 Solution to problem 13.45

Problem 13.46 A right regular pentagonal pyramid, edge of base 25 mm and height 60 mm, rests on its base on HP such that one of its base edges is parallel to the VP. A section plane perpendicular to the HP and inclined to the VP at 30° cuts the pyramid and is 12 mm away from the axis. Draw its top view, sectional front view and true shape of the section.

Solution: The interpretation of the solution is left to the student. See Fig. 13.48.

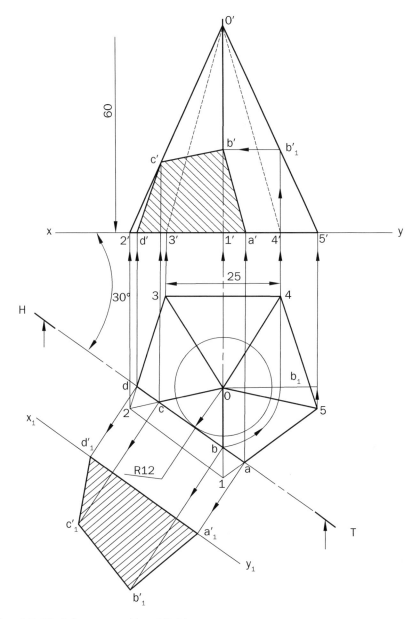

Fig. 13.48 Solution to problem 13.46

Problem 13.47 A hexagonal pyramid, side of base 30 mm and height 60 mm, has its base on the HP and an edge of the base makes an angle of 30° to the VP. It is cut by a plane perpendicular to the HP and inclined at 30° to VP at a distance of 18 mm from the base along the axis. Draw the sectional elevation and true shape of the section.

Or

A right regular hexagonal pyramid, edge of base 30 mm and height 60 mm, rests on its base in HP, with one of its base edges perpendicular to VP. The pyramid is cut by a section plane inclined to VP at 30° and perpendicular to HP and it is at a distance of 18 mm from the axis. Draw its top view, sectional front view and true shape of the section.

(PTU, Jalandhar December 2005)

Solution: The interpretation of the solution is left to the reader. See Fig. 13.49.

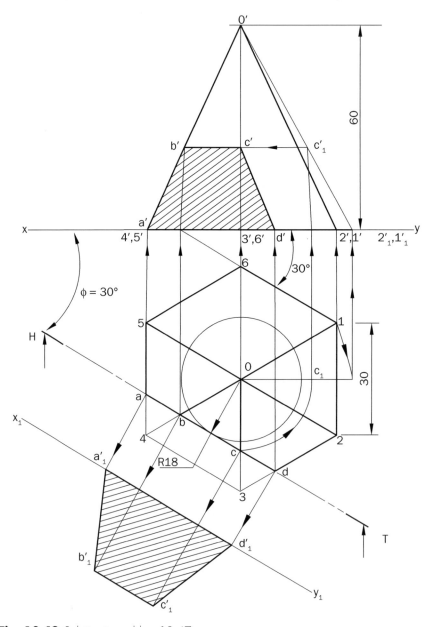

Fig. 13.49 Solution to problem 13.47

Problem 13.48 A hexagonal pyramid, side of base 25 mm and height 65 mm, is lying on HP on one of its triangular faces with its axis parallel to VP. It is cut by an auxiliary vertical plane (AVP) inclined to VP by 30° and passing though a point on the axis, 20 mm from the apex. Draw its projections and true shape of the section.

(PTU, Jalandhar December 2005, 2010)

Solution: The solution of this problem is self-explanatory. See Fig. 13.50.

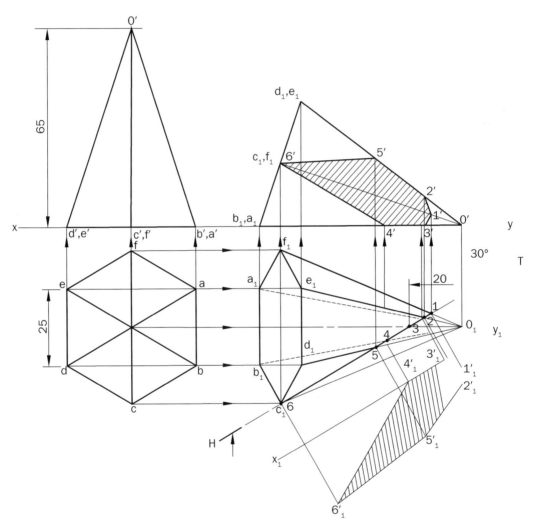

Fig. 13.50 Solution to problem 13.48

Problem 13.49 A right regular hexagonal prism, edge of base 30 mm and height 80 mm long, has an edge of its base in HP with its axis inclined at 60° to the HP and parallel to the VP. A section plane inclined to the VP at 60° and perpendicular to the HP cuts the axis at a distance of 60 mm from its bottoms end. Draw its top view and sectional front view.

Solution: All the construction lines are retained to make the solution self-explanatory see Fig. 13.51.

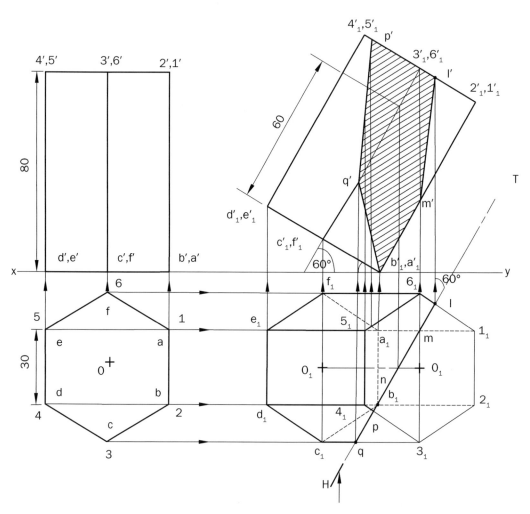

Fig. 13.51 Solution to problem 13.49

13.10 Section Plane Perpendicular to Both HP and VP

When a section plane is perpendicular to both HP and VP, its HT and VT become collinear. The projection of such a section is an edge view, i.e., a line in both front and top views, whereas the projection of the section in the profile plane gives true shape of the section.

Problem 13.50 A right circular cone, diameter of base 50 mm, axis 50 mm long, rests on its base on HP. A section plane perpendicular to both HP and VP cuts the cone and 10 mm away from the axis. Draw its front view, top view and sectional left side view.

Solution:

(i)   Draw the projections of the cone in the given position and name the points on it.
(ii) Since the section plane is perpendicular to both HP and VP, so the section plane line is seen as an edge view or a line both in front and top views. The side view will show the true shape of the section.
(iii) The cutting plane line, i.e., HT and VT both, cuts the various elements as shown in Fig. 13.52.
(iv) Project all the points of intersection in the left side view to their corresponding elements. Join these points in proper order and draw section lines in it.

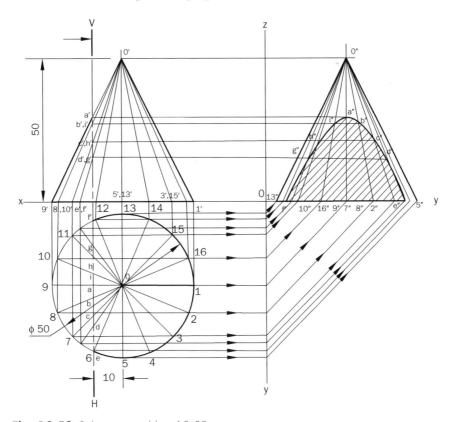

Fig. 13.52 Solution to problem 13.50

Problem 13.51 A right circular cone, diameter of base 50 mm and height 60 mm, rests on its base rim on HP, such that its axis is parallel to VP and inclined at 45° to the HP. A section plane perpendicular to both HP and VP cuts the axis of the cone at a distance of 25 mm from its vertex. Draw its front view, top view and sectional side view.

Solution: The interpretation of the solution is left to the student. See Fig. 13.53.

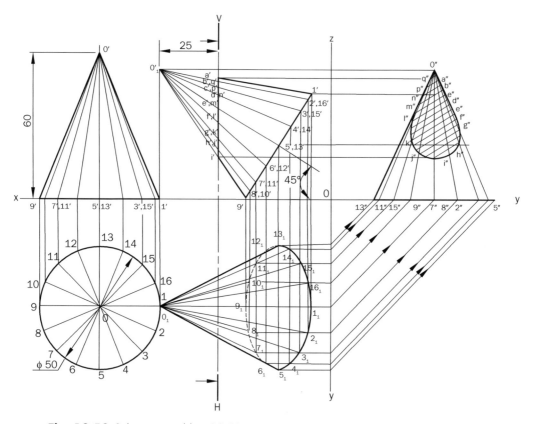

Fig. 13.53 Solution to problem 13.51

Problem 13.52 A right regular pentagonal prism, side of base 25 mm, height 50 mm, lies on one of its rectangular faces on HP, such that its axis is inclined at 30° to the VP and parallel to the HP. A section plane perpendicular to both HP and VP cuts the prism, meeting its axis at a distance of 10 mm from the base which is away from the VP. Draw its front view, top view and sectional side view.

Solution:

(i)   Draw the projections of the prism in the given position and label it.

(ii)   As the section plane is perpendicular to both HP and VP, hence its section plane line is seen as an edge or a line both in front and top views. The side view will show the true shape of the section.

(iii) The cutting plane line cuts the various elements as shown in Fig. 13.54.

(iv) Project all the points of intersection in the left side view to their corresponding edges (elements). Join these points in proper order and draw section lines in it.

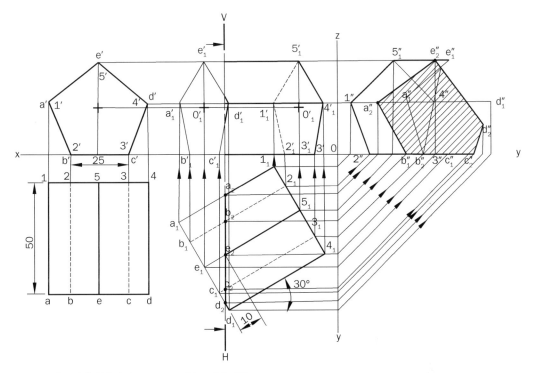

Fig. 13.54 Solution to problem 13.52

Exercises

Section Plane Parallel to the HP

13.1 A right regular pentagonal pyramid, side of base 25 mm and height 50 mm, rests on its base on HP such that one of its base edges is parallel to the VP. A section plane parallel to the HP cuts the pyramid bisecting its axis. Draw its front view and sectional top view.

13.2 A square pyramid, edge of base 25 mm and height 50 mm, rests on its base on HP such that one of its base edges makes an angle of 30° with the VP. A section plane parallel to the HP and perpendicular to the VP cuts the pyramid bisecting its axis. Draw its front view and sectional top view.

13.3 A right circular cone, diameter of base 50 mm and height 60 mm, rests on HP on its base rim such that its axis is parallel to the VP and inclined to the HP at 45°. A section plane parallel to the HP and perpendicular to the VP cuts the cone bisecting its axis. Draw its front view and sectional top view.

13.4 A right regular pentagonal prism, side of base 25 mm and axis 50 mm, lies on one of its rectangular faces on HP such that its axis is parallel to the both HP and VP. A section plane parallel to and 15 mm above the HP cuts the prism. Draw its front view and sectional top view.

Section Plane Parallel to the VP

13.5   A cube of 35 mm edge rests on its base on HP such that one of its faces is inclined at 30°
       to the VP. A section plane parallel to the VP cuts the cube at a distance of 10 mm from the
       axis. Draw its top view and sectional front view.

13.6   A right regular pentagonal prism, side of base 25 mm and length of the axis 60 mm, lies
       on one of its rectangular faces on HP with its axis inclined at 45° to VP. A section plane
       parallel to the VP and perpendicular to the HP cuts the axis of the prism at a distance of
       10 mm from the end face away from the VP. Draw its top view and sectional front view.

13.7   A right circular cone diameter of base 50 mm and height 60 mm rests on HP on its base
       rim such that its axis is parallel to the VP and inclined to the HP at 45°. A section plane
       parallel to the VP and perpendicular to the HP at a distance of 10 mm away from its vertex
       cuts the cone. Draw its top view and sectional front view.

13.8   A right regular hexagonal pyramid, edge of base 30 mm and height 55 mm, rests on its
       base on HP, such that one of its base edges is perpendicular to the VP. A section plane
       parallel to the VP cuts the pyramid at a distance of 10 mm from the axis. Draw its top view
       and sectional front view.

13.9   A right circular cylinder, diameter of base 40 mm and height 60 mm, is lying on HP on
       one of its elements, such that its axis is inclined at 30° to the VP. A section plane parallel
       to VP cuts the cylinder at a distance of 15 mm from its end face meeting its axis. Draw its
       sectional front view and top view.

Section Plane Perpendicular to the VP and Inclined to the HP

13.10 A right regular pentagonal pyramid, edge of base 25 mm and height 50 mm, rests on
       its base on HP such that one of its base edges is parallel to the VP. A section plane
       perpendicular to the VP and inclined to the HP at 45° cuts the pyramid, bisecting its axis.
       Draw its front view, sectional top view and true shape of the section.

13.11 A right circular cone, diameter of base 50 mm and height 60 mm, rests on its base on
       HP. A section plane perpendicular to the VP and inclined to the HP at 45° cuts the cone
       bisecting its axis. Draw its front view, sectional top view and true shape of the section.

13.12 A right circular cylinder diameter of base 50 mm and height 60 mm rests on its base
       on HP. A section plane perpendicular to the VP and inclined to the HP at 45° cuts the
       cylinder bisecting its axis. Draw its front view, sectional top view and true shape of the
       section.

13.13 A right circular cylinder, diameter of base 30 mm and height 60 mm long, lies on its base
       in HP. A section plane perpendicular to the VP and inclined at 45° to the HP cuts the axis
       at a point 20 mm from its top end. Draw its front view, sectional top view and true shape
       of the section.

Section Plane Perpendicular to the HP and Inclined to the VP

13.14 A right regular hexagonal pyramid, edge of base 25 mm and height 50 mm, lies on HP
       on one of its triangular faces such that its axis is parallel to the VP. A section plane
       perpendicular to the HP and inclined to the VP at 30° cuts the pyramid, bisecting its axis.
       Draw its top view, sectional front view and true shape of the section.

13.15 A right regular pentagonal pyramid, edge of base 25 mm and height 50 mm, rests on its base on HP such that one of its base edges is perpendicular to VP. A section plane perpendicular to the HP and inclined to the VP at 45° cuts the pyramid and is 10 mm in front of its axis. Draw its top view, sectional front view and true shape of the section.

13.16 A right circular cone, diameter of base 50 mm and height 65 mm, rests on its base in HP. A section plane perpendicular to the HP and inclined to the VP at 30° cuts the cone and is 15 mm away from the axis. Draw its top view, sectional front view and true shape of the section.

Section Plane Perpendicular to Both HP and VP

13.17 A right regular hexagonal pyramid, edge of base 25 mm and height 50 mm, rests on HP on one of its base edges such that the edge is perpendicular to the VP and its axis makes an angle of 45° to the HP. A section plane perpendicular to both HP and VP cuts the pyramid, bisecting its axis. Draw its front view, top view, sectional right side view and true shape of the section.

13.18 A right regular pentagonal prism, side of base 25 mm and height 50 mm, lies on one of its rectangular faces on HP, such that its axis is inclined to the VP at 45°. A section plane perpendicular to both HP and VP cuts the prism, meeting its axis at a distance of 10 mm from the end face which is away from the VP. Draw its front view, top view and true shape of the section.

13.19 A right circular cone, diameter of base 60 mm, axis 50 mm long, rests on is base on HP. A section plane perpendicular to both HP and VP cuts the cone and 12 mm away from the axis. Draw its front view, top view and sectional left side view.

13.20 A right circular cone, diameter of base 60 mm and height 60 mm, rests on its base rim on HP, such that its axis is parallel to VP and inclined at 45° to the HP. A section plane perpendicular to both HP and VP cuts the axis of the cone at a distance of 30 mm from its vertex. Draw its front view, top view and sectional side view.

Objective Questions

13.1 Distinguish between frustum of a solid and truncated of a solid.

13.2 Why are the solids sectioned?

13.3 What is the difference between apparent section and true section?

13.4 The sectional views are used to see the .... details of the objects.

13.5 When a section plane is parallel to the HP, the true shape of the section is viewed in the ....

13.6 The projection obtained on a VP of a cut solid is called sectional ....

13.7 The true shape of the section is an .... when a cylinder is cut by a section plane inclined to the axis.

13.8 The section planes are represented by its .... on HP and VP.

13.9 What is a cutting plane?

13.10 For obtaining a sectional view of a solid, the part of the object between the section plane and .... is assumed to be removed.

13.11 What is the principle of sectioning?

13.12 Name different types of sectioning methods.

13.13 What do you mean by sectional view?

13.14 Where and why is a cutting plane drawn in a drawing?

13.15 What is the true shape of the section obtained by cutting a cone parallel to one of the generators?

13.16 A cone is cut by a plane in such a way that the cutting plane passes through the apex. What is the true shape of the section?

13.17 Section portion is represented by .... lines.

13.18 Explain the procedure for working out true shape of surface inclined to HP.

13.19 The true shape of a section will be seen in the front view of an object when the section plane is .... to the VP.

13.20 Section lines are inclined at .... angle to ....line.

13.21 What do you mean by VT and HT of section plane?

Answers

13.4 Interior	13.7 Ellipse	13.17 Continuous thin
13.5 Top view	13.8 Traces	13.19 Parallel
13.6 Front view	13.10 Observer	13.20 45°, horizontal

Chapter 14

Development of Surfaces

14.1 Introduction

The complete surface of an object laid out on a plane is called the development of the surface or flat pattern of the object. 'Development' is a term frequently used in sheet metal work where it means the unfolding or unrolling of a detailed object into a flat sheet, called pattern (see Fig. 14.1). The development of geometrical surfaces is important in the fabrication of not only small and simple shapes made of thin sheet metal, but also for sophisticated pieces of hardware.

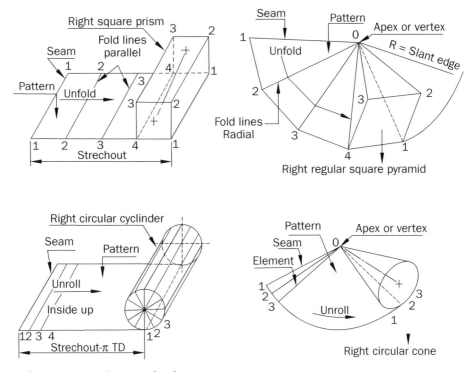

Fig. 14.1 Development of surfaces

In the development of a geometric surface, the opening should be determined first. Every line used in making the development must represent the true length of that line on an actual surface.

The knowledge of development of surfaces is essential in many industries such as automobile, ship building, packaging, sheet metal work, etc.

14.2 Methods of Development

There are four methods of development. These are:

(i) **Parallel Line Method:** It is employed in case of prisms and cylinders in which stretch-out line principle is used. Stretch-out is given by the perimeter of the object measured in a plane at right angles to the axis.

(ii) **Radial Line Method:** It is applied for the development of solids with slant edges like pyramids and cones. Here the true length of the slant edge of the cone or pyramid is taken as the radius of the arc of development.

(iii) **Triangulation Method:** It is used to develop transition pieces. It is a method of dividing a surface into a number of triangles and transforming them into the development.

(iv) **Approximate Method:** It is used to develop objects of double-curved or warped surfaces as sphere, paraboloid and hyperboloid.

Only the lateral surfaces of the solids have been developed and shown as presented here, omitting the bases or ends of solids.

14.3 Parallel Line Method

The surfaces of right prisms, oblique prisms, right cylinders and oblique prisms are developed by this method. In this method right section and stretch-out line principle is used. Parallel lines, parallel to the axis of the detail, are shown on a view which shows them as their true lengths.

(a) **Development of right prisms:** Development of the lateral surface of a right prism consists of the same number of rectangles in contact as the number of the sides of the base of the prism. One side of the rectangle is equal to the length of the axis and the other side equal to the length of the side of the base.

Problem 14.1. A right regular pentagonal prism, side of base 30 mm and height 50 mm, is truncated at the top and cut away from below as shown in Fig. 14.2 (a). Develop the lateral surface of the remaining prism.

(PTU, Jalandhar May 2009)

Solution: For its solution, see Fig. 14.2 (*b*), which is self-explanatory.

1. Draw the given front and top views.
2. The edge view of the right section appears in front view which gives the lateral corner edges as true lengths.
3. Draw the stretch-out line parallel to the edge view of the right section. Transfer along it the distance 1–2, 2–3, 3–4, 4–5, 5–1 measuring from top view.
4. Fix the ends of the lateral corner edges by taking measurements from front view.
5. Vertical position of these points is established by projecting them from front view as shown.
6. Observe that the quarter circular cut in front view is through the prism. As such its development will comprise of two symmetrical curves, one at entrance and other at exit. See Fig. 14.2 (a) and (b).

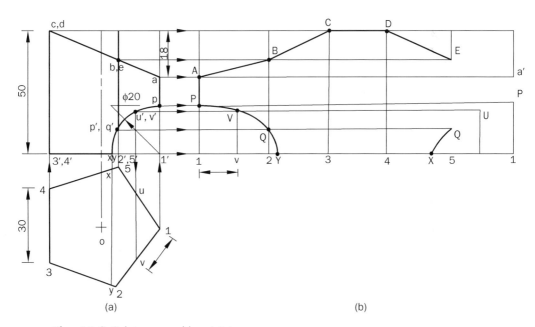

(a)   (b)

Fig. 14.2 Solution to problem 14.1

Problem 14.2 Develop the part pentagonal sheet metal detail as shown in Fig. 14.3 (a).

Solution:

(i)   Draw the given views of the pentagonal prism.
(ii) Complete the development of the full pentagonal prism along the stretch-out line.
(iii) Locate the intersection points on the edges of the prism.
(iv) Transfer the above points to the development by projecting them.
(v)   Draw the lines joining these points and complete the development as shown in Fig. 14.3(b).

Problem 14.3 A right regular hexagonal prism, side of base 30 mm and height 60 mm, is truncated at the top as shown in Fig. 14.4 (a). Develop the lateral surface of the remaining prism.

(*PTU, Jalandhar May 2011*)

Solution: For its solution, see Fig. 14.4 (*b*), which is itself self-explanatory.

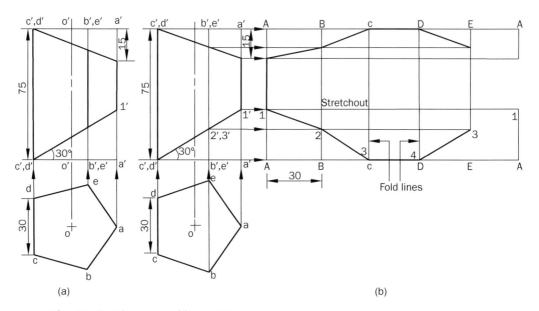

(a)   (b)

Fig. 14.3 Solution to problem 14.2

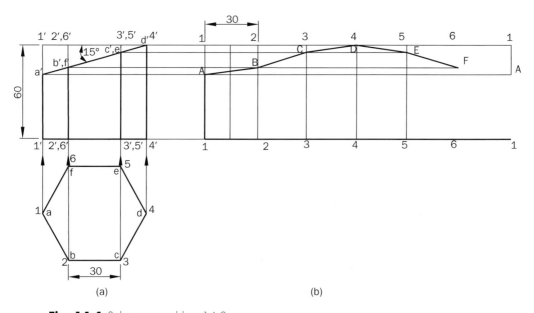

(a)   (b)

Fig. 14.4 Solution to problem 14.3

Problem 14.4 A right regular hexagonal prism, side of base 30 mm and height 60 mm, is truncated at the top and cut away from below as shown in Fig. 14.5 (a). Develop the lateral surface of the remaining prism.

Solution:

    (i)   Draw the given views of the hexagonal prism.

    (ii)   Complete the development of the full hexagonal prism along its stretch-out line.

    (iii) Draw the stretch-out line parallel to the edge view of the right section. Transfer along it the distance 1-2, 2-3, 3-4, 4-5, 5-6 and 6-1, measuring from the top view.

    (iv) On the quarter circle in front view, take some points, say, g', h', i', j', etc. Mark their corresponding top views g, h, i, j, etc., too. Distances of each of these points is measured and transfered from the top view to the development of the hexagonal prism by projecting them.

    (v)   Draw the lines joining these points and complete the development as shown in Fig. 14.5 (b).

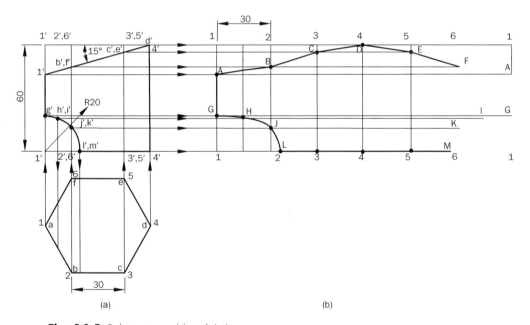

(a)   (b)

Fig. 14.5 Solution to problem 14.4

Problem 14.5 A pentagonal prism of 25 mm base edge and 50 mm length is resting on its base with an edge of base at 45° to VP. The prism is cut by a sectional plane inclined at 30° to HP and passes through a point 25 mm from the base along its axis. Develop the truncated prism.

(PTU, Jalandhar December 2003)

Solution: All the construction lines are retained to make the solution self-explanatory. See Fig. 14.6.

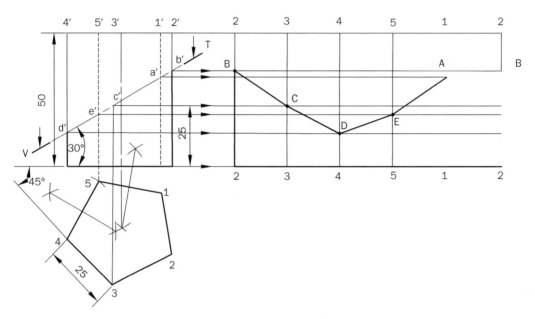

Fig. 14.6 Solution to problem 14.5

Problem 14.6 A right regular square prism, side of base 25 mm and height 50 mm, rests on its base on HP such that its vertical faces are equally inclined to the VP. A horizontal circular hole of diameter 30 mm is drilled centrally through it such that the axis of the hole cuts the diagonally opposite vertical edges. Develop its lateral surfaces.

(PTU, Jalandhar December 2009, May 2009)

Solution:

(i)   Draw the front and top views of the prism satisfying all the condition.

(ii)   Develop the lateral surface of the complete square prism.

(iii)   Divide the circle in the front view into sixteen equal parts and name these points as a', b', c', etc. and then project these points into top view as a, b, c, etc.

(iv)   Transfer the distances of these points in top view from lateral corner points. For example, from point 2 towards 3, to the stretch-out line of the development, such that $2g = 2g$, etc. Erect perpendiculars to the stretch-out line through these points, to cut the corresponding horizontal projections, taken into the development, from the front views a', b', c', etc., of these points a, b, c, etc.

(v)   Join these points by a smooth curve in the front view as shown in Fig. 14.7.

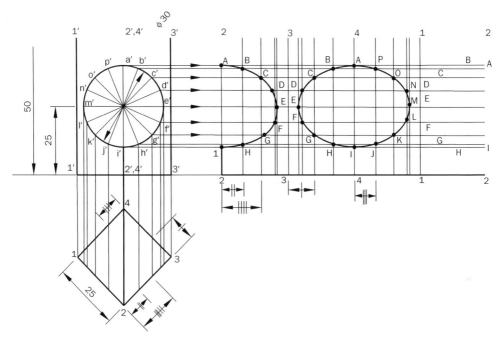

Fig. 14.7 Solution to problem 14.6

Problem 14.7 Front and top views of a right regular square prism are shown in Fig. 14.8 (a). Develop its lateral surface.

Solution: For its solution, see Fig. 14.8 (b), which is itself self-explanatory.

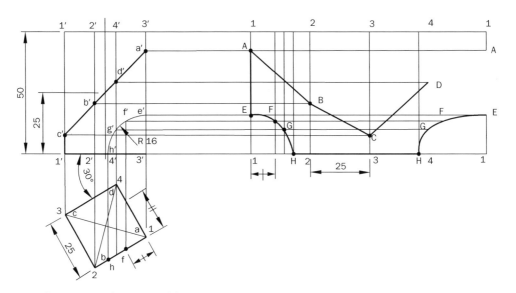

Fig. 14.8 Solution to problem 14.7

Problem 14.8 A right regular hexagonal prism, side of base 30 mm and height 60 mm, rests on its base on HP with one of its base side parallel to VP. A horizontal circular hole of diameter 40 mm is drilled centrally through it, such that the axis of the hole is perpendicular to it. Develop its lateral surface.

Solution: All the construction lines are retained to make the solution self-explanatory. See Fig. 14.9.

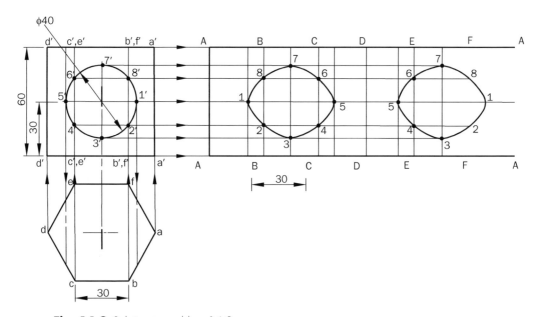

Fig. 14.9 Solution to problem 14.8

Problem 14.9 Front and top views of a right regular square prism are shown in Fig. 14.10 (a). Develop its lateral surface.

Solution: All the construction lines are retained to make the solution self-explanatory. See Fig. 14.10 (b).

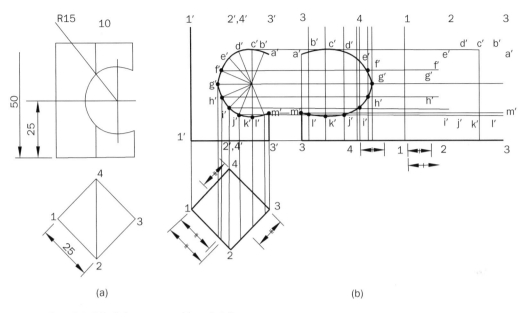

Fig. 14.10 Solution to problem 14.9

(b) **Development of oblique prisms:** When the axis of a prism is not at right angle to its base, the solid is known as oblique prism. The lateral surface of an oblique prism is developed by the same parallel line method as used for right prisms.

Problem 14.10 Front and top views of an oblique hexagonal prism are shown in Fig. 14.11 (a). Develop its lateral surface.

Solution:

   (i)   Draw the given views of an oblique hexagonal prism.
   (ii)   As the axis of the prism is inclined to the HP and parallel to the VP. The true length of the lateral edges are shown in the front view. The right section view can be drawn as an auxiliary view, as shown in Fig. 14.11 (*b*).
  (iii) The development can be completed by the same method as used for the right prisms.

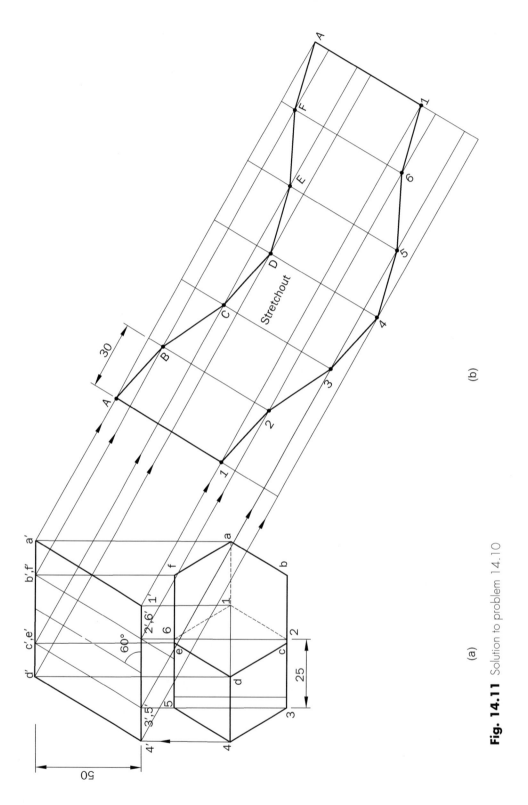

30

50

60°

25

a'   b',f'   c',e'   d'

a
f   b
2',6' 1'
6
e   c
d

1'
1
2

3',5'
5
4'
3
4
5

A   B   C   D   E   F   A

Stretchout

1   2   3   4   5   6   1

(a)

(b)

Fig. 14.11 Solution to problem 14.10

Problem 14.11 Front and top views of an oblique square prism are shown in Fig. 14.12. Develop its lateral surface.

Solution:

    (i)   Draw the given views of an oblique square prism.

   (ii)   As the axis of the prism is inclined to the HP and parallel to the VP, the true lengths of the lateral edges are shown in the front view. The right section view can be drawn as an auxiliary view, as shown in Fig. 14.12.

  (iii)   The development can be completed by the same method as used for the right prism.

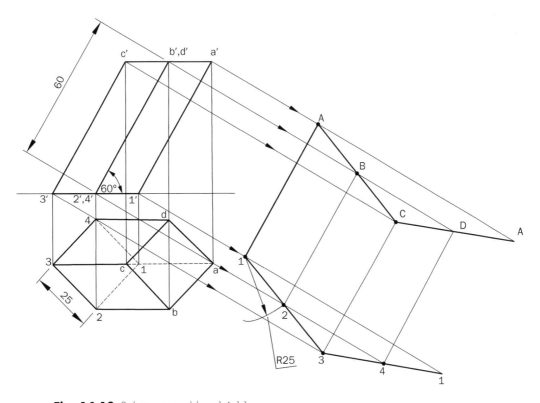

Fig. 14.12 Solution to problem 14.11

(c) Development of right cylinders: The development of the surface of a right circular cylinder is a rectangle, having one side equal to its circumference of its base circle and the other side equal to the height of the cylinder.

Since the surface of a cylinder is smooth and curved, it means that a cylindrical surface has no fold lines. However, a series of elements may be obtained by dividing its base circle into sixteen equal parts, is drawn as the surface of a cylinder.

Problem 14.12 A truncated cylinder is shown in Fig. 14.13 (a). Develop its lateral surface.

Or

A cylinder of base 40 mm and height 60 is cut by a section plane which makes 45° with the HP at a distance of 40 mm from lower base. Draw the development for the lower part of the cylinder.

(PTU, Jalandhar May 2006, December 2006)

Solution:

    (i)   Draw the given views of the truncated cylinder.

   (ii)   Complete the development of the full cylinder along its stretch-out line.

  (iii)   Mark the points of intersection a', b', c', etc., between the generators and truncated zone of the cylinder.

  (iv)   Locate the points a', b', c', etc., on the generator 1, 2, 3, etc., by projecting from the front view.

   (v)   Join these points by a smooth curve as shown in Fig. 14. 13 (b).

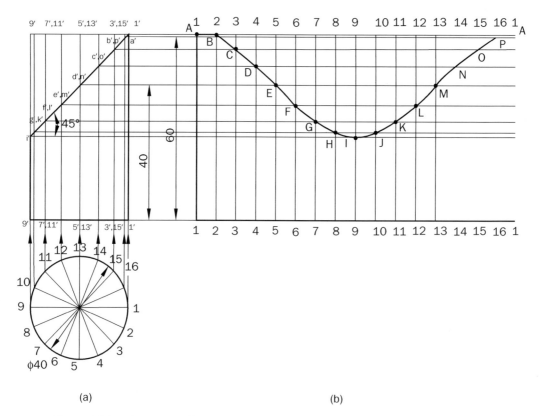

(a)   (b)

Fig. 14.13 Solution to problem 14.12

Problem 14.13 Develop the lateral surface of the right circular cylinder cut at top and bottom, as shown in Fig. 14.14 (a).

Solution: For its solution, see Fig. 14.14 (*b*), which is self-explanatory.

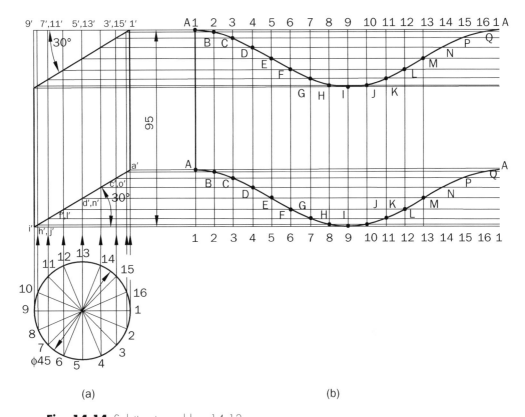

(a)   (b)

Fig. 14.14 Solution to problem 14.13

Problem 14.14 Develop the circular pattern of a right circular cylindrical pipe of 50 mm diameter and 60 mm height. It has a horizontal circular hole of diameter 30 mm drilled centrally through it such that the axes of the hole and the cylinder are mutually perpendicular to each other.

(*PTU, Jalandhar May 2000*)

Or

A cylinder of diameter 50 mm and height 60 mm has a hole of diameter 30 mm drilled in it such that its axis intersects that of the cylinder at the middle at right angle. Draw the development.

(*PTU, Jalandhar December 2002*)

Solution:

(i)   Draw the top and front views of the given cylindrical pipe.

(ii) Develop the complete cylinder by taking a stretch-out line.

(iii) Name the points where the circle for the hole in the front view intersects the elements of the cylindrical pipe.

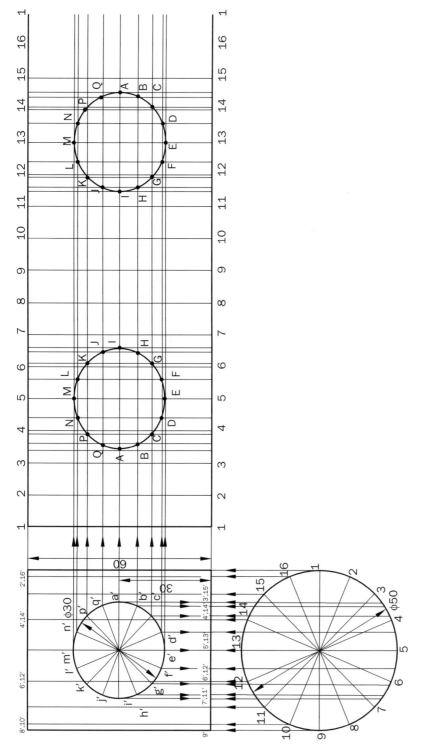

Fig. 14.15 Solution to problem 14.14

(iv) Project these points in the top view as *a*, *b*, *c*, and so on.

(v) Transfer the distances of these points in the top view from elements of the cylindrical pipe. For example, from point 3 towards 4, to the stretch-out line of the development, such that 3*a* = 3A, etc. Erect perpendiculars to the stretch-out line through these points, to cut the horizontal projections, taken into the development, from the front views *a'*, *b'*, *c'*, etc.

(vi) Join these points by a smooth curve in the front view as shown in Fig. 14.15.

Problem 14.15 Develop the lateral surface of the right circular cylinder as shown in Fig. 14.15 (a).

Solution: All the construction lines are retained to make the figure self-explanatory. See Fig. 14.161 (*b*).

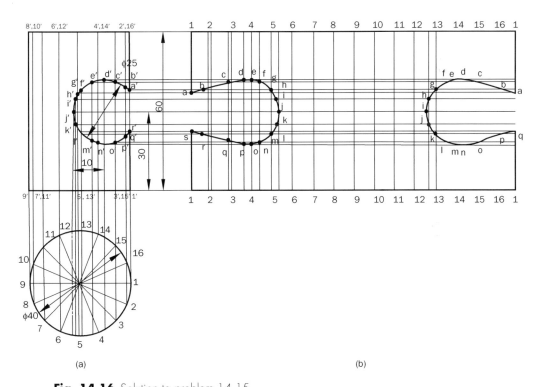

(a)   (b)

Fig. 14.16 Solution to problem 14.15

Problem 14.16 Develop the lateral surface of the right circular cylinder as shown in Fig. 14.17 (a).

Solution: All the construction lines are retained to make the figure self-explanatory. See Fig. 14.17 (*b*).

Problem 14.17 Draw the development of the lateral surface of the cylinder cut by the plances as shown in Fig. 14.18 (a).

Solution: The interpretation of the solution is left to the student. See Fig. 14.18 (*b*).

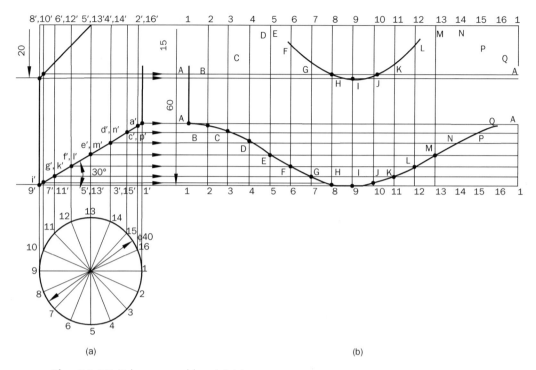

Fig. 14.17 Solution to problem 14.16

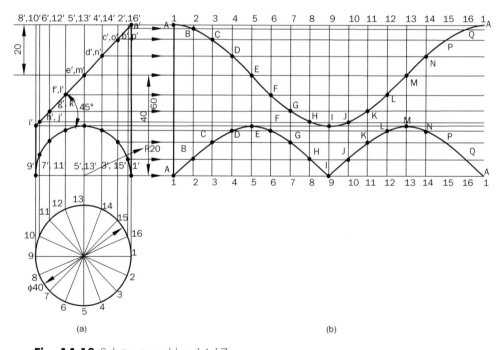

Fig. 14.18 Solution to problem 14.17

Problem 14.18 Draw the development of the lateral surface of the right circular cylinder as shown in Fig. 14.19 (a).

Solution: For its solution, see Fig. 14.19 (b), which is self-explanatory.

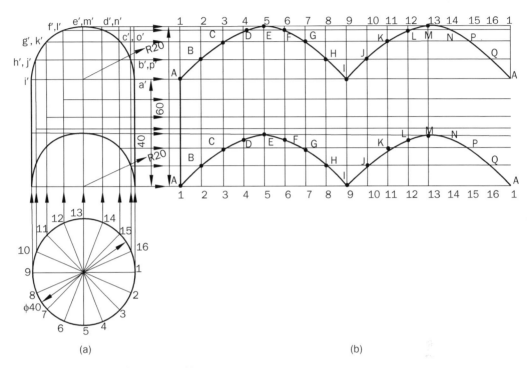

Fig. 14.19 Solution to problem 14.18

Problem 14.19 A cylinder of 45 mm diameter and 60 mm length is resting on one of its bases on HP. It is cut by a section plane inclined at 60° with HP and perpendicular to VP passing through a point on the axis 15 mm from its top end. Draw its front view, sectional top view and develop the lateral surface of the remaining solid.

Solution:

(i) Draw the projections of the cylinder in the given position and label it.
(ii) Draw the cutting plane line VT inclined at 60° to x-y and passing through a point on the axis 15 mm from the top end of the cylinder.
(iii) The cutting plane line VT cuts the various elements of the cylinder. Project the points of intersection in the top view to their corresponding elements. Join these points in proper order and draw section lines in it.
(iv) Complete the development of the remaining cylinder along its stretch-out line as shown in Fig. 14.20.

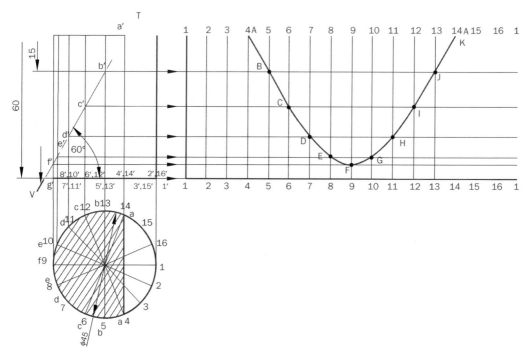

Fig. 14.20 Solution to problem 14.19

Problem 14.20 Three cylindrical pipes of diameter 30 mm form a Y-piece as shown in Fig. 14.21 (a). Draw the development of the lateral surface of each pipe.

Solution:

(i) Draw the given view. Draw a semi-circle on the base of the pipe A as diameter and divide it into sixteen equal parts as shown in Fig. 14.21 (*b*).

(ii) Develop these three pipes independently, using the principle of parallel line method.

(iii) Since pipes B and C are similar, hence their developments are also similar. Therefore, development of pipe B is done here.

(d) **Development of oblique cylinders.** When the axis of an oblique cylinder is not at right angle to its base, the solid is known as oblique cylinder. Therefore, its cross section at right angles to the axis is elliptical. An oblique cylinder may be considered of as a regular oblique prism having infinite number of elements. Therefore, the development of an oblique cylinder may be constructed by using the same method as already been described for oblique prisms.

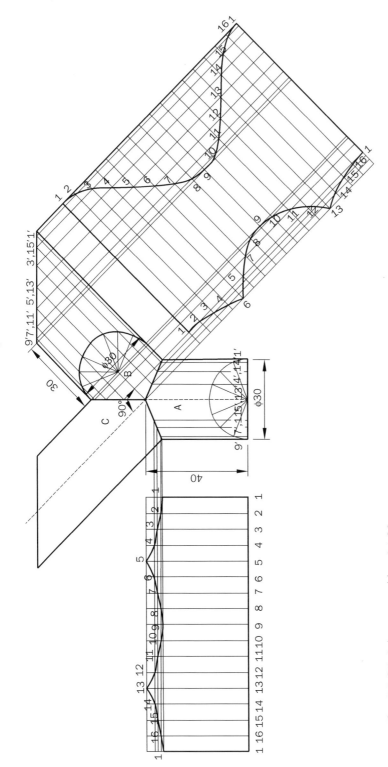

Fig. 14.21 Solution to problem 14.20

Problem 14.21 Draw the inside pattern of an oblique cylinder of 40 mm base diameter and 60 mm vertical height, with its axis inclined to its base at 60° as shown in Fig. 14.22 (a).

Solution:

(i)   Draw the given views of an oblique cylinder.
(ii) Divide the circle into sixteen equal parts in the top view and draw surface lines for a right cylinder in the front view.
(iii) In this case, the stretch-out line of the pattern will not be equal to the circumference of the circular top view.
(iv) Project the end points of the surface to the pattern as shown in Fig. 14.22 (*b*).

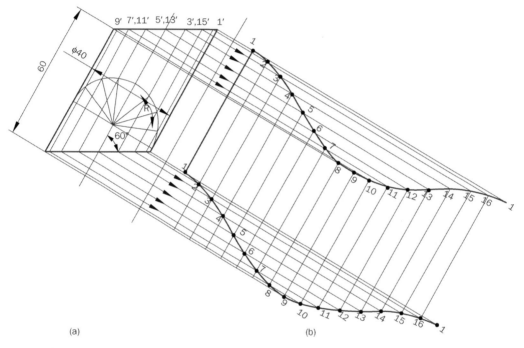

(a)   (b)

Fig. 14.22 Solution to problem 14.21

Problem 14.22 Develop the full size inside pattern of an oblique cylindrical piece made of sheet metal, shown in Fig. 14.23 (b).

Solution: All the construction lines are retained to make the figure self-explanatory. See Fig. 14.23 (*b*).

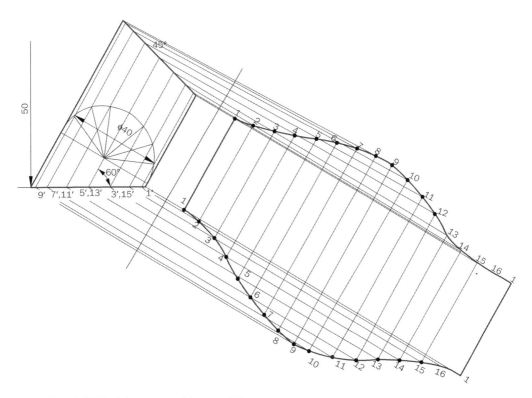

Fig. 14.23 Solution to problem 14.22

14.4 Radial Line Method

The radial line method is used for the development of pyramids and cones in which apex or vertex is taken as centre and its slant edge or generator as the radius of its development.

(a) **Development of right pyramids:** The development of the lateral surface of a pyramid consists of a number of equal isosceles triangles in contact. The base and the sides of each triangle are equal to the edge of the base and the slant edge of the pyramid respectively. The method of developing the inside pattern of the right pyramids are explained in the following solved problems.

Problem 14.23 A square pyramid edge of base 30 mm and height 50 mm rests on its base in HP such that all of its base edges are equally inclined to VP. A section plane parallel to the HP cuts the pyramid bisecting its axis. Draw its front view, sectional top view and develop the lateral surface of the pyramid.

Solution:

(i) Draw a square 1234 in the top view, keeping its base edge 4-1 or 4-3 inclined to the VP at 45°. Project its corresponding front view 1'2'3'4'.

(ii) Here o'1' or o' 3' slant edges give the true length.

(iii) With o' as centre and radius equal to o'1', draw an arc as shown in the Fig. 14.24.

(iv) Select point 1 on this arc and then cut off four equal divisions of length equal to the length of an edge from the top view.

(v) Name these points as 2, 3, 4, and 1. Join 1 to 2, 2 to 3, 3 to 4 and 4 to 1 by straight lines and also join these points to points o' to complete the development.

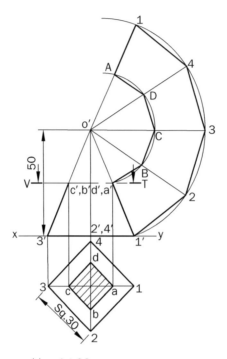

Fig. 14.24 Solution to problem 14.23

Problem 14.24 A right regular hexagonal pyramid, edge of 20 mm and height 40 mm, rests on its base in HP such that one of its base edge parallel to the VP. Draw its projections and develop its lateral surface.

(PTU, Jalandhar May 2007)

Solution:

(i) Draw a regular hexagon, 123456 in the top view, keeping one of its base edge say 5-6 or 3-2 parallel to the VP. Project its corresponding front view 1' 2' 3' 4' 5' 6'.

(ii) The true lengths of the slant edges of o' 1' or o' 4' of a regular hexagon pyramid are to be measured from the front view, as the top view of these edges 01 or 04 are parallel to the x-y.

(iii) So with o' as centre and radius equal to o' 1', draw an arc as shown in Fig. 14.25.

(iv) Select point 1 on this arc and then cut off six equal divisions of length equal to the length of an edge from the top view.

(v) Name these points as 2, 3, 4, 5, 6 and 1. Join 1 to 2, 2 to 3, 3 to 4, 4 to 5, 5 to 6 and 6 to 1 by straight lines and also join these points to point o' to complete the development.

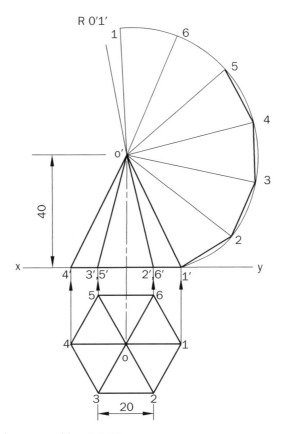

Fig. 14.25 Solution to problem 14.24

Problem 14.25 A square pyramid, edge of base 30 mm and height 50 mm, rests on its base on HP such that one of its base edges is parallel to the VP. Draw its projections and develop its lateral surface.

(*PTU, Jalandhar December 2004*)

Solution:

(i)   Draw a square 1234 in the top view, keeping its base edge 1-4 parallel to the VP. Project its corresponding front view $1'2'3'4'$.

(ii)   Here, none of the slant edges gives the true length. The true length of a slant edge $o'1_1'$ of a square pyramid is to be measured from the front view, as the top view of that edge is parallel to x-y.

(iii) As all the slant edges are of the same length for a square pyramid, so with $o'1_1'$ as radius and o' as centre, draw an arc as shown in Fig. 14.26.

(iv)   Select point 1 on this arc and then cut off four equal divisions of length equal to the length of an edge from the top view.

(v)   Name these points as 2, 3, 4 and 1. Join 1 to 2, 2 to 3, 3 to 4 and 4 to 1 by straight lines and also join these points to point o' to complete the development.

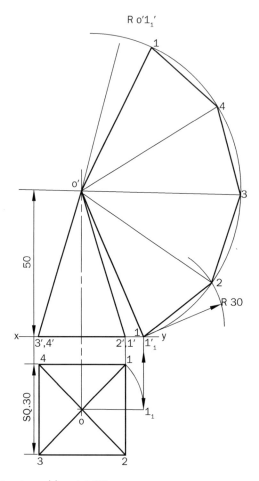

Fig. 14.26 Solution to problem 14.25

Problem 14.26 A square pyramid, edge of base 30 mm and height 50 mm, rests on its base in HP such that one of its base edges is parallel to the VP. A section plane perpendicular to the VP and inclined to the HP at 30° cuts the pyramid bisecting its axis. Draw its front view, sectional top view and develop the lateral surface of the truncated pyramid.

Solution:

(i)   The development is first drawn as a complete pyramid from the apex or vertex to the base.

(ii) The true lengths of cut away parts of the slant edges, from the apex o to the points a, b, c and d are found by projecting horizontally the point a', b', c' and d' in the front view to new positions on o'1$_1$' slant edge (true length).

(iii) These true length distances are measured along their respective radial lines from o' in the development at points A, B, C and D. These points are then joined by straight lines as shown in Fig. 14.27 to complete the development of the truncated pyramid.

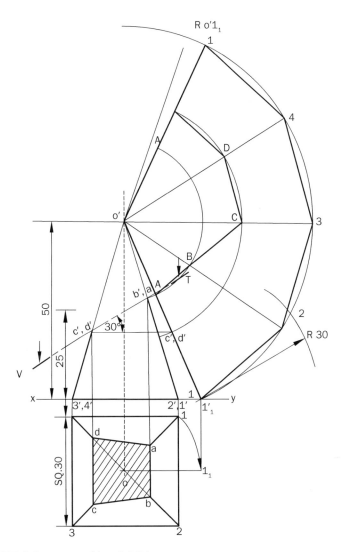

Fig. 14.27 Solution to problem 14.26

Problem 14.27 A square pyramid, edge of base 30 mm and height 50 mm, rests on its base, in HP such that all its base edges are equally inclined to the VP. Draw its projections and develop its lateral surface.

(*PTU, Jalandhar December 2004*)

Solution:

(i) Draw a square 1234 in the top view, keeping its base edges equally inclined to the VP. Project its corresponding front view, 1'2'3'4'.

(ii) The true lengths of the slant edges of o'1' and o'3' of a square pyramid are to be measured from the front view, as the top view of these edges o1 and o3, respectively, are parallel to the x-y.

(iii) As all the slant edges are of the same length for a square pyramid, so with o'1' as radius and o' as centre, draw an arc as shown in Fig. 14.28.

(iv) Select point 1 on this arc and then cut off four equal divisions of length equal to the length of an edge from the top view.

(v) Name these points as 2, 3, 4 and 1. Join 1 to 2, 2 to 3, 3 to 4 and 4 to 1 by straight lines and also join these points to point o' to complete the development.

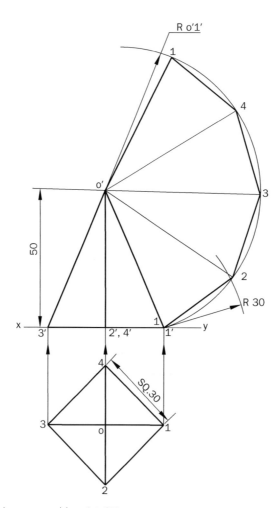

Fig. 14.28 Solution to problem 14.27

Problem 14.28 A square pyramid, edge of base 30 mm and height 50 mm, is resting on its base in HP such that all of its base edges are equally inclined to the VP. A section plane perpendicular to the VP and inclined to the HP at 30° cuts the pyramid, bisecting its axis. Draw its front view, sectional top view and develop the lateral surface of the truncated pyramid.

Solution: The solution to this problem is self-explanatory. See Fig. 14.29.

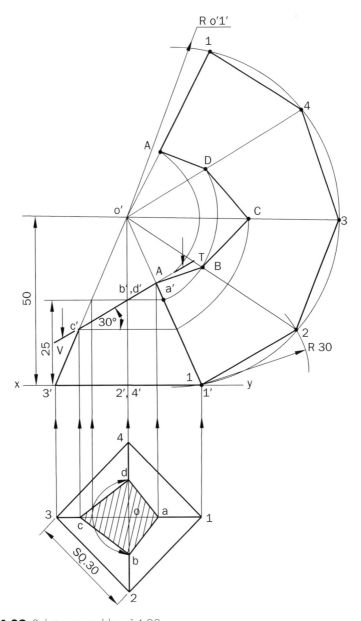

Fig. 14.29 Solution to problem 14.28

Problem 14.29 Draw the development of the lateral surface of the part P of the pyramid, the front view of which is shown in Fig. 14.30 (a). The pyramid is hexagonal, two sides of the base parallel to the VP.

(*PTU, Jalandhar December 2002*)

Solution: All the construction lines are retained to make the solution self-explanatory. See Fig. 14.30 (*b*).

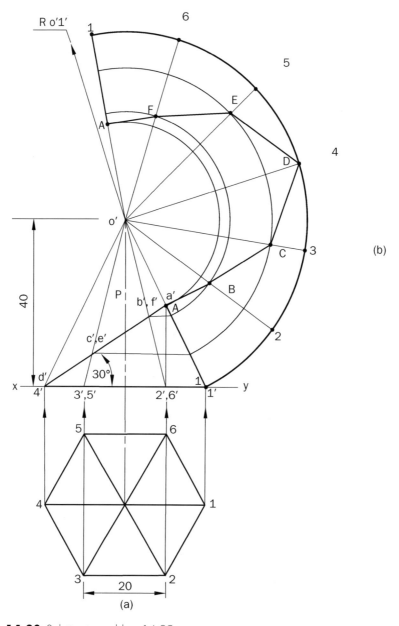

Fig. 14.30 Solution to problem 14.29

Problem 14.30 A right regular pentagonal pyramid, edge of base 25 mm and height 60 mm, is lying on one of its triangular faces in HP, with its axis parallel to the VP. A section plane perpendicular to the HP and inclined to the VP at 30° cuts pyramid, bisecting its axis. Draw its top view, sectional front view and develop the lateral surface of the remaining pyramid.

Solution:

(i) Draw the projections of the pentagonal pyramid in the given conditions.

(ii) Then cut the solid in the top view by a section plane perpendicular to the HP and inclined to the VP at 30°, fulfilling all the given conditions and then draw the sectional front view from it.

(iii) Develop the remaining part of the pyramid as discussed in the previous problems using radial line method. See Fig. 14.31.

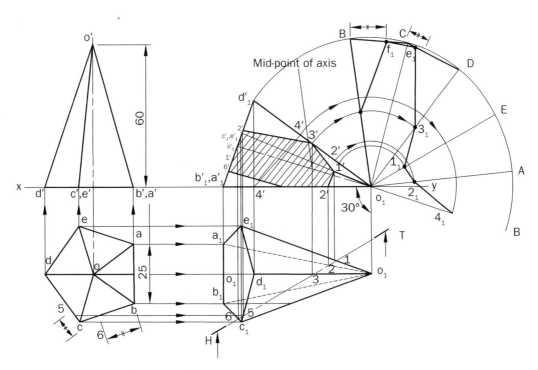

Fig. 14.31 Solution to problem 14.30

Problem 14.31 A right regular pentagonal pyramid, edge of base 30 mm and height 50 mm, rests on its base in HP with one of its base edges perpendicular to the VP. A section plane perpendicular to the VP and parallel to the HP cuts the pyramid, bisecting its axis. Draw its front view, sectional top view and develop the lateral surface of the remaining pyramid.

Solution:

(i)   Draw the projections of the pentagonal pyramid in the given conditions.

(ii)   Cut the pyramid in the front view by a section plane perpendicular to the VP and parallel to the HP, fulfilling all the given conditions and then draw the sectional top view.

(iii)   Develop the remaining part of the pyramid as discussed in the earlier problems using radial line method. See Fig. 14.32.

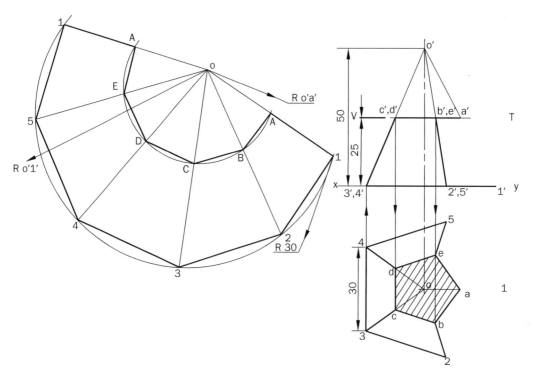

Fig. 14.32 Solution to problem 14.31

Problem 14.32 A right regular hexagonal pyramid, edge of base 25 mm and height 55 mm, rests on its base in HP with one of its base edges parallel to the VP. A section plane perpendicular to the VP and inclined to the HP at 30° cuts the pyramid, bisecting its axis. Draw its front view, sectional top view and develop the lateral surface of the remaining pyramid.

Solution: This problem consists of three distinct parts:

(i) First, draw the projections of the hexagonal pyramid in the given conditions.

(ii) Second, add the section plane to the front view to satisfy the given conditions and draw the sectional top view applying principles of section of the solids.

(iii) Finally, develop the remaining part of the pyramid as discussed in the previous problems using radial line method. See Fig. 14.33.

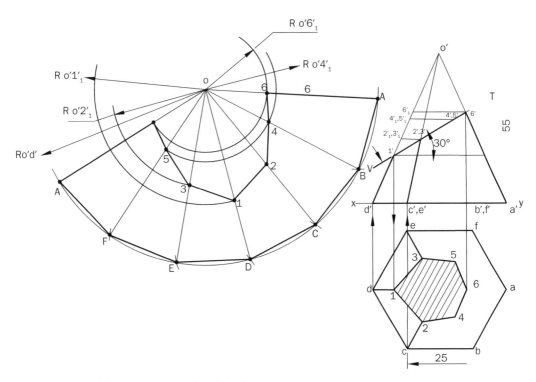

Fig. 14.33 Solution to problem 14.32

Problem 14.33 Develop the surface of a hexagonal pyramid side of base 28 mm and height 60 mm. The pyramid is resting on its base on the ground and the edge of the base is inclined at 20° to VP.

(*PTU, Jalandhar May 2005*)

Solution: All construction lines are shown to make the figure self-explanatory. See Fig. 14.34.

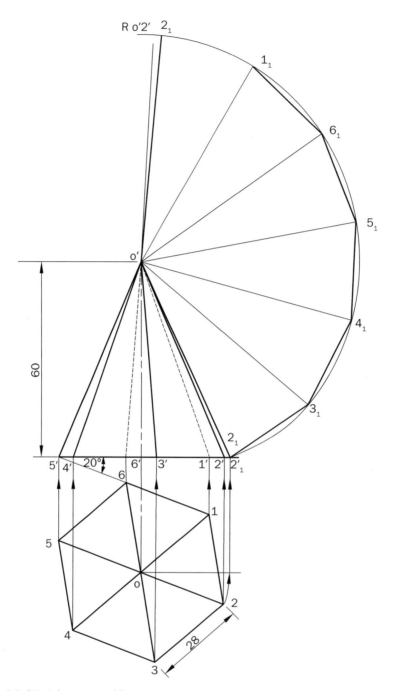

Fig. 14.34 Solution to problem 14.33

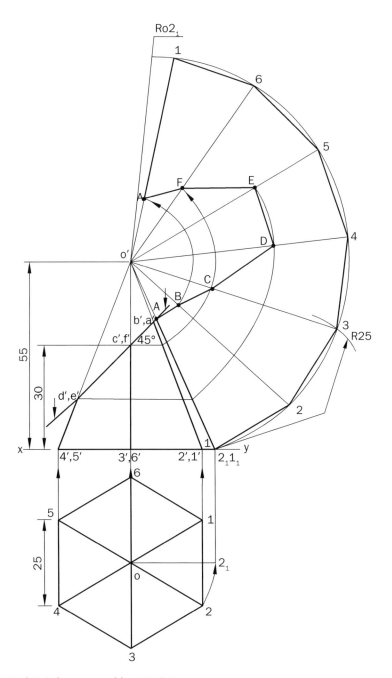

Fig. 14.35 Solution to problem 14.34

Problem 14.34 A right regular hexagonal pyramid edge of base 25 mm and height 55 mm is resting on its base in HP with one of its base edges perpendicular to the VP. It is cut by a plane, which is inclined at 45° to the HP and perpendicular to the VP, at a distance of 30 mm from its base. Draw its projections and develop its lateral surface.

Solution:

(i)   Draw a hexagon 123456 in the top view, keeping its base edge (12 or 45) perpendicular to the *VP*. Project its corresponding front view 1'2'3'4'5'6'.

(ii) Then cut the solid in the front view by a section plane perpendicular to the *VP* and inclined to the *HP* at 45°, fulfilling all the given conditions.

(iii) Here, none of the slant edges gives the true length. The true length of a slant edge $o'1_1'$ $(o'2_1')$ of a hexagonal pyramid is to be measured from the front view, as the top view of that edge is parallel to x-y.

(iv) As all the slant edges are of the same length for a hexagonal pyramid, so with $o'1_1'$ $(o'2_1')$ as radius and o' as centre, draw an *arc* as shown in Fig. 14.35.

(v) Develop the remaining part of the pyramid as described is the previous problems using radial line method.

Problem 14.35 A square pyramid edge of base 30 mm and height 50 mm is resting on HP on its base with an edge of base inclined at 30° to the VP. It is cut by a sectional plane which is inclined at 45° to the HP and perpendicular to the VP, bisecting its axis. Draw its front view, sectional top view and develop the lateral surface of the remaining pyramid.

Solution: The interpretation of the solution is left to the reader. See Fig. 14.36.

Problem 14.36 A right regular hexagonal pyramid edge of base 25 mm and height 50 mm is resting on its base is HP, with one of its base edges parallel to the VP. A string is wound round the surface of the pyramid from the right extreme point on the base and ending at the same point. Determine the shortest path required. Also show the path of the string in the front and top views.

Solution:

(i)   Draw the projections of the pyramid and develop its lateral surface as described in the earlier problems.

(ii) Draw the straight line *AA* in the development to get the shorter path of the string required.

(iii) Mark the points *BCDEF* on the development.

(iv) Proceed the reverse direction to get the points *b'c'd'e'f'* in the front view.

(v) Project the points *bcdef* and join these points in the top view. See Fig. 14.37.

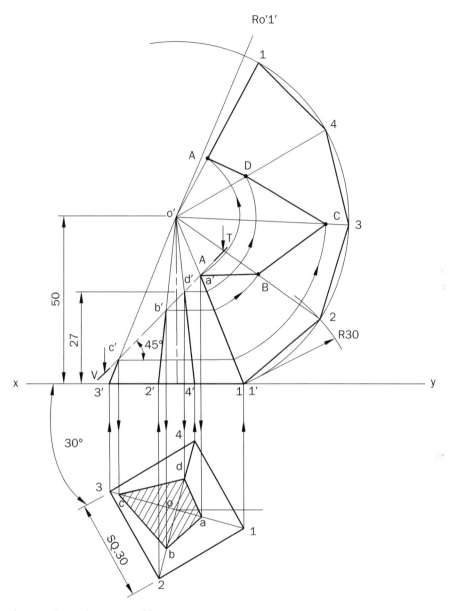

Fig. 14.36 Solution to problem 14.35

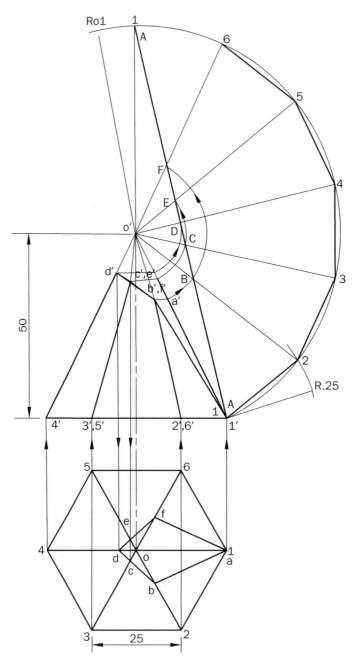

Fig. 14.37 Solution to problem 14.36

Problem 14.37 A square pyramid edge of base 40 mm and height 60 mm is lying on one of its triangular faces on HP, with its axis parallel to the VP. It is cut by a sectional plane which is perpendicular to the VP and parallel to the HP, bisecting its axis. Draw its front view, sectional top view and develop the lateral surface of the remaining pyramid.

Solution: All the construction lines are retained to make the solution self-explanatory. See Fig. 14.38.

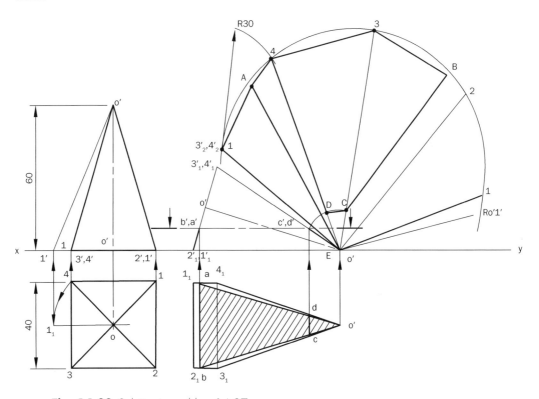

Fig. 14.38 Solution to problem 14.37

Problem 14.38 Develop the surface of a pentagonal pyramid having its base edge 30 mm and axis 60 mm long.

(PTU, Jalandhar December 2005)

Solution:

(i)   Draw a pentagon 12345 in the top view. Project its corresponding front view 1' 2' 3' 4' 5'.

(ii) As the top view of slant edge 01 is parallel to x-y, so o' 1' slant edge of a pentagenal pyramid gives the true length.

(iii) With o' as centre, radius equal to o' 1', draw an *arc* as shown in Fig. 14.39.

(iv) Select point 1 on this *arc* and then cut off five equal divisions of length equal to the length of an edge from top view.

(v)   Name these points as 2, 3, 4 and 5. Join 1 to 2, 2 to 3, 3 to 4, 4 to 5 and 5 to 1 by straight lines and also join these points to point o' to complete the development.

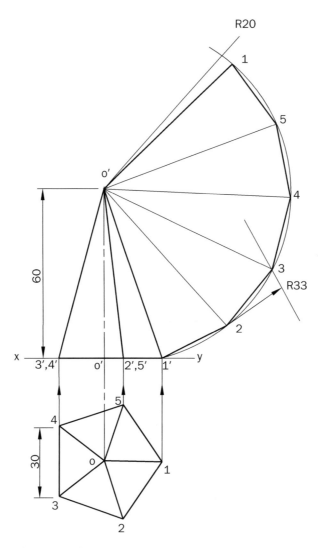

Fig. 14.39 Solution to problem 14.38

Problem 14.39 A right regular hexagonal pyramid, edge of base 20 mm and height 40 mm, rests on its base in HP, with one of its base edges perpendicular to VP. It is cut by a section plane which is perpendicular to VP and inclined to HP at 30° bisecting its axis. Draw its front view, sectional top view and develop the lateral surface of the truncated pyramid.

Solution: All the construction lines are retained to make the solution self-explanatory. See Fig. 14.40.

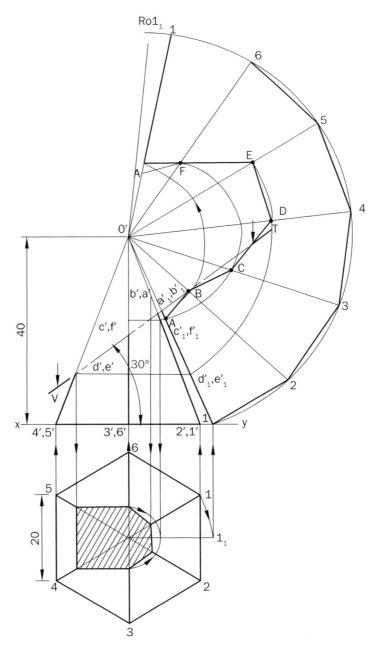

Fig. 14.40 Solution to problem 14.39

Problem 14.40 A right regular hexagonal pyramid, edge of base 30 mm and axis 65 mm long, rests on its base on HP with one of its base edges is parallel to the VP. A horizontal circular hole of diameter 30 mm is drilled at a distance of 20 mm from the base of the pyramid and to the axis perpendicular to the VP. Develop its lateral surfaces.

Solution: All the construction lines are retained to make the solution self-explanatory. See Fig. 14.41.

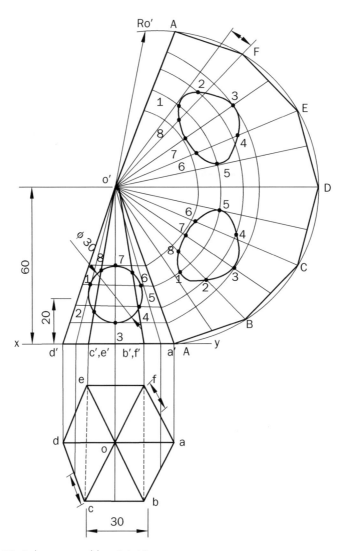

Fig. 14.41 Solution to problem 14.40

(b) **Development of oblique pyramids.** When the axis of an oblique pyramid is not at right angle to its base, then the solid is known as oblique pyramid. The lateral surfaces of these solids consist of triangles. It is needed to determine the true lengths of all the slant edges. Therefore, the development of an oblique pyramid may be constructed by using the radial line method as already been described for right pyramids. The procedure of developing the lateral surface of oblique pyramids is explained in the following problems.

Problem 14.41 Two views of an oblique square pyramid are shown in Fig. 14.42 (a). Develop its lateral surface.

Solution:

(i)   Draw the given views. Here the true lengths are different for the slant edges, whereas the base edges are of true length in the top view.

(ii)   Find out the true lengths for all the slant edges as shown in Fig. 14.42 (b).

(iii) Complete the problem as discussed in the right pyramids.

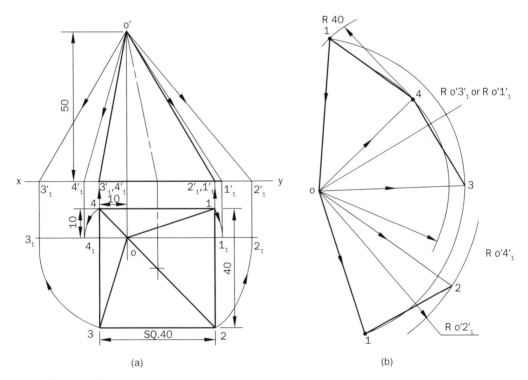

(a)   (b)

Fig. 14.42 Solution to problem 14.41

Problem 14.42 Two views of an oblique square pyramid are shown in Fig. 14.43 (a). Develop its lateral surface.

Solution: The interpretation of the solution is left to the reader. See Fig. 14.43 (b).

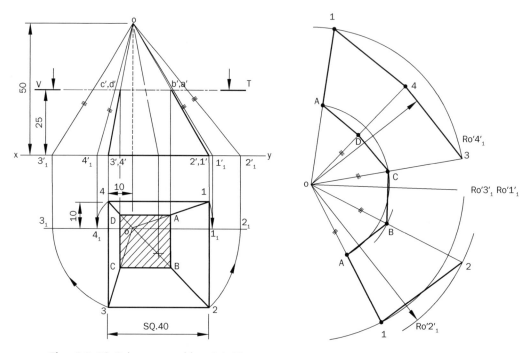

Fig. 14.43 Solution to problem 14.42

(c) **Development of right cones:** Development of the curved surface of a cone is a sector of a circle having its radius equal to the generator, i.e., slant height, and length of the arc is equal to the circumference of the base circle.

Problem 14.43 A right circular cone, diameter of base 50 mm and height 60 mm, rests on its base in HP. A section plane perpendicular to the VP and parallel to the HP cuts the cone, bisecting its axis. Draw its front view, sectional top view and develop the lateral surface of the remaining part of the cone.

(PTU, Jalandhar May 2010)

Solution:

(i) Draw the front and top views of the cone.
(ii) Divide the base circle into sixteen equal parts.

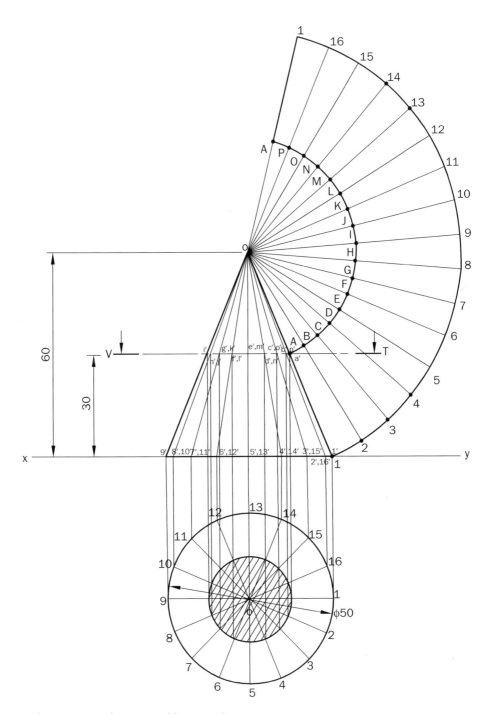

Fig. 14.44 Solution to problem 14.43

(iii) With centre o' and radius equal to o'1' or o'9', draw an arc of the circle. The length of this arc should be equal to the circumference of the base circle. This arc can be determined in two ways:

(a) Calculate the subtended angle θ by the formula

$$\theta = 360° \times \frac{\text{Radius of the base circle}}{\text{Slant height of the cone}}$$

Cut off the arc so that it makes the angle θ at the centre and then divide it into sixteen equal parts.

(b) By dividing the arc with a compass or a divider into sixteen equal parts or divisions, where each division must be equal to the one of the divisions of the base circle, this method will give an approximate length of the circumference.

(iv) Join the division points with o, thereby completing the development of the whole cone.

(v) The cut portion of the cone may be deducted from this development by making the positions of the points at which the generators are cut. The true lengths of these points on the cut away portion of the cone are found by projecting their points of intersection with the section plane in the front view horizontally on to true length o'1' or o'9'. These true length distances are then transferred to their respective elements as shown in Fig. 14.44.

(vi) Join all these points by a smooth curve to complete the development of the frustum.

(vii) Also draw its front view and sectional top view, as discussed in the sections of solids chapter 13.

Problem 14.44 A right circular cone, diameter of base 50 mm and height 60 mm, rests on its base in HP. A section plane perpendicular to the VP and inclined to the HP at 30° cuts the cone, bisecting its axis. Draw projections of the truncated cone and develop its lateral surface.

Solution: All construction lines are shown to make the figure self-explanatory. See Fig. 14.45.

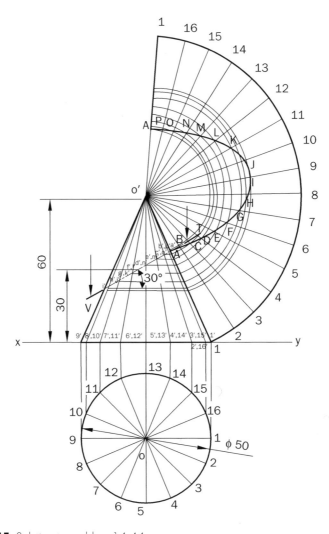

Fig. 14.45 Solution to problem 14.44

Problem 14.45 Develop the lateral surface of a right circular cone as shown in Fig. 14.46 (a).

Solution: All construction lines are retained to make the solution easy to understand. See Fig. 14.46 (b).

Problem 14.46 Develop the lateral surface of a right circular cone with an equilateral triangular hole as shown in Fig. 14.47 (a).

Solution:

  (i)   Draw the given front view and project the corresponding top view.
  (ii) Develop the lateral surface of the frustum of the cone by the radial line method.

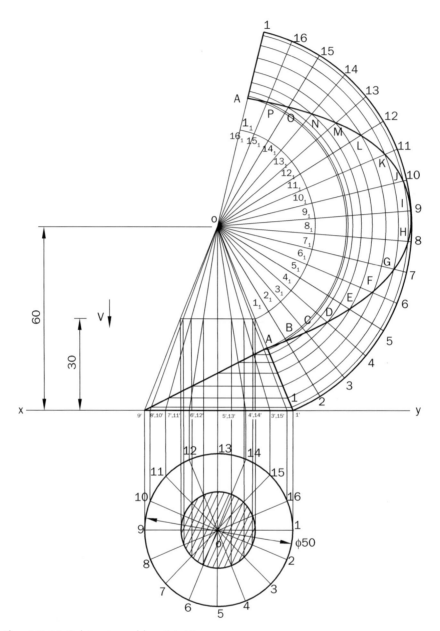

Fig. 14.46 Solution to problem 14.45

(iii) Take a number of points, e.g., *a′*, *b′*, *c′*, *d′*, *e′*, *f*, *g′*, on the triangular hole in the front view and transfer these points to their respective positions in the development.
(iv) Join these points by a smooth curve as shown in Fig. 14.47 (*b*).

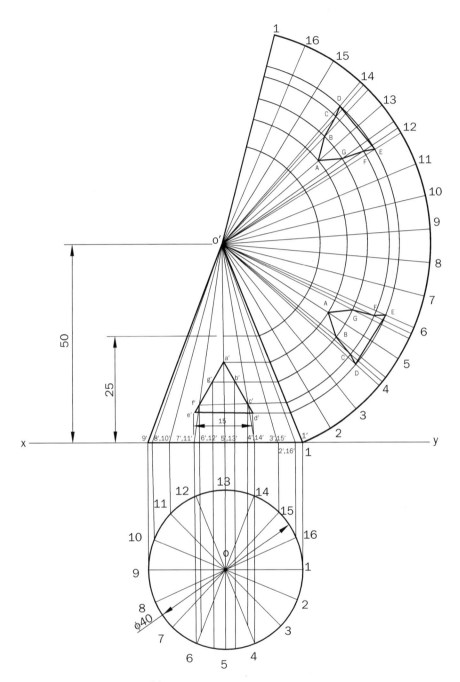

Fig. 14.47 Solution to problem 14.46

Problem 14.47 A right circular cone, diameter of base 50 mm and height 60 mm, is lying on one of its generators on HP with its axis parallel to the VP. A section plane parallel to the HP and perpendicular to the VP cuts the axis at a distance of 15 mm from its base. Draw front view, sectional top view and develop the lateral surface of the cut cone.

Solution: This problem consists of three different parts:

(i)   First of all, draw the projections of the cone in the given conditions.
(ii) Next add the section plane in the front view to satisfy the given conditions and draw the sectional top view, applying principles of section of solids.
(iii) Finally develop the lateral surface of the cut cone using radial line method.

All construction lines are retained to make the solution more understandable. See Fig. 14.48.

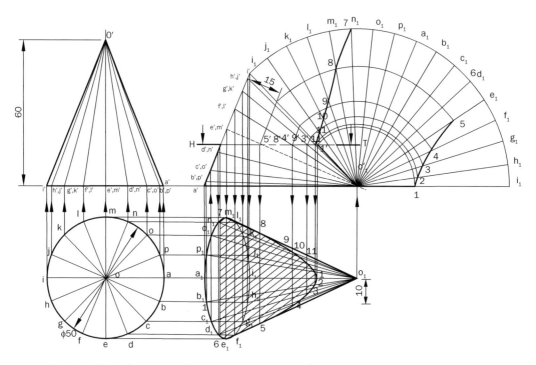

Fig. 14.48 Solution to problem 14.47

Problem 14.48 Develop the lateral surface of a funnel as shown in Fig. 14.49 (a).

Solution: This object consists of two parts: (*i*) frustum of a cone and (*ii*) a right cylinder. All construction lines are shown to make the figure self-explanatory. See Fig. 14.49 (*b*).

Problem 14.49 Develop the lateral surface of a funnel as shown in Fig. 14.50 (a).

(*PTU, Jalandhar May 2001, 2011*)

Solution: All construction lines are shown to make the figure self-explanatory. See Fig. 14.50 (*b*).

Problem 14.50 Draw the development of a given Fig. 14.51 (a).

(*PTU, Jalandhar December 2002*)

Solution: All the construction lines are retained to make the solution to the given figure self-explanatory. See Fig. 14.51 (*b*).

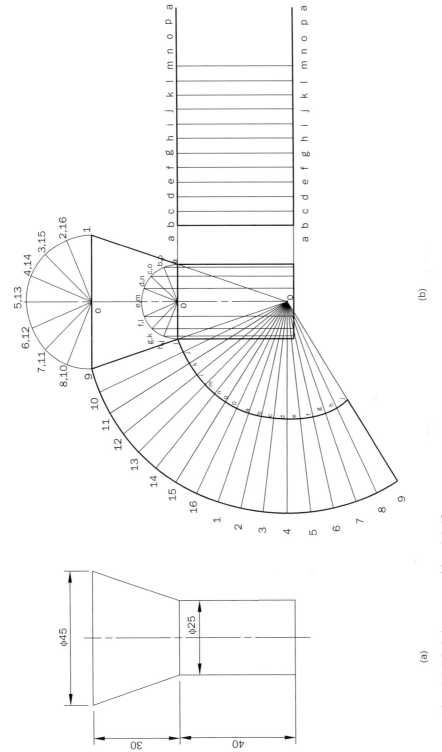

Fig. 14.49 Solution to problem 14.48

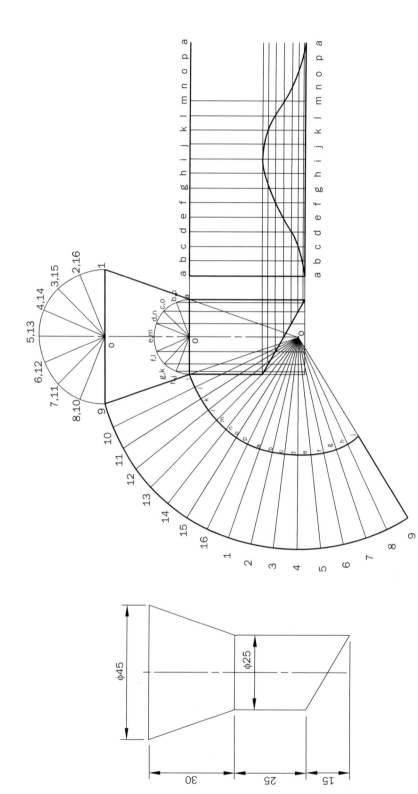

Fig. 14.50 Solution to problem 14.49

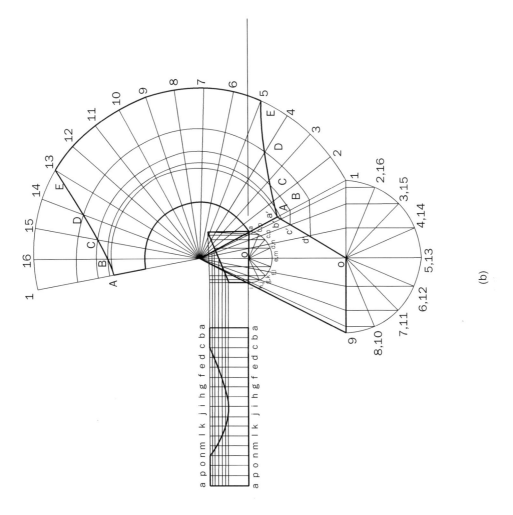

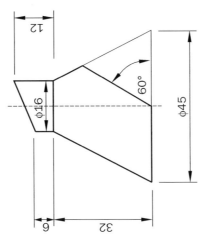

(a)

(b)

Fig. 14.51 Solution to problem 14.50

Problem 14.51 Draw the development of a conical object with base diameter as 50 mm and the top diameter as 35 mm. Height of the object is 35 mm.

Solution: All the construction lines are retained to make the solution to the given figure self-explanatory. See Fig. 14.52.

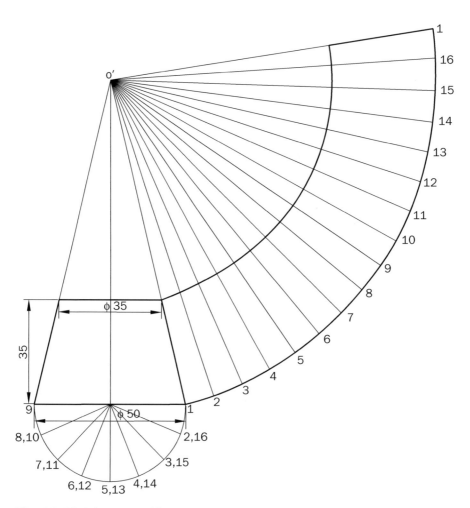

Fig. 14.52 Solution to problem 14.51

Problem 14.52 A right circular cylinder of 30 mm diameter and 35 mm height of axis is cut by sectional plane inclined at 30° to HP and passes 18 mm from the base along the axis. Draw the development of the truncated cylinder.

(PTU, Jalandhar May 2004)

Solution: The solution to this problem is self-explanatory. See Fig. 14.53.

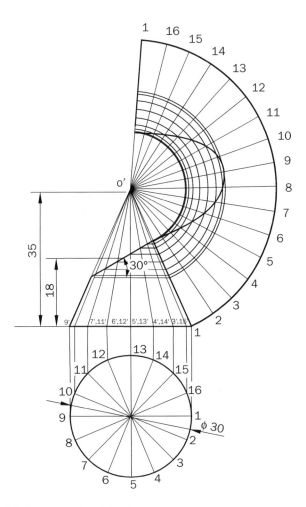

Fig. 14.53 Solution to problem 14.52

(d) **Development of oblique cones.** When the axis of an oblique cone is not at right angle to its base, then the solid is known as oblique cone. Therefore, its development is different from that of a right cone.

Problem 14.53 Develop the lateral surface of an oblique cone, diameter of base 40 mm and height 40 mm, having its axis inclined at 60° at its base.

Solution: All the construction lines are retained to make the solution self-explanatory. See Fig. 14.54.

14.5 Triangulation Method

This method is used for transition pieces. In most cases, transition pieces are composed of plane and conical surfaces. As transition pieces are made of different kinds of surfaces, so these are generally developed by triangulation method.

Development of transition pieces

The triangulation method is used for transition pieces. This method consists of dividing the surfaces into suitable triangles and placing these side by side after finding the true lengths of each side of the triangle. The procedure of development of a few transition pieces is illustrated in the following problems.

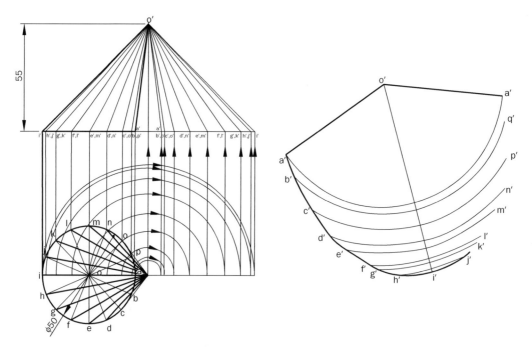

Fig. 14.54 Solution to problem 14.53

Problem 14.54 Figure 14.55 (a) shows a paper tray model. Develop the surface of a paper tray model.

Solution: For its solution, see Fig. 14.55 (*b*).

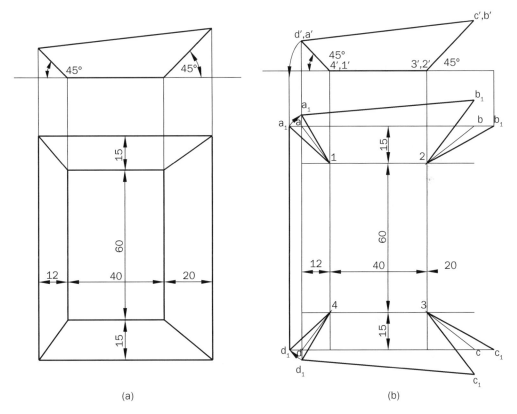

(a)   (b)

Fig. 14.55 Solution to problem 14.54

Problem 14.55 An air conditioning duct of a square cross section 50 mm × 50 mm connects a circular pipe of 30 mm diameter through the transition piece. Draw the projections and develop the lateral surface of the transition piece.

Solution:

(i)   Draw the front view and top view of the given object.
(ii)   Divide the top view of the circle into twelve equal parts.
(iii)   The transition piece is composed of four isosceles triangles and four conical surfaces.
(iv)   Begin the development along the line 1-A. The conical surfaces are developed by the triangulisation method as shown in Fig. 14.56. In the top view, join the division of the circle 1, 2, 3, etc., with the corner *a*, *b*, *c*, etc. Project them in the front view. Obtain the true length of the sides of each triangle.

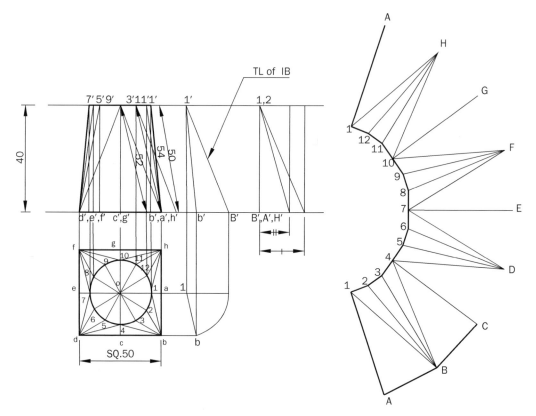

Fig. 14.56 Solution to problem 14.55

14.6 Approximate Method

The approximate method is used for double curved surfaces such as sphere, etc.

Development of sphere
The surface of a sphere can be approximately developed by dividing it into a number of parts. There are two methods in which a sphere can be developed:

   (a) Zone Method
   (b) Lune or Gore Method

   (a) **Zone Method**
  In this method, the sphere is cut into a number of zones and substituting each zone as a frustum of a right circular cone. Now develop each zone separately by radial line method as already discussed in Section 14.4.

Problem 14.56 Draw the development of a sphere of 50 mm diameter by zone method.

(*PTU, Jalandhar December 2004, May 2005*)

Solution:

(i)   Draw the front view and top view of the sphere and in the front view, draw a number of lines parallel to the diameter passing through the points A, B, C and D respectively.

(ii) Project the points of intersection of these parallel lines in the top view and draw circles through the points of intersection of projectors with the diameters of the top view.

(iii) Join points A and B, B and C, C and D.

(iv) Extend AB to meet the axis produced at O_1, BC to meet at O_2 and CD to meet at O_3. Point O_1, O_2 and O_3 are the apexes of cones of which each zone is a frustum. The top zone is regarded as the full cone with apex at O_4.

(v)   Now develop each zone assuming it to be the frustum of the cone, as shown in Fig. 14.57.

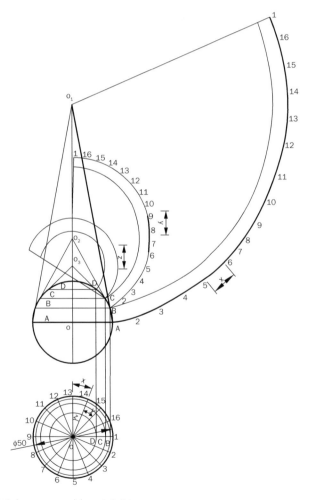

Fig. 14.57 Solution to problem 14.56

(b) **Lune or Gore Method**

In this method, the sphere is cut into a number of equal meridian sections called lunes. A meridian section of the sphere so sectioned is developed assuming it to be an approximate cylinder. As all the meridian sections are equal, other sections are duplicated with the first developed section used as template.

Problem 14.57 Draw the development of a sphere of 50 mm diameter by Lune method.

Solution:

(i)   Draw the top and front views of the sphere. Divide the front view of the sphere into sixteen equal parts.

(ii)   Divide the half of the semi-circle in a number of equal parts (say eight)

(iii) Develop the one-sixteenth of a sector of the sphere as shown in Fig. 14.58.

(iv)   Draw the centre line of the sector *OBAO* and project points a', b', c', d', e', f, g and h' on this line.

(v)   Draw arcs a' - b', c' - d', e' - f' and g' - h' with centre O and radius equal to the points of intersection of projection of points a', c', e' and g'.

(vi)   Draw a stretch-out line 13 - 5 equal to the length of an arc of a semi-circle. Divide it into eight equal parts and draw parallel lines through each division marks.

(vii) Join the points of intersections of the parallel lines with respective projectors by a smooth curve. This is the development of one sixteenth sector of the sphere.

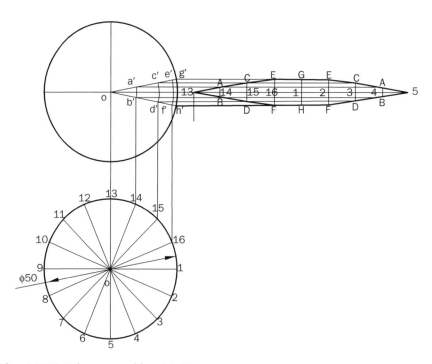

Fig. 14.58 Solution to problem 14.57

Exercises

Parallel Line Method

14.1 A right regular pentagonal prism, side of base 25 mm and height 50 mm, rests on its base in HP with one of its base edges perpendicular to the VP. A section plane perpendicular to the VP and inclined to the HP at 45° cuts the prism, bisecting its axis. Develop the lateral surface of the truncated prism.

14.2 A square prism is cut at top and bottom ends, as shown in Fig. 14.59. Develop its lateral surface.

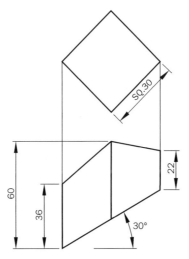

Fig. 14.59

14.3 A right circular cylinder, diameter of base 40 mm and height 60 mm, has a cut of 30 mm radius at the top end as shown in Fig. 14.60. Develop its lateral surface.

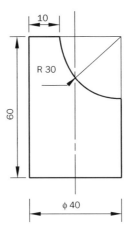

Fig. 14.60

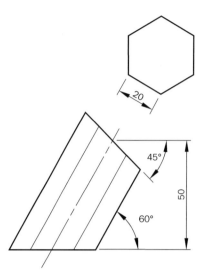

Fig. 14.61

14.4   Fig. 14.61 shows two views of an oblique prism. Develop its lateral surface.

14.5   Develop the lateral surface of an oblique cylinder as shown in Fig. 14.62.

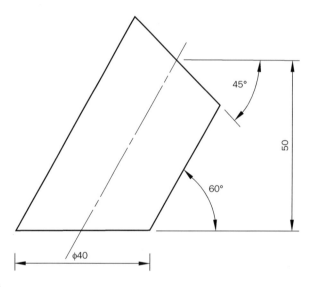

Fig. 14.62

14.6   Develop the lateral surface of the Y-piece formed by the three cylindrical pipes of 30 mm diameter, as shown in Fig. 14.63.

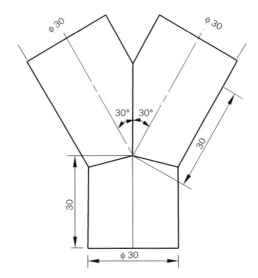

Fig. 14.63

14.7 A right regular square prism, side of base 30 mm and height 50 mm, rests on its base on HP such that its vertical faces are equally inclined to the VP. A horizontal circular hole of diameter 30 mm is drilled centrally through it such that the axis of the hole cuts the diagonally opposite vertical edges. Develop its lateral surfaces.

Radial Line Method

14.8 A right circular cone, diameter of base 40 mm and height 60 mm, rests on its base in HP. The front view is cut by a plane passing through the mid-point of the axis at an angle of 45° to the HP. Draw the development of the truncated cone.

14.9 Develop the lateral surface of the conical part as shown in Fig. 14.64.

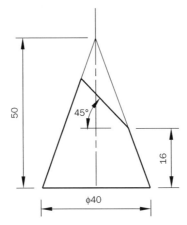

Fig. 14.64

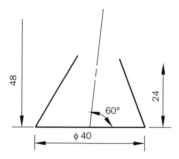

Fig. 14.65

14.10 Develop the lateral surface of the oblique cone as shown in Fig. 14.65.

14.11 Develop the lateral surface of the cone, having a circular hole cut in it as shown in Fig. 14.66.

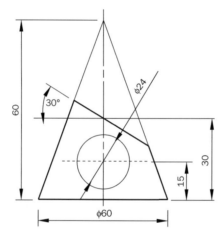

Fig. 14.66

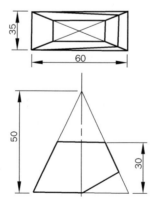

Fig. 14.67

14.12 Develop the lateral surface of the object as shown in Fig. 14.67.

14.13 A right regular pentagonal pyramid, edge of base 25 mm and height 50 mm, rests on its base in HP with one of its base edges perpendicular to the VP. Draw its projections and develop its lateral surface.

14.14 A right rectangular pyramid of base 50 mm × 30 mm and height 60 mm rests on its base in HP with one of its base side parallel to the VP. A section plane perpendicular to the VP and inclined to the HP at 30° cuts the pyramid, bisecting its axis. Develop the lateral surface of the truncated pyramid.

14.15 A right regular hexagonal pyramid, edge of base 25 mm and height 50 mm rests on its base in HP. A hole of 15 mm diameter is drilled through the pyramid at right angles to its axis and at a height of 15 mm from the base. Develop the lateral surface of the pyramid.

14.16 A right regular hexagonal pyramid, edge of base 25 mm and height 50 mm, is lying on one of its triangular faces on HP with its axis parallel to the VP. A section plane perpendicular to the HP and inclined to the VP at 30° cuts the pyramid bisecting its axis. Develop the lateral surface of the truncated pyramid.

14.17 Develop the lateral surface of an oblique cone, diameter of base 40 mm and height 40 mm, having its axis inclined at 60° to its base.

14.18 A right rectangular pyramid of base 45 mm × 35 mm height 60 mm rests on its base in HP with one of its base sides parallel to VP. A section plane perpendicular to the VP and inclined at 30° to the HP cuts the pyramid, bisecting its axis. Develop the lateral surface of the truncated pyramid.

14.19 A right regular hexagonal pyramid, edge of base 25 mm, height 60 mm, rests on its base on HP with one of its base edges parallel to VP. A section plane perpendicular to HP and inclined to the VP at 30° cuts the pyramid and is 10 mm away from the axis. Draw its top view, sectional front view and true shape of the section. Also develop the lateral surface of the truncated pyramid.

14.20 A right circular cone, diameter of base 50 mm and height 65 mm, rests on its base in HP. A section plane perpendicular to the HP and inclined to the VP at 45° cuts the cone and is 12 mm away from the axis. Draw its top view, sectional front view and true shape of the section. Also develop the lateral surface of the truncated cone.

14.21 A right regular pentagonal prism, side of base 25 mm and height 60 mm, rests on an edge of its base on HP, such that one of its base corners lies on HP and its axis is inclined at 45° to the HP and parallel to the VP. A section plane perpendicular to the HP and inclined to the VP at 45° cuts the prism bisecting its axis. Draw its top view and sectional front view. Also develop the lateral surface of the truncated prism.

Triangulation Method

14.22 An air conditioning duct of a square cross section 60 mm × 60 mm connects a circular pipe of 30 mm diameter through the transition piece. Draw the projections and develop the lateral surface of the transition piece.

14.23 Develop the lateral surface of a transition piece to join coaxial circular and hexagonal duct holes 55 mm apart. Given the diameter of the circular hole as 40 mm and side of the hexagonal hole as 40 mm.

14.24 Develop the inside pattern of a sheet metal tray as shown in Fig. 14.68.

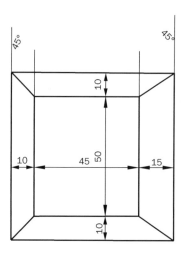

Fig. 14.68

Approximate Method

14.25 Draw the development of a sphere of 60 mm diameter by (*a*) Lune method and (*b*) Zone method.

Objective Questions

14.1 The complete layout of a 3 - D object on a 2- D sheet is known as..................... of surfaces.

14.2 The general application of development is used in the............................ metal work.

14.3 Radial line method is used for developing...................... and.................... objects.

14.4 Every line on the development must show the............................ length of the corresponding line on the surface to be developed.

14.5 What do you mean by development of surfaces?

14.6 What are different methods of development of surfaces?

14.7 Give practical examples of development of surfaces.

14.8 .... method is used for the development of sphere.

14.9 Distinguish between single curved surface and double curved surface.

14.10 What is the importance of development of surface is a manufacturing industry?

14.11 What is the shape obtained by development of a cone?

14.12 Explain the method used for development of a right cone.

Answers

14.1 Development

14.2 Sheet

14.3 Cones, pyramids

14.4 True

14.8 Approximate

Chapter 15

Intersection of Surfaces

15.1 Introduction

The term 'intersection of surfaces' is used when the surfaces of two solids intersect each other, such as of cylinders, cones, prisms, pyramids, etc. The term 'interpenetration of solids' is used when three-dimensional solids such as cylinders, cones, prisms, pyramids, etc., intersect each other. The term 'intersection of surfaces' is used for all situations, irrespective of the fact that the solids interpenetrating each other have plane or curved surfaces.

When two solids interpenetrate each other, their surfaces meet in a common line called the line or curve of intersection. This line of intersection of two surfaces may be joined by determining key points and joining these points in the correct order. The resulting line may be straight or curved depending upon the nature of the intersecting surfaces. Determination of the line of intersection between two solids is important, as their development into flat patterns is otherwise not possible. The term 'development' refers to the unfolded sheet from which the required objects can be formed without stretching the material, i.e., hoppers, pipe joints, etc. In view of the above mentioned applications, intersection of surfaces may be classified into three categories:

(i) Intersection of two plane surfaces, *viz.* prism and prism, prism and pyramid.
(ii) Intersection of two curved surfaces, *viz.* cone and cylinders, cylinder and cylinder, cone and cone.
(iii) Intersection of a plane surface and a curved surface, *viz.* prism and cylinder, pyramid and cylinder, etc.

15.2 Methods of Determining Line of Intersection

The following two methods may be used to find the line of intersection:

(i) Line Method or Piercing Point Method
(ii) Cutting Plane Method

(i) **Line Method or Piercing Point Method:** When two solids meet each other, elements on one solid intersect with the surface of the other solid. Points of intersection of these solids with the surface of the other solid are then located. These points are the vertices of the line of intersection. These points are more easily located from the view in which the lateral surface of the solid appears edgewise, i.e., a line. Then, these points from one view are projected on to the other view. The curve joining these points will be a line of intersection.

(ii) **Cutting Plane Method:** This is the general method of finding the line of intersection of any two surfaces. In this method, the solids are assumed to be cut by a series of cutting planes. These planes may be perpendicular to HP or perpendicular to VP or oblique. Certain important points where the line or curve of intersection changes shape, known as key points, have to be located. Due care should be given in the selection of key points.

Each method is explained in detail while solving problems. Sound knowledge of projections of solids and sections of solids is quite essential while dealing with these problems.

Guidelines for the selection of method
A key to solving problems of intersection of surfaces with care and efficiency lies in the choice of method. The following points are to be considered regarding the choice of the method:

- When at least one of the intersecting surfaces is a plane surface, the line method is more useful.
- When it is difficult to obtain side view of any of the solids, the cutting plane method proves to be more convenient, irrespective of whether or not the surface is plane.
- When both the intersecting surfaces are curved, the cutting plane method is more convenient. However, if one of the surfaces is a cone, the line method may be preferred.

15.3 Intersection of Two Prisms

Prisms have plane lateral surfaces. The line of intersection between two plane surfaces is determined by locating the points at which the edges of one surface intersect the other surface and, then, joining these points by a straight line in a correct sequence. These points are called vertices.

Problem 15.1 A vertical prism, edge of base 30 mm and height 60 mm, is resting on its base in HP. It is completely penetrated by another horizontal square prism, edge of base 20 mm, such that the axes of the two prisms bisect each other at right angles, while the faces of the two prisms are equally inclined to the VP. Draw the projections of the solids, showing lines of intersection. Assume any suitable length of the penetrating prism.

(PTU, Jalandhar May 2011)

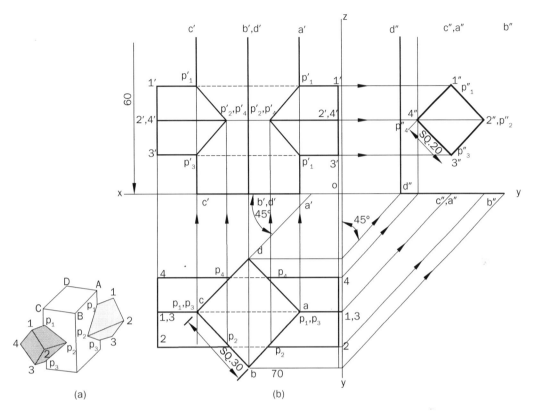

Fig. 15.1 Solution to problem 15.1

Solution:

(i) Draw the projections of the prism in the required position and label the various corner points.

(ii) The faces of the vertical prism are seen as edgewise in the top view, where it will be convenient to locate the points of intersection of the both prisms.

(iii) Lines 1-1 and 3-3 intersect the edges of the vertical prism in points p_1 and p_3, coinciding with a and c. Lines 2-2 and 4-4 meet the faces of the prism at p_2 and p_4 respectively.

(iv) Project these points to the corresponding lines in the front view. For example p_2 is projected to $p_2{}'$ on the line 2'-2' in the front view and so on. Note that point $p_4{}'$ coincides with $p_2{}'$.

(v) Join these points in a proper sequence, taking proper care of visible and invisible lines as shown in Fig. 15.1.

(vi) Lines $p_1{}'p_4{}'$ and $p_3{}'p_4{}'$, which have been projected as hidden lines in front view, are overlapped by $p_1{}'p_2{}'$ and $p_2{}'p_3{}'$ visible lines in the front view. These lines show the line of intersection.

(vii) The edges of the penetrating prism, which are overlapped by the vertical prism (i.e., penetrated) are shown as dashed (invisible) lines. The portions of the edges of the vertical prism, which don't exist, need not to be shown or kept fainter.

Problem 15.2 A vertical square prism, edge of base 30 mm and height 60 mm, is resting on its base in HP. It is completely penetrated by another horizontal square prism, edge of base 20 mm, such that the axis of the penetrating prism is perpendicular to and 5 mm in front of the axis of the vertical prism, while the faces of the two prisms are equally inclined to the VP. Draw the projections of the solids, showing lines of intersection. Assume any suitable length of the penetrating prism.

(PTU, Jalandhar December 2010)

Solution:

(i)   Draw the projections of the prism in the required position and label the various corner points.

(ii)   The points p_1', p_2', p_3', p_4' at which the edges of the horizontal prism intersect the faces of the vertical prism are to be located from the top view. In addition to these points, it is also necessary to locate the points $(p_5'$ and $p_6')$ at which the edges of the vertical prism are intersected by the horizontal prism. These points can easily be located in the side view when the horizontal prism appears as an edge view.

(iii)   These points are numbered as p_5' and p_6' in the left side view. Project these two points to their respective edges in front view as p_5' and p_6'.

(iv)   Join these points in the correct sequence as shown in Fig. 15.2. Proper care should be taken to determine the visible and invisible lines.

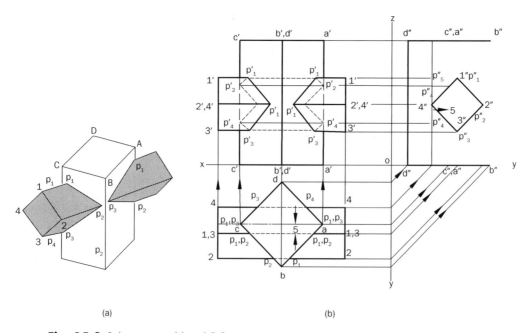

(a)   (b)

Fig. 15.2 Solution to problem 15.2

Problem 15.3 A vertical square prism, edge of base 30 mm and height 60 mm, has a face inclined at 30° to the VP. It is completely penetrated by another square prism, edge of base 20 mm and 70 mm long axis, faces of which are equally inclined to the VP. The axes of the two prisms are parallel to the VP and bisect each other at right angles. Draw the projections of the solids showing lines of intersection.

Solution:

(i)   Draw the projections of the prism in the required position and label the various corner points.

(ii) The faces of the vertical prism are projected edgewise in the top view, where it is easy to locate the points of intersection of the horizontal prism with the vertical prism. For example, the edge 1-1 of the horizontal prism pierces the vertical prism at p_1 in the top view. This point p_1 is projected to p_1' on the line 1'-1' in the front view. Similarly, plot the other points of intersection.

(iii) In addition to these points, it is necessary to find the points at which the edges of the vertical prism are cut. These points can easily be located in the side view, where the horizontal prism appears as an edge view.

(iv) Mark these points as p_2'', p_4'', p_6'', p_8'' in the left side view. Project these points to their respective edges in the front view as p_2', p_4', p_6' and p_8', respectively.

(v)   Join these points in proper sequence as shown in Fig. 15.3. Proper care should be taken to determine the visible and invisible lines.

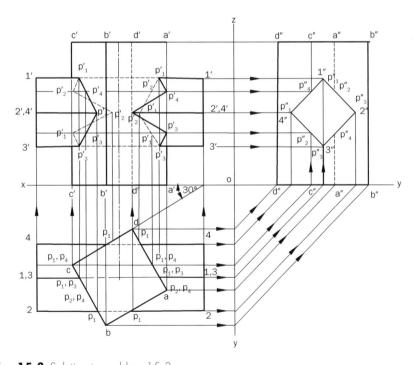

Fig. 15.3 Solution to problem 15.3

15.4 Intersection of Cylinder and Cylinder

As the cylinders have their lateral surfaces curved, the curve of intersection of two intersecting cylinders may be located by any one of the two methods. Draw a number of lines on the surface of the penetrating cylinder or by passing a series of cutting planes, parallel to the axis of the penetrating cylinder. For plotting an accurate curve, certain 'key points' at which the shape of curve changes must be located. These are the points placed at outermost or extreme lines of the each cylinder pierce the surface of the other cylinder.

Problem 15.4 A vertical cylinder diameter of base 40 mm and height 60 mm rests on its base in HP is completely penetrated by another cylinder of the same dimensions. The axes of the two cylinders bisect each other at right angles and are parallel to VP. Draw the projections of the solids showing curve of intersection.

(*PTU, Jalandhar December 2004, May 2007, December 2007, May 2008, May 2010*)

Solution:
Line Method

  (i)   Draw the projections of the cylinders in the required position.
  (ii)   Divide the side view of the horizontal cylinder into sixteen equal parts and complete the front and top views corresponding to these points.
  (iii)   Mark the points of intersection from p_1 to p_{16} in the top view and then project these points to their corresponding elements in the front view from p_1' to p_{16}'.
  (iv)   Join all the points in a proper sequence by free hand as shown in Fig 15.4

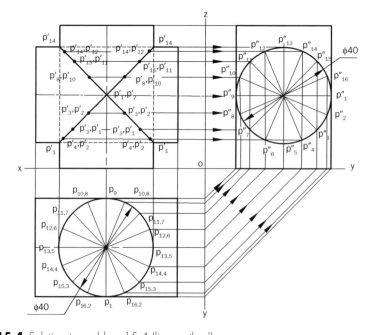

Fig. 15.4 Solution to problem 15.4 (line method)

Cutting Method

(i) Draw the projections of the cylinders in the required position.

(ii) Consider a series of horizontal cutting planes, say, nine (CP_1 to CP_9), passing through the horizontally and vertically placed cylinders. The section of the horizontal cylinder is a rectangle and the section shape of the vertical cylinder is a circle of diameter equal to the cylinder diameter. Points at which sides of the rectangles intersect the circle are being placed on the curve of intersection. For example, consider a horizontal section plane passing through points 12 and 14. In the front view, it is seen as a line coinciding with the line 12'-12' or 14'-14'. Points p_{12} and p_{14}, which the sides 12-12 and 14-14 of the rectangle cut the circle, lie on the curve. First of all mark these points in the top view and then project to points p_{12}' and p_{14}' on the lines 12'-12' and 14'-14', respectively, and so on.

(iii) Join all the points in a proper sequence by free hand as shown in Fig. 15.5.

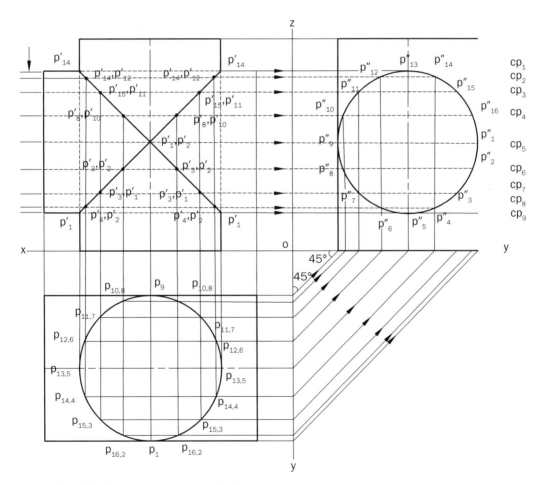

Fig. 15.5 Solution to problem 15.4 (cutting plane method)

Problem 15.5 A vertical cylinder, diameter of base 55 and height 70 mm, resting on its base on HP is completely penetrated by another cylinder of the diameter 40 mm and 65 mm length, such that their axes bisect each other at right angles and are parallel to the VP. Draw the projections of the solids showing curve of intersection.

(PTU, Jalandhar May 2005, December 2006, December 2007, May 2010)

Solution:
Line Method

(i) Draw the projections of the cylinders in the required position.

(ii) Divide the side view of the horizontal cylinder into sixteen equal parts and complete the front and top views corresponding to these points.

(iii) Mark the points of intersection (p_1 to p_{16}) in the top view and then project these points to their corresponding elements in the front view.

(iv) Join all the points in a proper sequence by free hand as shown in Fig. 15.6

Cutting Method

The interpretation of the solution is left to the reader. See Fig. 15.7

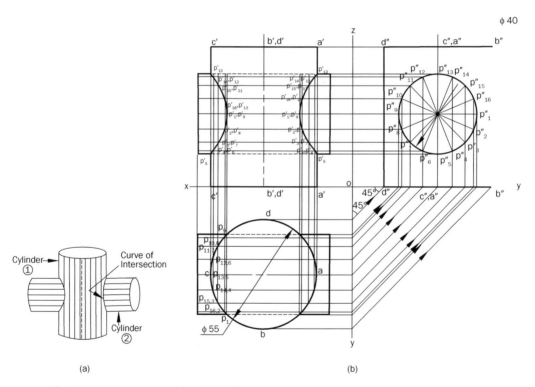

(a)   (b)

Fig. 15.6 Solution to problem 15.5 (line method)

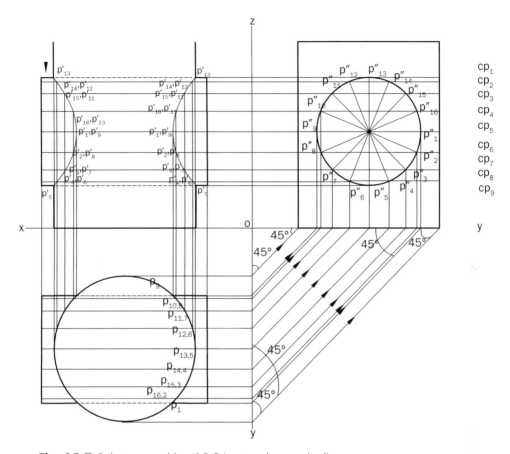

Fig. 15.7 Solution to problem 15.5 (cutting plane method)

Problem 15.6 A vertical cylinder, diamter of base 55 mm and height 70 mm, is resting on its base in HP. It is intersected by another cylinder of diameter 40 mm and 65 mm length, such that the axes of the two cylinders are mutually perpendicular and 10 mm in front of the axis of the vertical cylinder. Draw the projections of the solids showing the curve of intersection.

Solution:

(i)   Draw the projections of the solids in the required position.

(ii) The whole of penetrating cylinder is not contained by the vertical cylinder as shown in Fig. 15.9. Divide the side view of the horizontal cylinder into sixteen equal parts and complete the front and top views corresponding to these points. The points p_{16}'', p_1'' and p_2'' lie on the part of the cylinder which projects outside the vertical cylinder. Therefore, there will be no points p_{16}', p_1' and p_2' in the front view on the curve of intersection.

(iii) Mark rest of the points by the line method.

(iv) Join these points in the proper order and care should be taken to determine the visible and invisible lines.

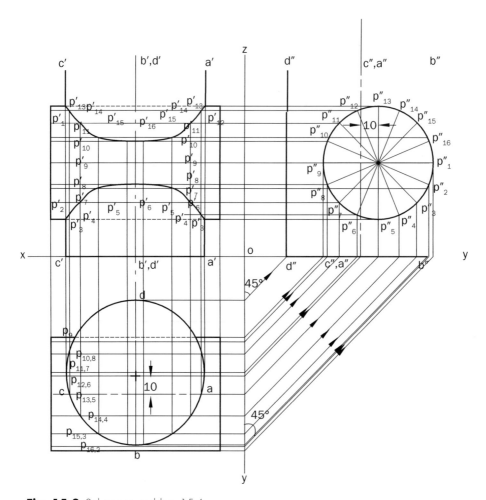

Fig. 15.8 Solution to problem 15.6

Problem 15.7 A cylinder of 45 mm diameter and 70 mm length standing on its base on HP is penetrated by another cylinder of same size. The axis of penetrating cylinder is parallel to both HP and VP and 10 mm away from the axis of the vertical cylinder. Draw the projections of the solid showing curves of interpenetration.

(PTU, Jalandhar December 2003)

Solution: For its solution, see Fig. 15.9.

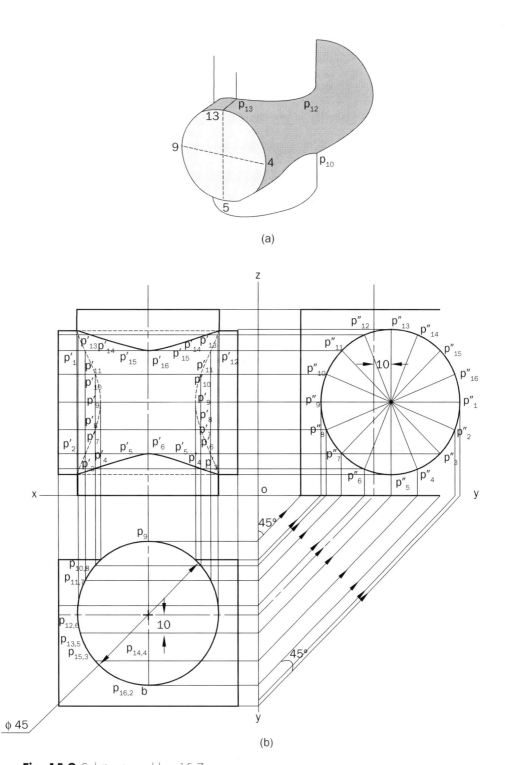

(a)

(b)

Fig. 15.9 Solution to problem 15.7

Problem 15.8 A right circular cylinder, base diameter 55 mm and height 85 mm, is resting on its base in HP. It is intersected by another cylinder of diameter of base 40 mm (Fig. 15.10). The axis of the penetrating cylinder is inclined to HP at 45° and is parallel to the VP and cuts the axis of the vertical cylinder at a distance of 20 mm from its base. Draw the projections of the solids showing curve of intersection.

(PTU, Jalandhar May 2005)

Solution:

(i)   Draw the top and front views of the vertically placed cylinder.

(ii)   In the front view, mark a point along its axis 20 mm above its base. Through this point, draw a centre line for the penetrating cylinder, inclined at 45° to the x-y.

(iii) Draw a line $x_1 \, y_1$ perpendicular to the centre line of the penetrating cylinder at a suitable distance as shown in Fig. 15.11 and cut half the diameter of the penetrating cylinder on this centre line on each side.

(iv)   Divide this circle into sixteen equal parts and complete the front and top views of the penetrating cylinder.

(v)   Mark the points of intersection in front and top views of the vertically placed cylinder.

(vi)   Join these points in proper sequence.

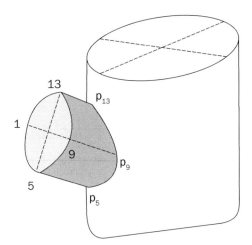

Fig. 15.10 Solution to problem 15.8

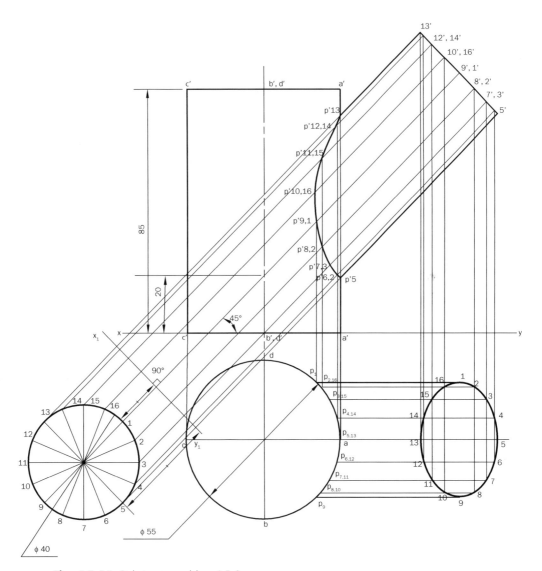

Fig. 15.11 Solution to problem 15.8

Problem 15.9 A cylinder of 60 mm diameter and 80 mm height stands on its base on HP. It is penetrated centrally by a cylinder, 40 mm diameter and 140 mm long, whose axis is parallel to the HP but inclined at 30° to the VP. Draw the projections showing curve of intersection.

(PTU, Jalandhar June 2003)

Solution: For its solution see Fig. 15.12.

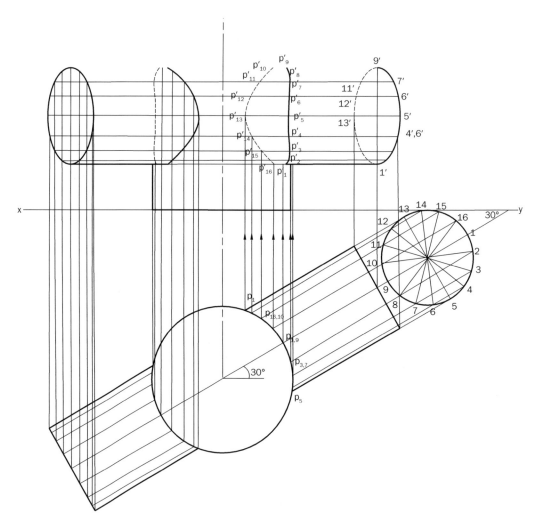

Fig. 15.12 Solution to problem 15.9

15.5 Intersection of Cylinder and Prism

As one of the intersecting solids is a solid of revolution, therefore the lines of intersection will be composed of curves. The various points plotted on the curves of intersection by either of the two methods are to be joined by means of a smooth curve.

Problem 15.10 A right circular cylinder of diameter of base 60 mm and length 70 mm resting on its base in HP is penetrated by a square prism of 30 mm base edge, such that the axes of the solids bisect each other at right angles. The faces of the prism are equally inclined to the VP. Draw the projections of the solids showing curve of intersection.

Solution:

(i) Draw the projections of the solids in the required position and name the corner points of the prism.

(ii) The edges of square prism intersect the edge view of the cylinder in the top view at points p_1, p_2, p_3, p_4. Project these points of intersection in the front view as p_1', p_2', p_3' and p_4'.

(iii) Divide the circle into sixteen equal parts. Project the key points of the cylinder too. These are the points of intersection of the cylinder with the surface of the prism, *viz.* p_5, p_6, p_7 and p_8 in the top view as shown in Fig. 15.13.

(iv) Project these points of intersection in the front view as p_5', p_6', p_7' and p_8'.

(v) Join all the points in the correct order and take proper care to determine the visible and invisible lines.

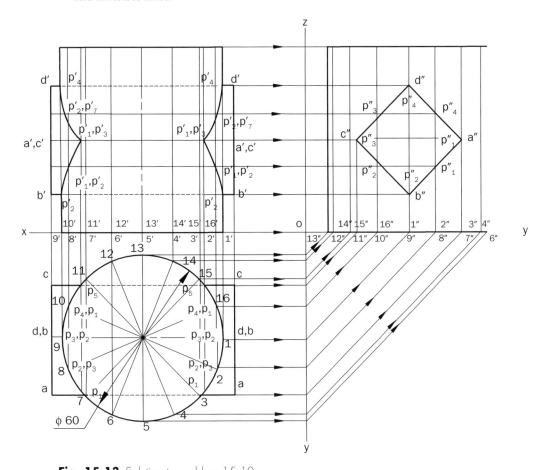

Fig. 15.13 Solution to problem 15.10

Problem 15.11 A square prism, edge of base 40 mm and height 70 mm, rests on its base in HP with one of its faces inclined at 30° to the VP. It is penetrated by a horizontal cylinder of diameter 45 mm and 80 mm length such that the axis of the cylinder is parallel to both HP and VP and is 5 mm in front of the axis of the prism. Draw the projections of the solids showing curve of intersection.

Solution:

(i) Draw the projections of the solids in the given position and name the corner points of the prism.

(ii) Divide the circle into sixteen equal parts project all the key points as shown in Fig. 15.14.

(iii) Draw a curve joining the points in a correct sequence. Proper care should be taken to determine the visible and invisible lines.

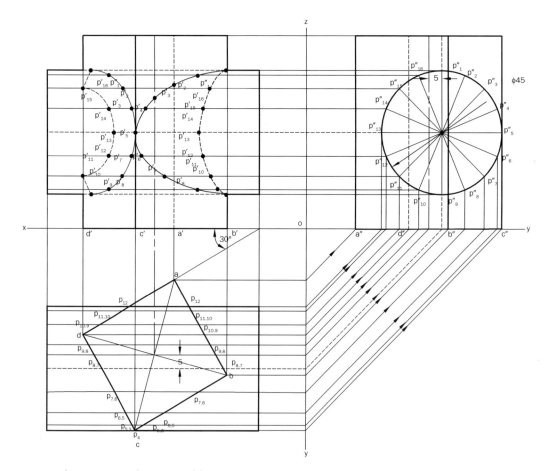

Fig. 15.14 Solution to problem 15.11

Problem 15.12 A square prism, edge of base 40 mm and height 70 mm, rests on its base in HP with one of its faces inclined at 30° to the VP. It has a hole of φ 45 mm drilled centrally through it. The centre line of the hole is parallel to both HP and VP and is 5 mm away from the axis of the prism. Draw the projections of the solids showing curve of intersection.

(PTU, Jalandhar December 2008)

Solution: For its solution, see Fig. 15.15.

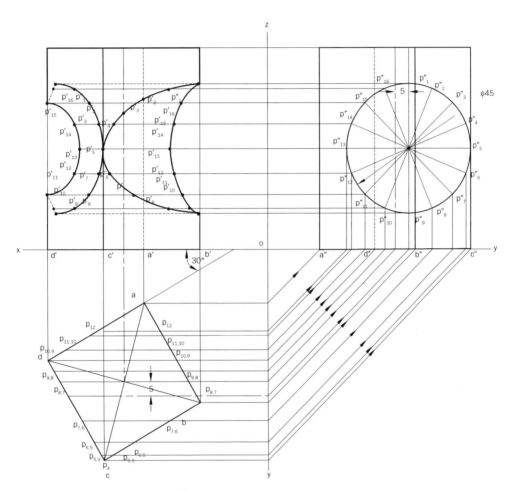

Fig. 15.15 Solution to problem 15.12

15.6 Intersection of Cylinder and Cone

Problem 15.13 A right circular cone, diameter of base 50 mm and height 60 mm, resting on its base in HP is completely penetrated by a cylinder of diameter 25 mm and 70 mm length. The axis of the penetrating cylinder is parallel to both HP and VP and intersects the axis of the cone at a distance of 20 mm from its base. Draw the projection of the solids showing curve of intersection.

(PTU, Jalandhar May 2004, December 2009)

Solution:

(i)   Draw the projections of the solids in the required position.

(ii) Divide the circle in the edge view into sixteen equal parts and name these parts as 1",
      2", 3", and so on.

(iii) Project these points in the front view and top view.

(iv) To mark the required points in the top view, draw circles with centre as o and diameters
      as p_5', p_5', p_{13}', p_{13}', etc., thereby cutting the lines through the corresponding points of
      front view at the required points as p_5', p_{13}', etc.

(v)   Project the points p_1, p_2, p_3, etc., from the top view to intersect the corresponding lines
      in the front view so as to mark the points as p_1', p_2', p_2', etc.

(vi) Join these points in the correct sequence as shown in Fig. 15.16.

Problem 15.14 A right circular cone, diameter of base 60 mm and height 70 mm, resting on its
base in HP is penetrated by a cylinder of diameter 30 mm and 90 mm length such that its axis
is parallel to both HP and VP and is 20 mm from its base and 5 mm in front of the axis of the
cone. Draw the projections of the solids showing curves of intersection.

Solution:

(i)   Draw the projections of the solids in the required position.

(ii) Divide the edge view of the circle (cone) into twelve equal parts and name these points
      as 1", 2", 3" and so on.

(iii) Draw the lines of the surface of the cylinder through these points in front and top
      views.

(iv) Assume a horizontal cutting plane through all these lines, the intersection of these
      planes are to be plotted in the top view. For example (see previous case problem 15.4
      cutting plane method), draw a circle in the top view with centre o and diameter p_9'
      p_9', $p_3'p_3'$, etc., thereby cutting the lines through corresponding front view points at the
      required points as p_9', p_3', etc.

(v)   Project the points p_1, p_2, p_3, etc., from the top view to intersect the corresponding lines
      in the front view so as to mark the points as p_1', p_2', p_3', etc.

(vi) Join these points in the correct sequence as shown in Fig. 15.17.

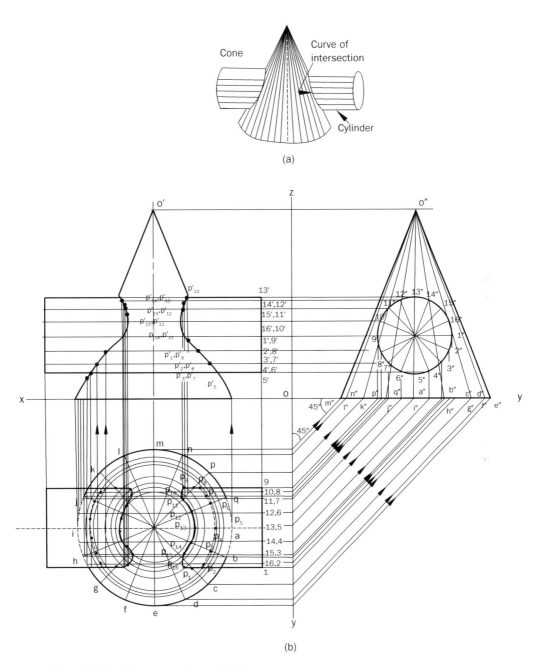

Fig. 15.16 Solution to problem 15.13

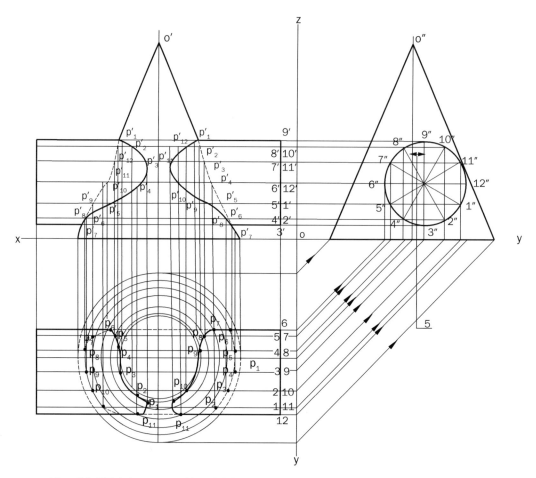

Fig. 15.17 Solution to problem 15.14

Problem 15.15 A vertical cone, diameter of base 70 mm and height 80 mm, resting on its base in HP is penetrated by a cylinder of diameter 30 mm and 90 mm length such that its axis is parallel to both HP and VP and is 20 mm from its base and 12 mm in front of the axis of the cone. Draw the projection of the solids showing curve of intersection.

Solution:

(i)   Draw the projections of the solids in the required position.

(ii)   The axis of the cylinder is so placed that a position of the cylinder remains outside the cone. So the curve changes the direction at point a'' and b'' in the left side view and becomes a single continuous curve as shown in Fig. 15.18.

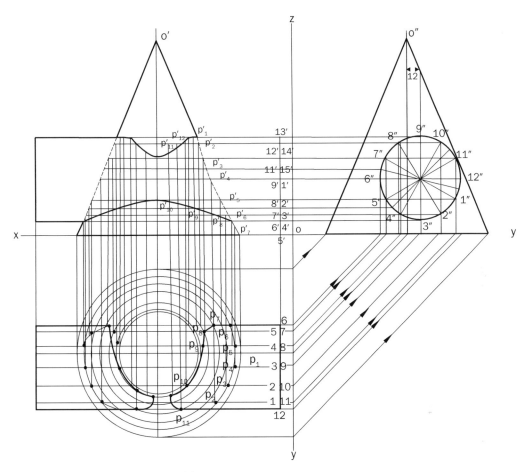

Fig. 15.18 Solution to problem 15.15

Problem 15.16 A vertical cone, diameter of base 50 mm and height 60 mm, resting on its base in HP is penetrated by a cylinder of diameter 30 mm, the axis of which is parallel to and 10 mm away from the axis of the cone. Draw the projection of the solids showing curves of intersection when (a) the plane containing the two axes is parallel to the VP and (b) the plane containing the two axes is inclined at 45° to the VP.

Solution:

(a) The plane containing the two axes is parallel to the VP.

(i) Draw the projection of the solids in the given positions.

(ii) Divide the circle of the cone into sixteen equal parts in the top view and name these parts as 1, 2, 3, etc.

(iii) Project these points in the front and left side views as 1', 2', 3', etc., and 1", 2", 3", etc., respectively.

(iv) The surface of the cylinder is seen as a circle in the top view. It intersects the lines drawn on the surface of the cone in the top view at points p_1, p_2, p_3 and so on. corresponding lines in the front and left side views respectively.

(v) Join these points in the correct sequence as shown in Fig. 15.19.

(b) The plane containing the two axes is inclined at 45° to the VP. Draw the projections of the solids in the given position. Adopt the same method as in the part (a). See Fig. 15.20.

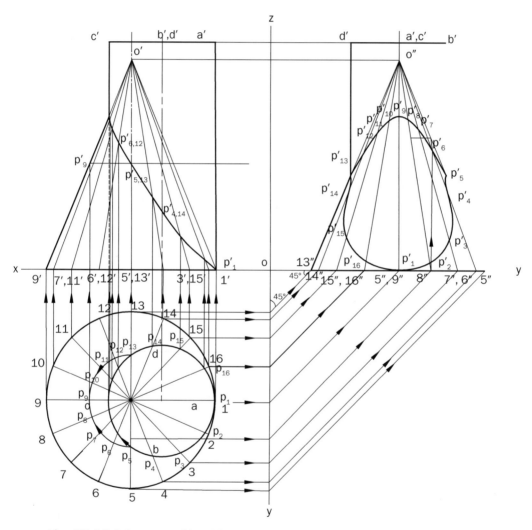

Fig. 15.19 Solution to problem 15.16 part (a)

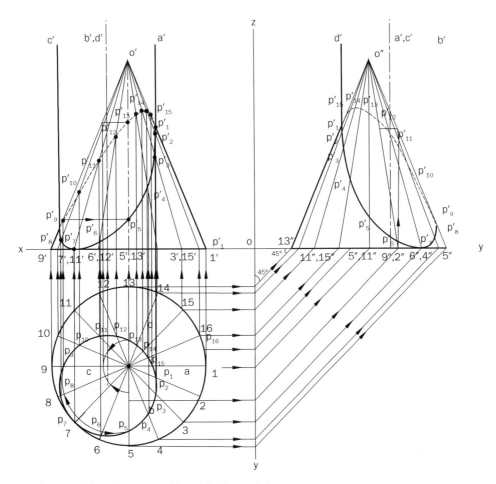

Fig. 15.20 Solution to problem 15.16 part (b)

15.7 Intersection of Cone and Prism

Problem 15.17 A right circular cone, diameter of base 55 mm and height 70 mm, resting on its base in HP is penetrated by a square prism of base 30 mm side in such a way that their axes are collinear. Draw three views of the solids showing curve of intersection. When (a) its base side is parallel to the VP and (b) its faces are equally inclined to the VP.

(PTU, Jalandhar December 2004)

Solution:

(a) Base side is parallel to the VP.

(i) Draw the projections of the solids in the given position.

(ii) Divide the circle of the cone into sixteen equal parts in the top view and label them as 1, 2, 3, etc.

(iii) Project these points in the front and left side views too.

(iv) The faces of the vertical prism are projected edgewise in the top view, where it is easy to locate the points of intersection with the vertical cone. For example (see Problem 15.4, cutting plane method) the points p_1, p_5, p_9, p_{13} are located at the base edge of the square prism. Since the surface of the cone is curved, hence other points of intersection on the square prism are located as p_2, p_3, p_4, etc.

(v) Project these points of intersection in the front view and side view as shown in the Fig. 15.21.

(vi) Join these points in the correct sequence.

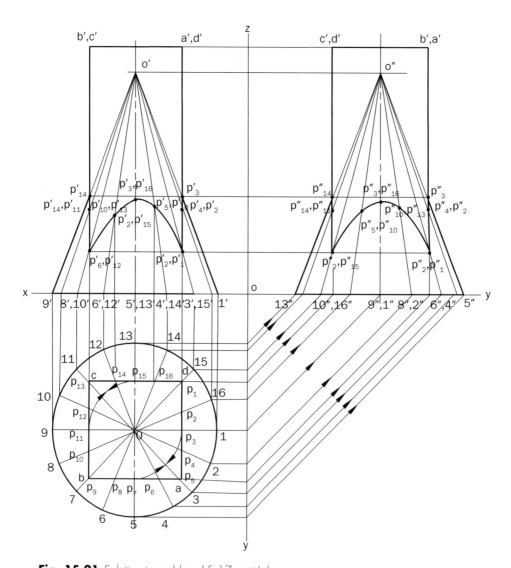

Fig. 15.21 Solution to problem 15.17 part (a)

(b) Its faces are equally inclined to the VP.

Draw the projections of the solids in the given position. Adopt the same method as in the case (*a*). See Fig. 15.22.

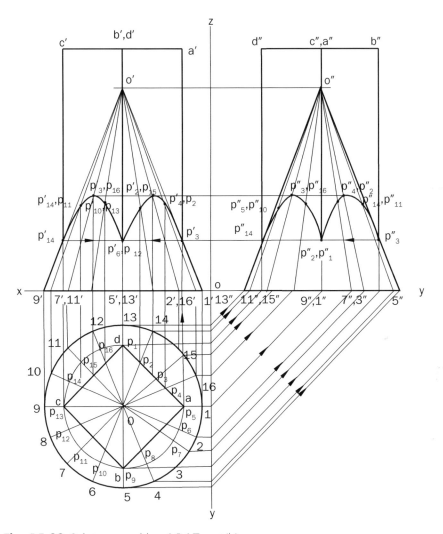

Fig. 15.22 Solution to problem 15.17 part (b)

Problem 15.18 A vertical cone, diameter of base 55 mm and height 70 mm, resting on its base in HP is penetrated by a square prism of side 25 mm. The axis of the prism is parallel to and 10 mm away from the axis of the cone. Draw the three views of the solids showing curve of intersection when the plane containing the two axes is parallel to the VP.

Solution: Adopt the same method as shown in problem 15.18. All construction lines are retained to make the solution easy to understand. See Fig. 15.23.

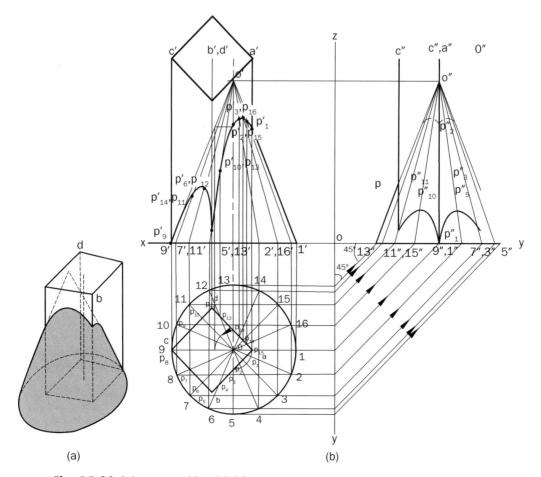

Fig. 15.23 Solution to problem 15.18

Problem 15.19 A vertical cone, diameter of base 55 mm and height 60 mm, resting on its base in HP is penetrated by a square prism of 20 mm side and 70 mm length such that the axes of the two solids intersect at right angles. The axis of the penetrating prism is parallel to both HP and VP and is 20 mm from the base of the cone. The faces of the penetrating prism are equally inclined to the VP. Draw the projections of the solids showing curve of intersection.

Solution:

(i)   Draw the projections of the solids in the given position.

(ii) Draw a horizontal straight line at a distance of 20 mm from the base of the cone in the front view, to represent the axis of the prism.

(iii) Draw a square of 20 mm side in the left side view, keeping its faces equally inclined to the VP.

(iv) Project the front and top views of the penetrating prism.

(v)   Divide the circle of the cone in the top view into sixteen equal parts.

(vi)   Project the points of intersection from the end view to the corresponding elements in the top and front views as shown in Fig. 15.24

(vii) Join all the points in the correct sequence and proper care should be taken to determine the visible and invisible lines.

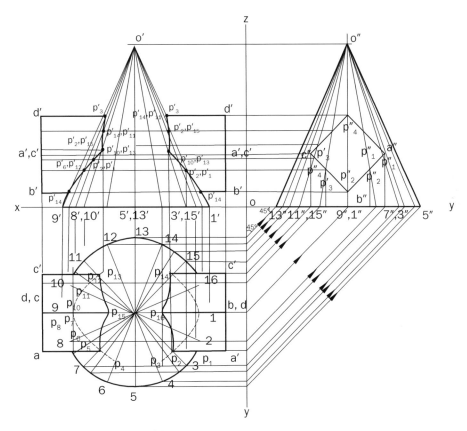

Fig. 15.24 Solution to problem 15.19

Exercises

Intersection of Prism and Prism

15.1   A vertical square prism, edge of base 40 mm and height 65 mm, rests on its base in HP. It is completely penetrated by another square prism, edge of base 25 mm, such that the axes of the two prisms bisect each other at right angles. The rectangular faces of the two prisms are equally inclined to the VP. Draw the projections of the solids, showing line of intersection. Assume any suitable length of the penetrating prism.

15.2   A vertical square prism, edge of base 40 mm and height 65 mm, rests on its base in HP. It is penetrated by a horizontal square prism, edge of base 25 mm, such that its axis is perpendicular to and is 6 mm in front of the axis of the vertical prism. The rectangular faces of both the prisms are equally inclined to the VP. Draw the projections of the solids, showing line of intersection. Assume any suitable length of the penetrating prism.

15.3 A vertical square prism, edge of base 40 mm and height 65 mm, is resting on its base in HP, with a vertical face inclined at 30° to the VP. It is penetrated by a horizontal square prism, edge of base 25 mm and 70 mm long, having one of its rectangular faces inclined at 30° to the VP. The axes of the two prisms bisect each other. Draw the projections of the solids showing line of intersection.

Intersection of Cylinder and Cylinder

15.4 A vertical cylinder of φ 50 mm and height 60 mm resting on its base in HP is completely penetrated by another cylinder of the same dimensions. The axes of the two cylinders bisect each other at right angles. Draw the projections of the solids showing curve of the intersection.

15.5 A vertical cylinder, diameter of base 60 mm and height 70 mm, resting on its base in HP is completely penetrated by another cylinder of the diameter 40 mm and 60 mm long, such that the axes of the two cylinders
   (a)   intersect each other at right angles and are parallel to VP.
   (b)   are mutually perpendicular but the axis of the cylinder and are both parallel to the VP.
   Draw the projection of the solids showing curves of intersection.

15.6 A vertical cylinder, diameter of base 60 mm and height 70 mm, resting on its base in HP is penetrated by another cylinder of the same dimensions. The axes of the two cylinders bisect each other at right angles. The axis of the penetrating cylinder is 10 mm in front of the axis of the vertical cylinder. Draw the projections of the solids showing curves of intersection.

15.7 A right circular cylinder, diameter of base 50 mm and height 70 mm, rests on its base on HP. It is intersected by another cylinder of 30 mm base diameter such that the axis of the penetrating cylinder is:
   (a)   inclined at 45° to the HP and parallel to the VP,
   (b)   intersects the axis of the vertical cylinder at a distance of 20 mm from its base, and
   (c)   is 50 mm in front of the axis of the vertical cylinder.
   Draw the projections of the solids curve of intersection.

15.8 A cylinder of 50 mm diameter and 60 mm height stands on its base on HP. It is penetrated centrally by a cylinder, 30 mm diameter and 100 mm long, whose axis is parallel to the HP but inclined at 30° to the VP. Draw the projections showing curve of intersection.

15.9 A right circular cylinder, diameter of base 55 mm and height 70 mm, is resting on its base in HP. It is intersected by another cylinder of diameter 40 mm and 65 mm long, such that the axes of the two cylinder are mutually perpendicular and 5 mm in front of the axis of the vertical cylinder. Draw the projections of the solids showing curve of intersection.

Intersection of Cylinder and Prism

15.10 A right circular cylinder of diameter of base 50 mm and 60 mm length is resting on its base in HP. It is penetrated by a square prism of 25 mm base edge, such that the axes of the solids bisect each other at right angles. The faces of the prism are equally inclined to the VP. Draw the projections of the solids showing curve of intersection.

15.11 A square prism, edge of base 40 mm and height 70 mm, resting on its base in HP with one of its faces inclined at 30° to the VP is penetrated by a horizontal cylinder of diameter 45 mm and 80 mm long such that the axis of the cylinder is parallel to both HP and VP and is 8 mm in front of the axis of the prism. Draw the projections of the solids showing curve of intersection.

Intersection of Cylinder and Cone

15.12 A right circular cone, diameter of base 50 mm and height 70 mm, resting on its base in HP is completely penetrated by a cylinder of diameter 25 mm and 60 mm long. The axis of the penetrating cylinder is parallel to both HP and VP and intersects the axis of the cone at a distance of 25 mm from its base. Draw the projections of the solids showing curve of intersection.

15.13 A right circular cone, diameter of base 50 mm and height 60 mm, resting on its base in HP is penetrated by a cylinder of diameter 30 mm and 70 mm length such that its axis is parallel to both HP and VP and is 20 mm from its base and 10 mm in front of the axis of the cone. Draw the projections of the solids showing curve of intersection.

15.14 A cone, base φ 50 mm and height 60 mm, rests on its base on HP. A right circular cylinder, base φ 25 mm, penetrates the cone such that its axis is parallel to and 6 mm away form the axis of the cone. Draw the projections of the solids showing curves of intersection when the plane containing the two axes is
(a) parallel to the VP and
(b) inclined to the VP at 45°

Intersection of Cone and Prism

15.15 A right circular cone, diameter of base 50 mm and height 60 mm, resting on its base in HP is penetrated by a square prism of base 25 mm side, in such a way that their axes are collinear. Draw three vies of the solids showing curves of intersection. When (a) its base side is parallel to the VP and (b) its faces are equally inclined to the VP.

15.16 A vertical cone, diameter of base 55 mm and height 60 mm, resting on its base in HP is penetrated by a square prism of 20 mm side and 70 mm long such that the axes of the two solids intersect at right angles. The axis of the penetrating prism is parallel to both HP and VP and is 25 mm from the base of the cone. The faces of the penetrating prism are equally inclined to the VP. Draw the projections of the solids showing curve of intersection.

Objective Questions

15.1 Differentiate between intersection of surfaces and interpenetration of solids

15.2 Give the practical applications of intersection of surfaces or interpenetration of solids.

15.3 Name the methods for finding the lines or curves of intersection.

15.4 The line of intersection between two prisms consists of ....

15.5 The line of intersection between a prism and a pyramid consists of ....

15.6 The line of intersection between a cylinder and a prism consists of..........................

15.7 The intersection between a solid resting on HP and a plane inclined to the HP is a .... in the front view.

15.8 The line of intersection depends upon the .... of the solids.

Answers

15.4 Straight lines	15.6 Curved lines	15.8 Nature or type
15.5 Straight lines	15.7 Line	

Chapter 16

Isometric Projection

16.1 Introduction

In engineering practice, it is usual to draw two or more than two orthographic projections to show the true shape and size of an object. Such drawings can be correctly interpreted only by those persons who have a sound knowledge of the principles of orthographic projections. To make the drawing more understandable, several forms of one plane projection drawings are used to supplement the orthographic drawings. These are called pictorial drawings, which can be easily understood by everybody without any formal training. Pictorial drawings are mainly used to show complicated shapes such as aircraft, ships, buildings, etc.

16.2 Classification of Pictorial Drawings

Pictorial drawings are classified into three categories:

(a) Axonometric Projection
(b) Oblique Projection
(c) Perspective Projection

Axonometric Projection: It is a three-dimensional projection of an object, as shown in Fig. 16.1 (*a*). It is a form of orthographic projection, since the projectors are perpendicular to the plane of projection and parallel to each other.

Oblique Projection: It is a three-dimensional projection of an object on a projection plane, as shown in Fig. 16.1 (*b*). In this, the projectors are parallel to each other, but are oblique to the picture plane.

Perspective Projection: It is the most realistic projection. In this, the projectors converge towards the viewer's eye, making different angles to the picture plane, as shown in Fig. 16.1 (*c*).

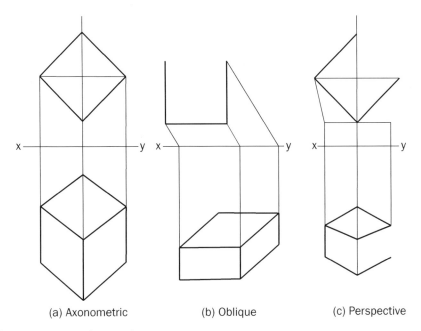

(a) Axonometric   (b) Oblique   (c) Perspective

Fig. 16.1 Types of pictorial projections

16.3 Axonometric Projection

It is a form of orthographic projection and is obtained by projecting an object placed in an oblique position to the plane of projection. Axonometric projection is classified as:

- Isometric Projection
- Dimetric Projection
- Trimetric Projection

Out of these, only isometric projection is dealt with here.

16.4 Isometric Projection

It is a type of pictorial drawing in which the three dimensions of a solid are not only shown in one view, but their actual sizes can be measured directly from it.

If a cube is resting on one of its corners on the HP, with its solid diagonal perpendicular to the VP, the front view of the cube is its isometric projection as shown in Fig. 16.2.

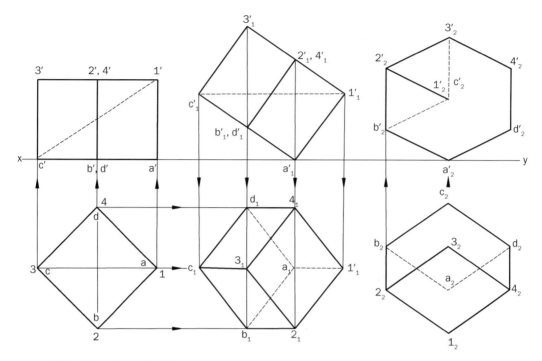

Fig. 16.2 Principles of isometric projection

Figure 16.3 shows the front view of a cube in the above position; the following conclusions are drawn from it:

- All the faces are equally inclined to the VP.
- The three edges *CB*, *CD* and *CG* make equal angles of 120° with each other.
- All other edges are parallel to any of these three edges. Isometric means equal measurement, i.e., each of the three planes of a cube is equally foreshortened.

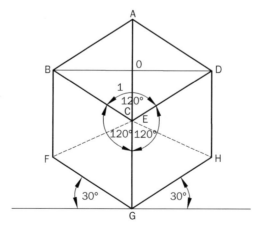

Fig. 16.3 Isometric projection of a cube

16.5 Terms Connected with Isometric Projection

The following are some important terms used in isometric projection. See Fig. 16.4

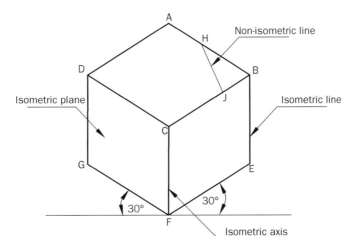

Fig. 16.4 Terms connected with isometric projection

- **Isometric Axes.** The three lines *CB*, *CD* and *CF* meeting at a point C and making 120° angles with each other are termed as isometric axes.
- **Isometric Lines.** The lines parallel to the isometric axes are called isometric lines.
- **Non-isometric Lines.** The lines which are not parallel to isometric axes are called non-isometric lines.
- **Isometric Planes.** The planes representing the faces of the cube as well as other planes parallel to these planes are called isometric planes.

16.6 Isometric Scale

Figure 16.5 (*a*) shows the isometric projection of a cube; as all the edges of the cube are equally foreshortened, the square faces appear as rhombuses. The rhombus a_1', b_1', c_1', d_1' shows the isometric projection of the top square face of the cube in which $b_1'\,d_1'$ is the true length of the diagonal. Construct a square $b_1'\,c_1''\,d_1'\,a_1''$ around $b_1'\,d_1'$ as a diagonal. The $b_1'\,a_1''$ shows the true length of $b_1'\,a_1'$.

$$In\ \Delta a_1'b_1'o, \quad \cos 30° = \frac{b_1'o}{b_1'a_1'}$$

$$\Rightarrow \qquad \frac{\sqrt{3}}{2} = \frac{b_1'o}{b_1'a_1'}$$

(i)

$In\ \Delta a_1"b_1'o,\quad \cos 45° = \dfrac{b_1'o}{b_1'a_1"}$

$\Rightarrow \qquad\qquad \dfrac{1}{\sqrt{2}} = \dfrac{b_1'o}{b_1'a_1"}$ (ii)

Divide (ii) by (i)

$$\dfrac{1}{\sqrt{2}} \times \dfrac{2}{\sqrt{3}} = \dfrac{b_1'o'}{b_1'a_1"} \times \dfrac{b_1'a_1'}{b_1'o}$$

$$\dfrac{\sqrt{2}}{\sqrt{3}} = \dfrac{b_1'a_1'}{b_1'a_1"}$$

$$\dfrac{b_1'a_1'}{b_1'a_1"} = 0.8165$$

$\Rightarrow \qquad \dfrac{\text{Isometric length}}{\text{True length}} = 0.8165$

Isometric length $= 0.8165 \times$ True length

While drawing an isometric projection, it is necessary to convert true lengths into isometric lengths. This is done easily by constructing and making use of an isometric scale. The method of constructing an isometric scale is as follows:

(i)   Draw a horizontal line of any length
(ii)   From any point on a horizontal line, draw two lines inclined at 30° and 45°, respectively.
(iii) Mark divisions of true lengths on 45° line.
(iv)   From each division point, draw vertical lines thereby meeting at 30° line at various points.
(v)   Then the corresponding scale projected on 30° line is called the isometric scale, as shown in Fig. 16.5 (b).

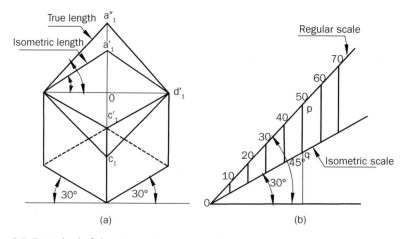

Fig. 16.5 Method of drawing an isometric scale

To obtain the isometric length of any dimensions, say, 50 mm, draw a vertical line from p, 50 mm division point, meeting the isometric scale at *q*. The length *oq* represents the isometric length.

16.7 Isometric Drawing

If the foreshortening of the isometric lines in an isometric projection is disregarded and, instead, the true lengths are marked, the view obtained will be exactly of the same shape but larger in proportion (about 22.5%) than that obtained by the use of the isometric scale, as shown in Fig. 16.6. To avoid this tedious construction of isometric projection, the true lengths are laid out along the isometric axes; the view obtained is called isometric drawing. It implies that the construction of an isometric drawing is much simpler compared to the isometric projection.

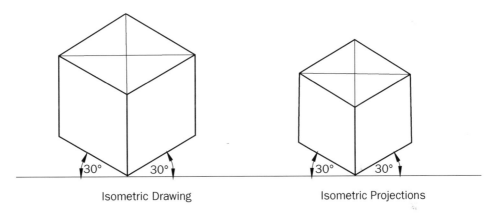

Isometric Drawing   Isometric Projections

Fig. 16.6 Isometric drawing is 22.5% larger than isometric projection

16.8 Isometric Dimensioning

The general rules for dimensioning have already been discussed in Chapter 3. All those rules hold good here too and in addition to those, the following rules must be taken care of.

- Extension lines, dimension lines and numerals for the isometric projection must be placed in the isometric planes of the faces as shown in Fig. 16.7.
- If possible, apply the dimensions to visible surfaces.

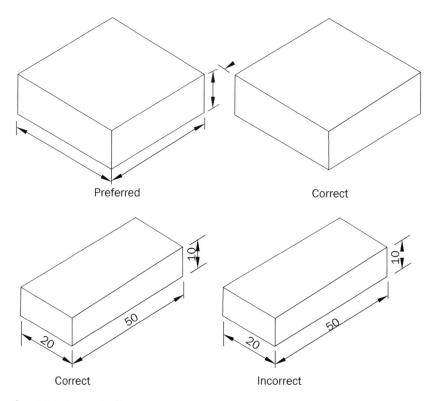

Fig. 16.7 Isometric dimensioning

16.9 Hidden and Centre Lines on an Isometric Projection

It is a usual practice to omit the hidden lines unless they are needed to make the drawing clearer. If an isometric projection is to be dimensioned and if it has holes, which much be located, centre lines must be drawn. The centre lines are placed on a plane in which the hole is shown and the dimensions are placed parallel to the planes. Fig. 16.8 shows the use of hidden and centre lines on an isometric projection.

16.10 Isometric Drawing or Projection of Plane Figures

Problem 16.1 Draw an isometric drawing of a square lamina of 30 mm side.

Solution:
Case I Vertical Plane

(i)   Draw a line at 30° to the horizontal and mark the length on it.
(ii) Draw verticals at the ends of the line and mark the length on these parallel lines.

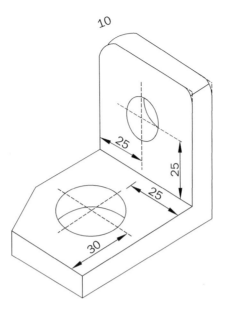

Fig. 16.8 Use of centre lines for dimensioning

(iii)   Join the ends of a straight line, which is also inclined at 30°.
As shown in Fig. 16.9, there are two possible positions for the plane.

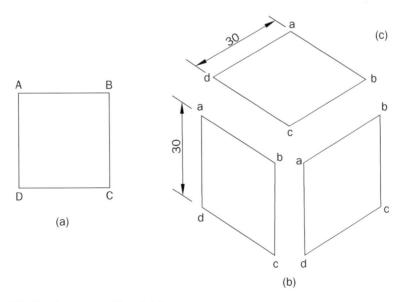

Fig. 16.9 Solution to problem 16.1

Case II Horizontal Plane

(i)   Draw two lines at 30° to the horizontal and mark the lengths along the same.

(ii)   Complete the figure by drawing 30° inclined lines at the ends until they intersect as shown in Fig. 16.9. The shape of an isometric drawing of a square lamina is a rhombus.

Problem 16.2 The front view of a triangle where surface is parallel to VP is shown in Fig. 16.10 (a). Draw its isometric drawing.

Solution:

(i)   Enclose the triangle in the rectangle and draw the isometric drawing of the rectangle.

(ii)   Mark a point *a* in *de* such that *da* = *DA*. Draw the triangle *abc*, which is the required isometric drawing. In an isometric drawing, angles do not increase or decrease in any fixed proportion. It can also be drawn in the other direction as shown in Fig. 16.10 (*b*).

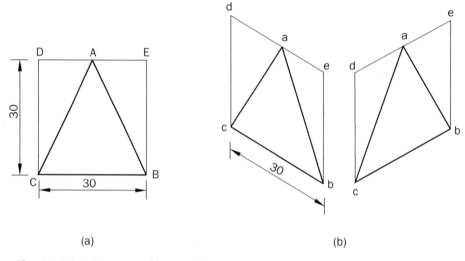

(a)   (b)

Fig. 16.10 Solution to problem 16.2

Problem 16.3 Draw an isometric drawing of a regular pentagon of 30 mm side.

Solution: The solution to this problem is self-explanatory. See Fig. 16.11.

Problem 16.4 The front view of a hexagon where surface is parallel to HP is shown in Fig. 16.12 (a). Draw its isometric drawing.

Solution: The solution to this problem is self-explanatory. See Fig. 16.12 (b).

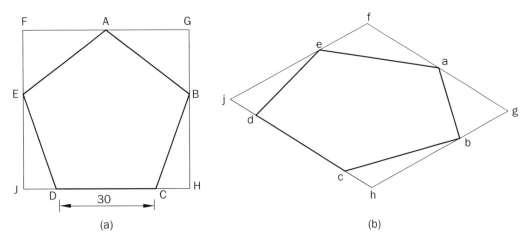

Fig. 16.11 Solution to problem 16.3

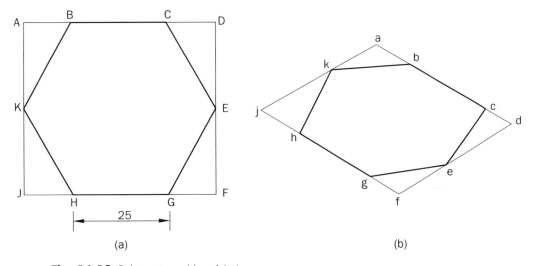

Fig. 16.12 Solution to problem 16.4

Problem 16.5 Draw the isometric drawing of a circle of 50 mm diameter.

Solution:

Offset Method

(i)   Divide the circle into convenient number of equal parts, after enclosing it in a square, as shown in Fig. 16.13 (*a*).

(ii) Determine the distances of the division points from the edges of the square, as shown in Fig. 16.13 (*b*).

(iii) Draw the isometric drawing of the square and mark the off-sets corresponding to the division points of the circle.

(iv) Join these points by a smooth curve.

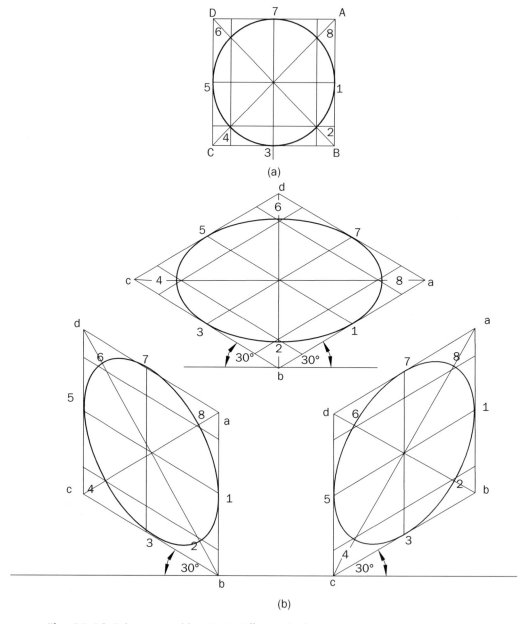

Fig. 16.13 Solution to problem 16.5 (Offset method)

Four Centre Method

(i)   Enclose the circle in a square, as shown in Fig. 16.13 (a).
(ii) Construct the isometric drawing of the square, which is a rhombus.
(iii) Locate the midpoints of the sides of the rhombus.
(iv) Join these midpoints to the nearest corners of the rhombus intersecting at c_1 and c_2.

(v)   With centres c_1 and c_2 and radius c_1 1 (c_2 2) draw two arcs as shown in Fig. 16.14.

(vi) With centres c_3 and c_4 and radius c_3 3 (c_4 4) draw two arcs meeting the above axes tangentially.

The ellipse obtained by the four centre method is not a true ellipse and it differs considerably in size and shape from the ellipse plotted by offset method. But owing to ease in construction and to avoid the labour of drawing free hand curves, this method is generally employed.

Problem 16.6 Figure 16.15 (a) shows the front view of a plane parallel to the VP. Draw its isometric drawing.

Solution: The solution to this problem is self-explanatory. See Fig. 16.15 (*b*).

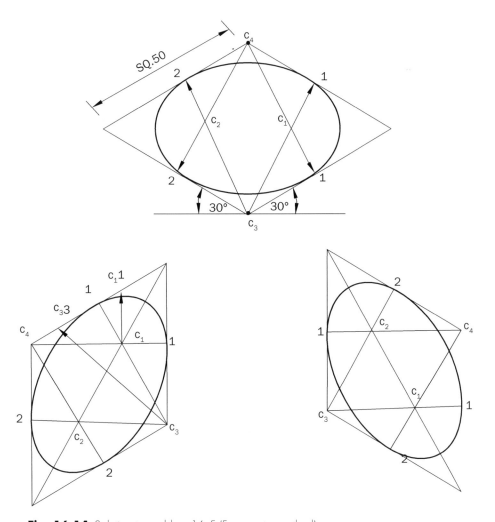

Fig. 16.14 Solution to problem 16.5 (Four centre method)

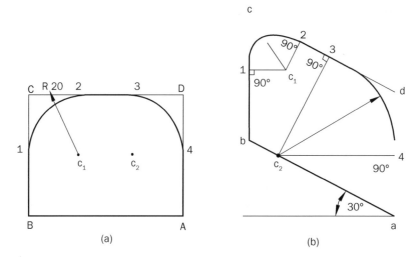

Fig. 16.15 Solution to problem 16.6

16.11 Isometric Drawings or Projections of Prisms, Pyramids, Cylinders and Cones

Problem 16.7 Two views of a block are given in Fig. 16.16 (a). Draw its isometric drawing.

Solution:

(i)   Draw the isometric drawing of the block.

(ii)   Provide the notch at the corner as shown in Fig. 16.16 (*b*).

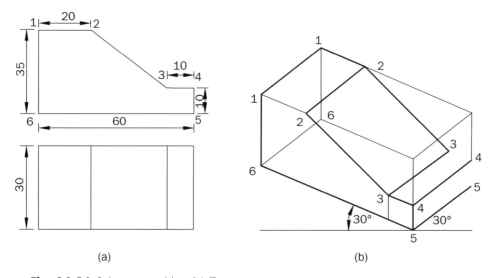

Fig. 16.16 Solution to problem 16.7

Problem 16.8 Draw an isometric projection of the frustum of a right regular hexagonal pyramid, side of base hexagon is 30 mm, side of top hexagon is 15 mm and height of the frustum is 40 mm.

Solution:

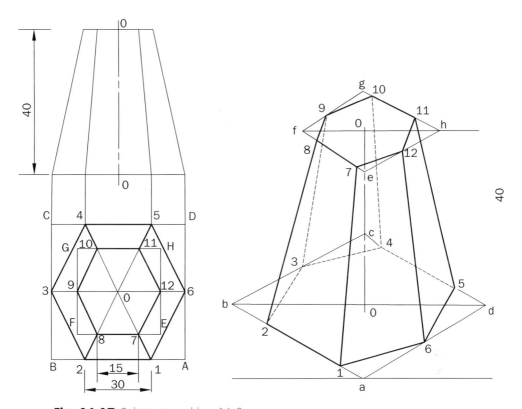

Fig. 16.17 Solution to problem 16.8

(i)   Draw front and top views of the hexagon (using isometric scale) and enclose it in rectangles as shown in Fig. 16.17 (*a*).
(ii) Draw the isometric projection of the enclosing boxes and locate the corners of the two hexagon bases.
(iii) Join the corners and compete the isometric projection as shown in Fig. 16.17 (*b*).

Problem 16.9 Front and top views of a right circular cylinder are shown in Fig. 16.18 (a). Draw the isometric drawing of the solids.

Solution: The solution to this problem is self-explanatory. See Fig. 16.18 (*b*).

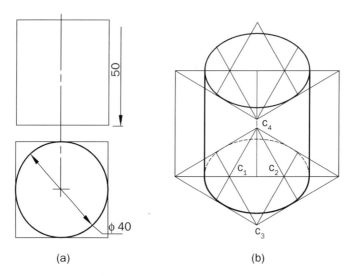

(a)   (b)

Fig. 16.18 Solution to problem 16.9

Problem 16.10 Draw the isometric projection of a cone, base 50 mm diameter and axis 60 mm long when its axis is vertical.

Solution:

    (i)   Draw an ellipse for the base (using isometric scale)
    (ii)   Determine the position of apex or vertex.
    (iii) Draw tangents to the ellipse from the apex as shown in Fig. 16.19.

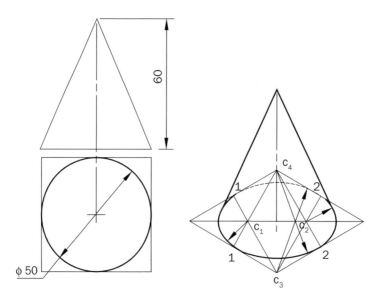

Fig. 16.19 Solution to problem 16.10

Problem 16.11 Draw the isometric drawing of the frustum of a cone as shown in Fig. 16.20 (a).

Solution: The solution to this problem is self-explanatory. See Fig. 16.20 (*b*).

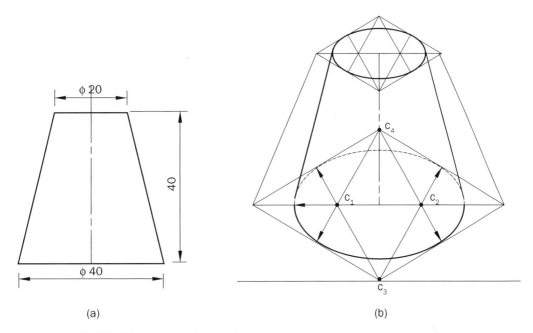

(a)   (b)

Fig. 16.20 Solution to problem 16.11

16.12 Isometric Projection of a Sphere

A sphere appears as a circle of diameter equal to the diameter of sphere when seen from any direction. Hence, the isometric projection of a sphere is also a circle of diameter equal to the diameter of the sphere.

The isometric projection of any curved surface is evidently the envelop of all lines which can be drawn on that surface. The great circles (circles cut by any plane passing through the centre) are the lines drawn on the surface of the sphere and these are nothing but ellipses having equal major axis. Therefore, the major axis is equal to the diameter of the sphere. Hence, the envelope is a circle whose diameter is equal to the diameter of the sphere as shown in Fig. 16.21.

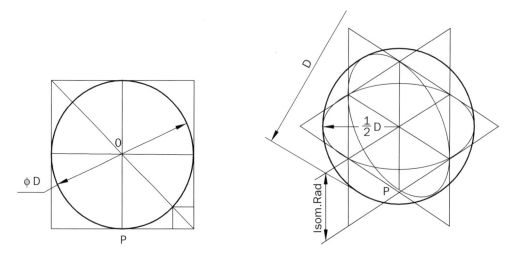

Fig. 16.21 Isometric projection of a sphere

Problem 16.12 Draw the isometric projection of a sphere resting centrally on the top of a square block, the front view of which is shown in Fig. 16.22 (a).

Solution:

(i)   Draw the isometric projection (using isometric scale) of the square block and locate the centre of its top surface.
(ii) Mark the centre of the sphere, such that it is equal to the isometric radius of the sphere.
(iii) With centre of the sphere and radius equal to the radius of the sphere, draw a circle which will be the isometric projection of the sphere, as shown in Fig. 16.22 (b).

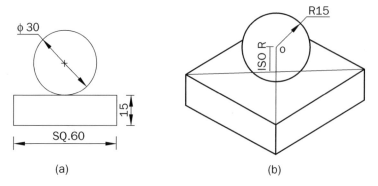

(a)   (b)

Fig. 16.22 Solution to problem 16.12

Additional Problems

Problem 16.13 A cube of 40 mm edge is placed centrally on the top of a square block of 60 mm edge and 20 mm thick. Draw the isometric projections of the two solids with the edges of the two block mutually parallel to each other.

(PTU, Jalandhar May 2004, May 2005, May 2007)

Solution: The solution to this problem is self-explanatory. See Fig. 16.23.

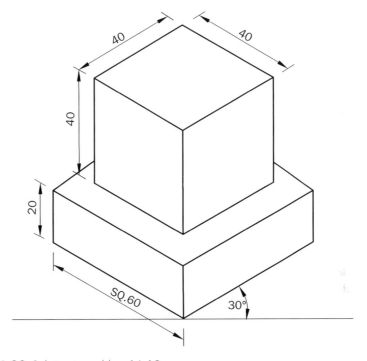

Fig. 16.23 Solution to problem 16.13

Problem 16.14 Draw the isometric projection of an object as shown in Fig. 16.24 (a).

Solution: The solution to this problem is self-explanatory. See Fig. 16.24 (b).

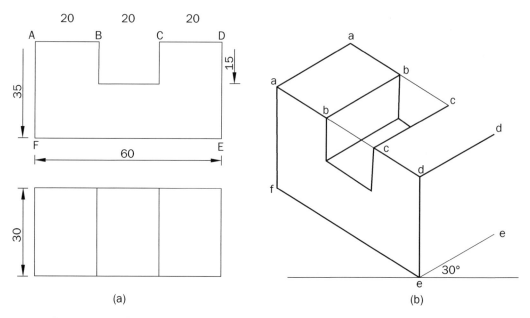

Fig. 16.24 Solution to problem 16.14

Problem 16.15 A right circular cone of φ 30 mm base and height 40 mm rests centrally on the top of a square block of 40 mm side and 15 mm thick. Draw the isometric projections of the solids.

Solution:

(i)   Draw the isometric projection of the block (using isometric scale).
(ii) Draw the rhombus systematically about the centre point of the top surface of the block as shown in Fig. 16.25.

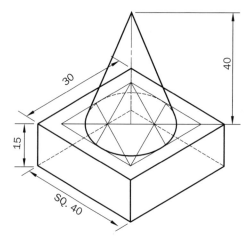

Fig. 16.25 Solution to problem 16.15

(iii)   Draw the ellipse for the cone base.
(iv)   Determine the position of apex or vertex.
(v)   Draw tangents to the ellipse from the apex or vertex.
(vi)   Complete the view by making the visible postions of the lines firm.

Problem 16.16 A sphere of φ 30 mm rests centrally on the top of a cube of 30 mm side. Draw the isometric projections of the solids.

(PTU, Jalandhar May 2011)

Solution: The solution to this problem is self-explanatory. See Fig. 16.26.

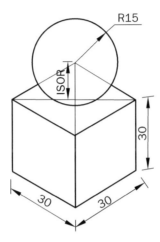

Fig. 16.26 Solution to problem 16.16

Problem 16.17 One view of an object is shown in Fig. 16.27 (a). Draw its isometric projection.

Solution: The solution to this problem is self-explanatory. See Fig. 16.27 (b).

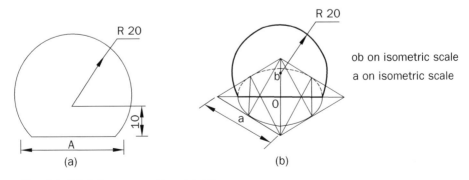

Fig. 16.27 Solution to problem 16.17

Problem 16.18 Three cubes of 40 mm, 30 mm and 20 mm are placed centrally such that the biggest cube at the bottom whereas the smallest on the top. Draw the isometric drawing of the solids.

(PTU, Jalandhar, June 2003, May 2011)

Solution: The solution to this problem is self-explanatory. See Fig. 16.28.

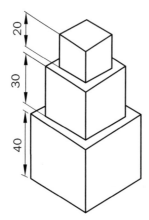

Fig. 16.28 Solution to problem 16.18

Problem 16.19 Two cubes of sides 20 mm and 30 mm are resting one upon another such that their vertical axes are in same line. Draw the isometric projection of the cubes assuming small cube is resting on bigger one.

(PTU, Jalandhar December 2004)

Solution: The solution to this problem is self-explanatory. See Fig. 16.29

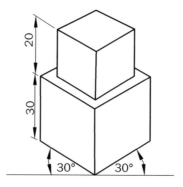

Fig. 16.29 Solution to problem 16.19

Problem 16.20 A cube of 30 mm side rests on the top of a cylindrical slab of 60 mm diameter and 25 mm thick. The axes of the solids are in same straight line. Draw an isometric projection of the solid.

(PTU, Jalandhar December 2004, May 2005, May 2010)

Solution:

(i)   Draw the isometric projection of the cylindrical slab using the isometric scale by four centre method.

(ii) Draw a cube of side 30 mm on the top of the cylindrical slab as shown in Fig. 16.30, such that the axes of the two solids are in a straight line.

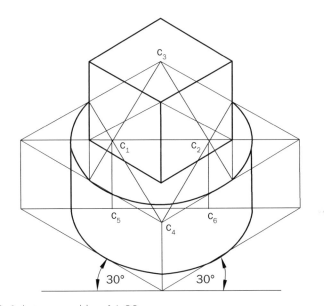

Fig. 16.30 Solution to problem 16.20

Problem 16.21 A cylindrical block of 45 mm diameter and 25 mm height is placed centrally on a cube of 45 mm side. The axes of the two solids are in the same straight line. Draw the isometric drawing of the solids.

(PTU, Jalandhar May 2004)

Solution: All construction lines are retained to help in understanding the solution. See Fig. 16.31.

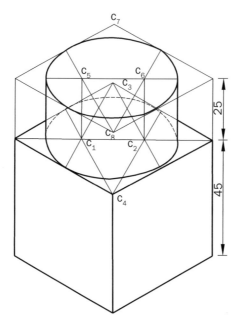

Fig. 16.31 Solution to problem 16.21

Problem 16.22 A hemi-sphere of 40 mm diameter rests on its circular base on the top of a cube of 40 mm. Draw the isometric projection of the solids.

(PTU, Jalandhar December 2003)

Solution:

(i)   Draw the isometric projection of the cube using the isometric scale.
(ii)   Draw the ellipse for the circular base of the hemisphere, using the four-centre method.
(iii)   Draw a circular arc of 20 mm radius tangential to the ellipse as shown in Fig. 16.32.

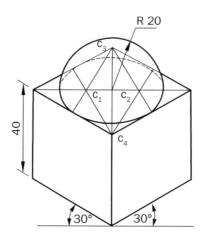

Fig. 16.32 Solution to problem 16.22

Problem 16.23 A square prism of side 30 mm and 40 mm high is resting on HP. A vertical square base of 10 mm side is cut through its face reaching other square face of the prism. Draw the isometric projection of the prism.

(*PTU, Jalandhar December 2003, December 2006, May 2008*)

Solution:

(i) Draw the isometric projection of the square prism using the isometric scale.

(ii) Draw a square base of 10 mm vertically as shown in Fig. 16.33. Complete the problem showing hidden edges by dotted lines.

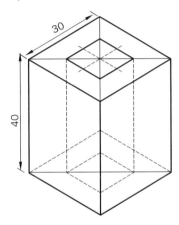

Fig. 16.33 Solution to problem 16.23

Problem 16.24 A right circular cone of φ 20 mm base and height 30 mm rests centrally on the top of a cube of 40 mm side. Draw the isometric projection of the two solids.

Solution: All the construction lines are retained to make the understanding of the solution easily. See Fig. 16.34.

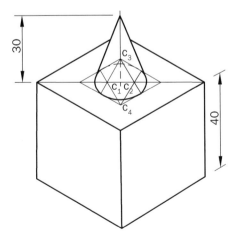

Fig. 16.34 Solution to problem 16.24

Problem 16.25 A cylindrical slab 60 mm diameter and 20 mm thick is surmounted by a cube of 28 mm edge. On the top of a cube, rests a square pyramid, attitude 30 mm and side of base 15 mm. The axes of the two solids are in the same straight line. Draw isometric projections of the solids.

(*PTU, Jalandhar December 2009*)

Solution: For its solution, see Fig. 16.35.

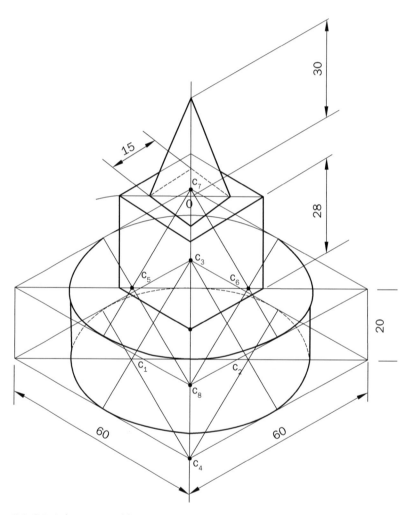

Fig. 16.35 Solution to problem 16.25

Problem 16.26 Draw the isometric projection of the three bricks of size 30 mm × 30 mm × 15 mm from the given front view and top view as shown in Fig. 16.36 (a).

Solution: For its solution, see Fig. 16.36 (*b*).

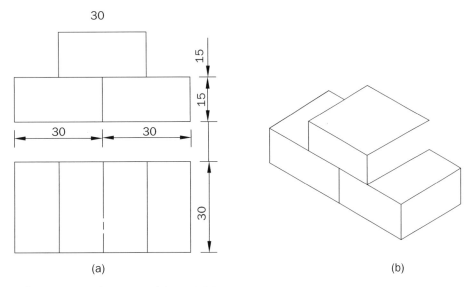

(a)   (b)

Fig. 16.36 Solution to problem 16.26

Problem 16.27 Front view of a given figure, 16.37 (a), is shown. Draw its isometric projection.

Solution: For its solution, see Fig. 16.37 (b).

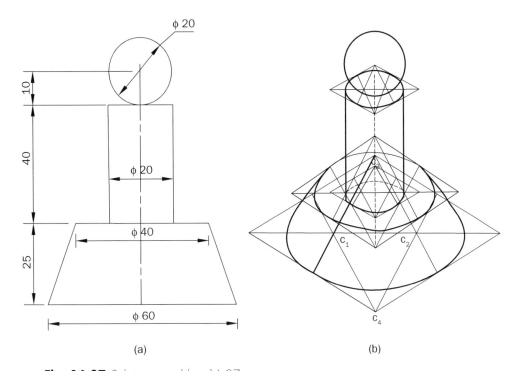

(a)   (b)

Fig. 16.37 Solution to problem 16.27

Problem 16.28 Figure 16.38 (a) shows three views of a cube cut by an oblique plane. Draw its isometric drawing.

Solution: For its solution, see Fig. 16.38 (*b*).

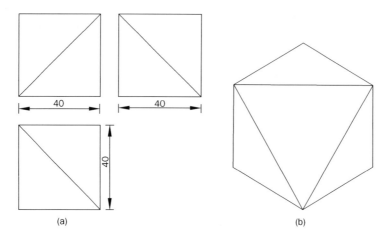

(a)   (b)

Fig. 16.38 Solution to problem 16.28

Problem 16.29 Figure 16.39 (a) shows the front view of a given figure. Draw its isometric projection.

Solution: For its solution, see Fig. 16.39 (*b*).

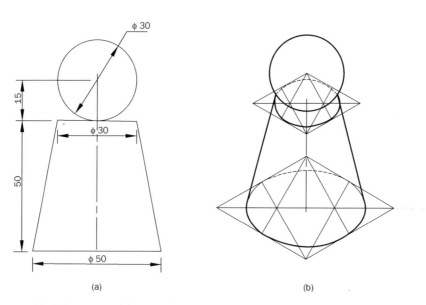

(a)   (b)

Fig. 16.39 Solution to problem 16.29

Problem 16.30 A sphere of 60 mm diameter is placed centrally on the top of a frustum of a square pyramid. The base of the frustum is 60 square, top 40 square and its height is 50 mm. Draw the isometric projection of the arrangement.

(PTU, Jalandhar May 2001)

Solution: For its solution, see Fig. 16.40.

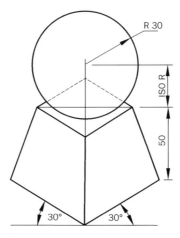

Fig. 16.40 Solution to problem 16.30

Problem 16.31 A right regular hexagonal prism, edge of base 20 mm and height 50 mm, has a circular hole of 20 mm diameter drilled centrally through it, along its axis. Draw its isometric projection.

(PTU, Jalandhar December 2003)

Solution: For its solution, see Fig. 16.41.

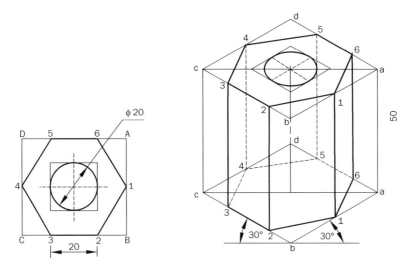

Fig. 16.41 Solution to problem 16.31

Problem 16.32 Draw isometric drawing of a pentagonal prism of base side 25 mm and axis 50 mm long resting on HP on one of its rectangular faces with its axis perpendicular to the VP.

Solution: The solution to this problem is self-explanatory. See Fig. 16.42.

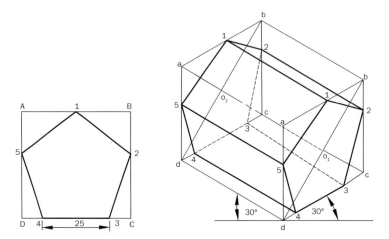

Fig. 16.42 Solution to problem 16.32

Problem 16.33 Draw isometric projection of a hexagonal prism of side of base 30 mm and height 50 mm surmounting a square pyramid of side 20 mm and height 45 mm such that the axes of the two solids are collinear and at least one of the edges of the two solids is parallel.

Solution: All the construction lines are retained to make the solution self–explanatory. See 16.43.

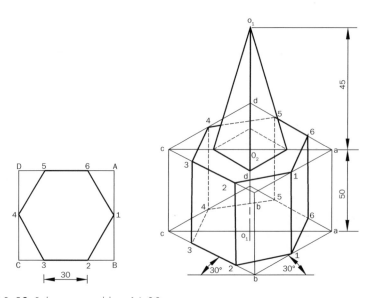

Fig. 16.43 Solution to problem 16.33

Problem 16.34 Draw isometric drawing of a funnel consisting of a cylinder and a frustum of a cone. The diameter of the cylinder is 30 mm and top diameter of the frustum is 50 mm. The height of the frustum of the cone and cylinder are both equal to 40 mm.

Solution:

  (i)   Draw isometric drawing of the cylinder.
  (ii) Draw a frustum of a cone as shown in Fig. 16.44. Complete the problem showing hidden edges by dotted lines.

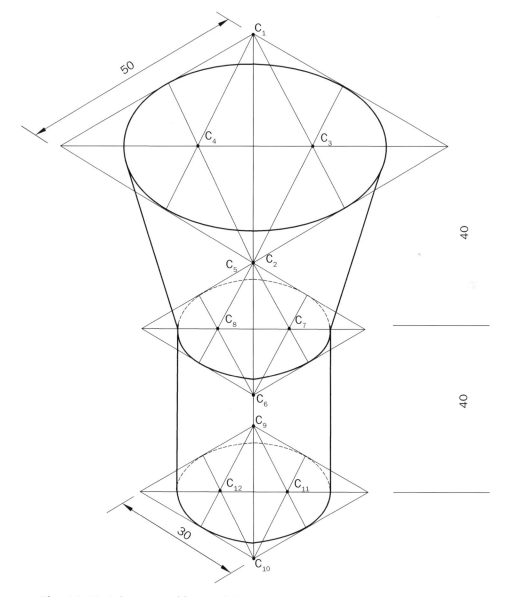

Fig. 16.44 Solution to problem 16.34

Problem 16.35 A right regular pentagonal prism, edge of base 30 mm and height 50 mm, has a circular hole of diameter 25 mm, drilled centrally through it, along its axis. Draw its isometric projection.

Solution:

    (i)   Draw the top view of the solid using the isometric scale.

    (ii)   Enclose the top view in a rectangular box.

    (iii) Locate the various positions of the corner points of the pentagon.

    (iv) Add the ellipse for the circular hole as shown in Fig. 16.45 and complete the isometric projection.

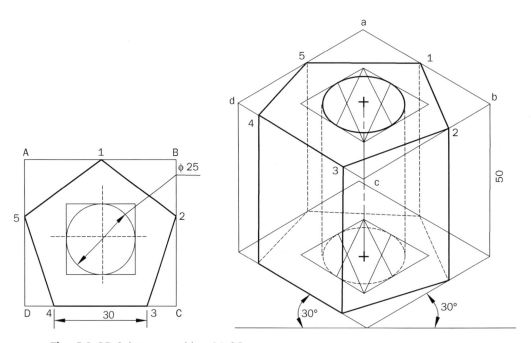

Fig. 16.45 Solution to problem 16.35

Problem 16.36 A square prism of side 40 mm and height 65 mm is resting on ground. A vertical hole of diameter 20 mm is cut through from top face reaching bottom face of the prism. Draw the isometric projection of the prism.

(PTU, Jalandhar December 2005)

Solution: The interpretation of the solution is left to the reader. See Fig. 16.46.

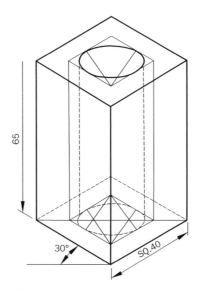

Fig. 16.46 Solution to problem 16.36

Problem 16.37 Draw the isometric projection of a sphere of radius 30 mm resting centrally on the top of the square prism of side 40 mm and height 60 mm.

(PTU, Jalandhar May 2006)

Solution: The interpretation of the solution is left to the reader. See Fig. 16.47.

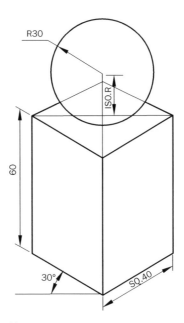

Fig. 16.47 Solution to problem 16.37

Problem 16.38 A pentagonal prism is placed on the square slab 60 mm and 20 mm height. The side of prism is 25 mm and height 50 mm. Draw its isometric projections.

Solution: All the construction lines are retained to make the understanding of the solution easily. See Fig. 16.48.

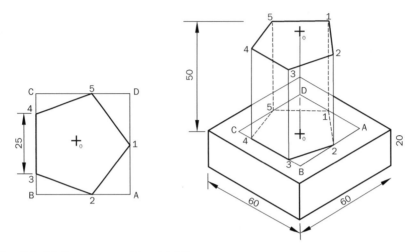

Fig. 16.48 Solution to problem 16.38

Problem 16.39 Draw isometric projections of a cylindrical block of 50 mm diameter and 20 mm thicknesses having a cube of 25 mm side resting centrally on top of it, which in turn is having a sphere of 25 mm diameter resting centrally on top of it.

(PTU, Jalandhar December 2007)

Solution: All construction lines are retained to make the solution self-explanatory. See Fig. 16.49.

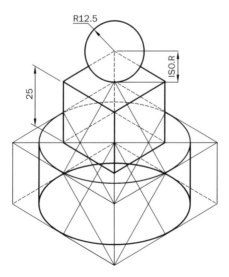

Fig. 16.49 Solution to problem 16.39

Problem 16.40 Draw the isometric projection from the orthographic projection of the block as shown in Fig. 16.2 (a).

Solution: The interpretation of the solution is left to the reader. See Fig. 16.50.

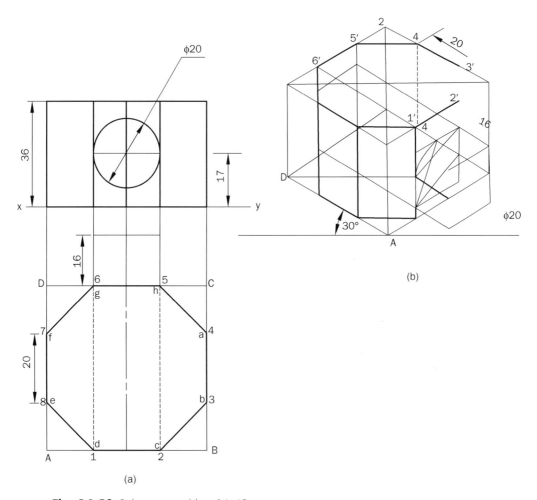

Fig. 16.50 Solution to problem 16.40

Problem 16.41 Draw isometric drawing of a hexagonal prism of base side 25 mm and axis 50 mm long (a) resting on HP on one of its rectangular faces with its axis perpendicular to the VP and (b) when its axis is kept perpendicular to HP.

Solution: All the construction lines are retained to make the solution self-explanatory. See Fig. 16.51.

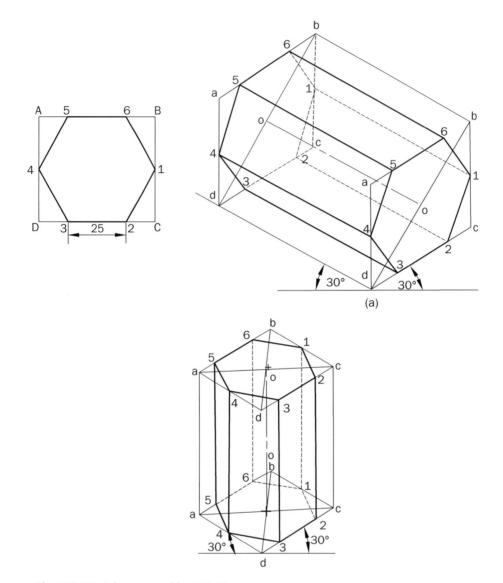

Fig. 16.51 Solution to problem 16.41

Problem 16.42 Draw isometric drawing of a hexagonal pyramid of base side 30 mm and axis 60 mm long with its axis kept perpendicular to HP.

Solution: The interpretation of the solution is left to the reader. See Fig. 16.52.

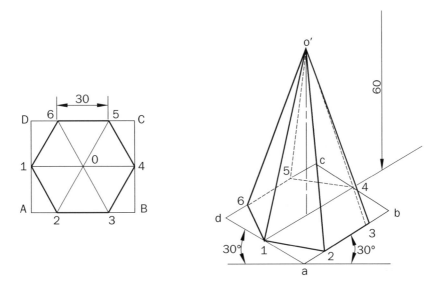

Fig. 16.52 Solution to problem 16.42

Problem 16.43 Draw isometric drawing of a frustum of a square pyramid; side of lower base is 40 mm, side of upper base is 20 mm and height of the frustum is 40 mm.

Solution: All the construction lines are retained to make the solution self-explanatory. See Fig. 16.53.

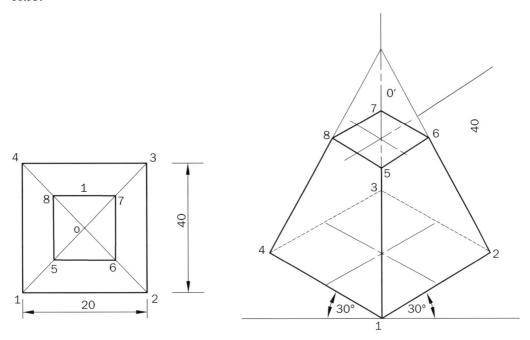

Fig. 16.53 Solution to problem 16.43

Problem 16.44 Draw the isometric drawing of a given frustum of a pentagonal pyramid as shown in Fig. 16.54 (a).

Solution: The solution to this problem is self-explanatory. See Fig. 16.54 (b)

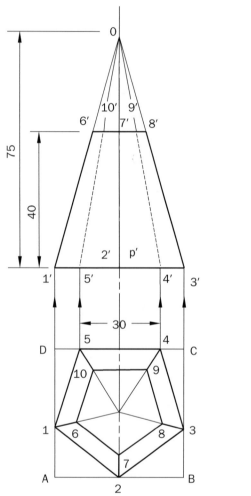

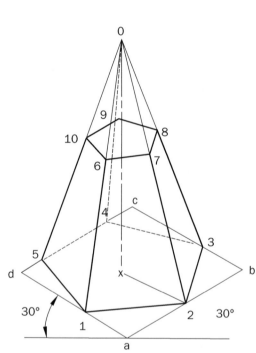

Fig. 16.54 Solution to problem 16.44

Exercises

Isometric Projections of Plane Figures

16.1 Draw the isometric projection of a regular pentagon of 25 mm side.

16.2 Draw the isometric projection of a circle of 60 mm diameter by both the methods.

16.3 Draw the isometric projection of a regular hexagonal of 25 mm side.

Isometric Projections of Prism, Pyramid, Cylinder and Cone

16.4 Draw an isometric drawing of the frustum of a right regular pentagonal pyramid; side of base hexagon is 25 mm, side of top hexagon is 15 mm and height of the frustum is 50 mm.

16.5 Draw the isometric drawing of a cylinder, base diameter 50 mm and height 50 mm long when its axis is (a) horizontal and (b) vertical.

16.6 Draw an isometric projection of the frustum of a right circular cone; base diameter 50 mm, top diameter 25 mm and height of the frustum is 40 mm.

16.7 A cube of 30 mm edge is placed centrally on the top of a square block of 50 mm edge and 15 mm thick. Draw the isometric projections of the two solids.

16.8 A right circular cone, diameter of base 40 mm and height 50 mm, rests centrally on the top of a cube of 50 mm. Draw the isometric projections of the solids.

16.9 A square prism of side 40 mm and 50 mm high is resting on HP. A vertical circular hole of 20 mm diameter is cut through its base reaching other side of the prism. Draw the isometric drawing of the solids.

16.10 A right regular hexagonal prism, side of base 20 mm and 50 mm long, lies on its rectangular face on HP. A right circular cylinder, diameter of base 30 mm and 45 mm long, rests centrally on the top rectangular surface of the prism. Draw the isometric projections of the solids.

16.11 A right regular hexagonal prism, side of base 20 mm and height 50 mm, lies on one of its rectangular faces. A right circular cone, diameter of base 30 mm and height 40 mm, rests centrally on the top rectangular surface of the prism. Draw the isometric drawing of the solids.

Isometric Projection of a Sphere

16.12 A sphere of φ 40 mm rests centrally on the top of a square block of 50 mm side and 15 mm thick. Draw the isometric projections of the solids.

16.13 A hemi-sphere of 30 mm diameter rests on its circular base on the top of a square block of 30 mm side and 15 mm thick. Draw the isometric projections of the solids.

16.14 A sphere of 50 mm diameter is placed centrally on the top of a cylinder of 60 mm base diameter and 50 mm high. Draw the isometric projection of the solids.

Additional Problems

16.15 Figures 16.53 to 16.61 show the orthographic projections of centain objects. Draw the isometric drawing of each.

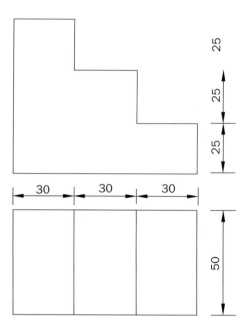

Fig. 16.55

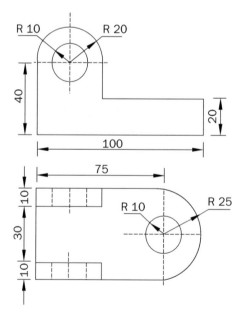

Fig. 16.56

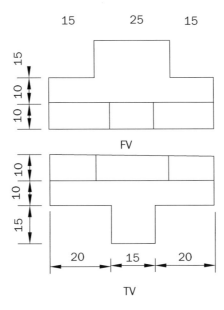

Fig. 16.57

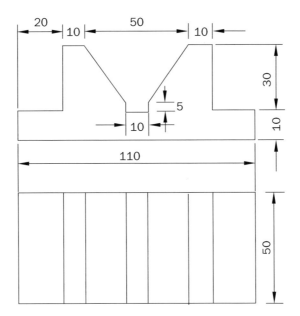

Fig. 16.58

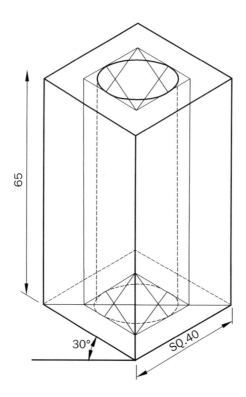

65

30°   SQ.40

Fig. 16.59

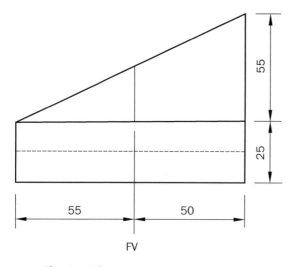

55

25

55   50

FV

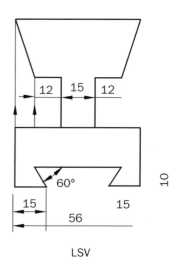

12   15   12

60°

15   15

56

10

LSV

Fig. 16.60

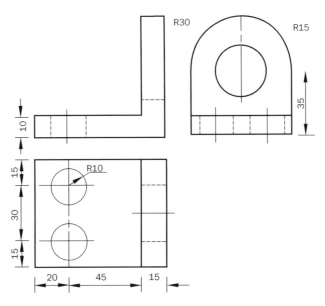

Fig. 16.61

Objective Questions

16.1 The isometric length is in the ratio of .... of the true length.

16.2 In isometric projections, dimensions lines are drawn parallel to ....

16.3 A circle in isometric projection appears as ....

16.4 Isometric projection of a sphere is a circle having a diameter .... of sphere.

16.5 The three forms of axonometric projection are dimetric, trimetric and .... projections.

16.6 What is the difference between isometric projection and isometric drawing?

16.7 Describe the four centre method of drawing isometric projection of a circle.

16.8 Define isometric axes, isometric lines and isometric planes.

16.9 Define isometric scale and how it is constructed.

16.10 What is the purpose of pictorial drawings?

16.11 Draw the isometric projections of (i) a circle and (ii) a square.

16.12 Give the various position of isometric axes.

16.13 How is isometric projection of an object obtained?

16.14 Draw a triangle of sides 40, 50 and 60 mm and draw its isometric projection considering it as top view.

16.15 Draw the isometric projection of a square considering it as (i) front view and (ii) top view.

Answers

16.1 $\sqrt{2} : \sqrt{3}$ 16.3 An ellipse   16.5 Isometric

16.2 Isometric axes   16.4 Equal to that

Chapter 17

Conversion of Pictorial Views into Orthographic Views

17.1 Introduction

Although a pictorial view helps in understanding the shape of the object, however it suffers from the drawback that it fails to convey the actual size and inner details of the object. This is because a pictorial view is drawn by seeing the object, as a three directional task. However, for design purpose, it requires the actual details of the object. For this purpose, the pictorial view of the object is converted into orthographic views by applying the principles of orthographic projections. Conversion of a pictorial view into the orthographic view requires sound knowledge of the principles of pictorial projection and some imagination.

17.2 Direction of Sight

For converting a pictorial view of an object into orthographic views, the direction from which the object is to be viewed for its front view is generally indicated by means of an arrow. The arrow must be parallel to the sloping axis. If there is no arrow, the direction for the front view may be decided to give the most prominent view; other views are drawn by looking in the directions perpendicular to the first direction.

17.3 Orthographic Views

Orthographic views can be drawn by two methods:

(i)   First-Angle Projection Method
(ii)   Third-Angle Projection Method

(i)   **First-Angle Projection Method**
In the first-angle projection method, the object lies in the first quadrant, i.e., above the HP and in front of the VP. The object lies in between the observer and the plane of projection. In this method, when the views are drawn in their relative positions, the top view is placed below the

front view and the left side view is placed to the right side of the front view. Thus in the first-angle projection method, either of the side views is so placed that it represents the side view of the object away from it, as shown in Fig. 17.1.

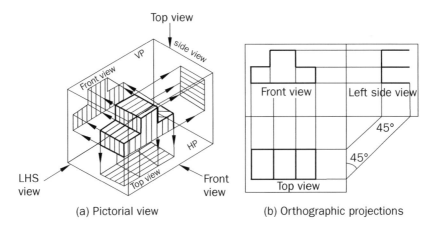

(a) Pictorial view   (b) Orthographic projections

Fig. 17.1 First-angle projection method (three views)

In the same way as discussed above, three more views may be obtained by placing the plane of projection on the front, top and left hand side of the object. The three views then obtained are called rear or back view, bottom view and right side view, respectively. The layout of all the six views on the drawing sheet is shown in Fig. 17.2.

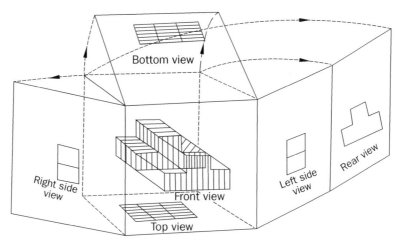

(a) Orthographic projections of unfolded non-transparent box

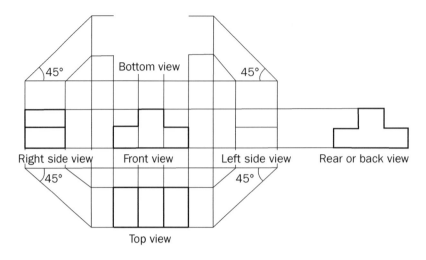

(b) Six views in first-angle projection method

Fig. 17.2 First-angle projection method (six views)

(ii)   Third-angle projection method

In the third-angle projection method, the object is assumed to be kept in the third quadrant, i.e., below the HP and behind the VP. The planes of projection are assumed to be transparent. In this case, the plane of projection lies in between the object and the observer. The three views, after the planes of projection are rotated, are shown in Fig. 17.3. Thus in third-angle projection method, either of the side view is so placed that it represents the side of the object nearer to it.

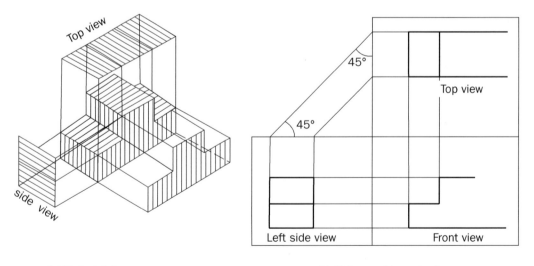

(a) Pictorial view   (b) Orthographic projections

Fig. 17.3 Third-angle projection method (three views)

In the same way as discussed above, three more views can be obtained by placing the picture plane on the back, bottom and on the right hand side of the object. The three views thus obtained are called rear or back view, bottom view and right side view, respectively. The layout of all the six views on the drawing sheet is shown in Fig. 17.4.

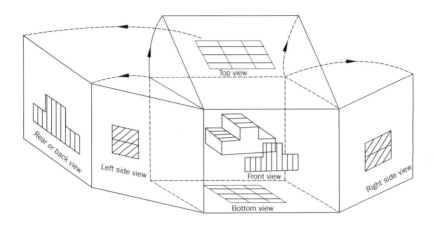

(a) Orthographic projections of unfolded transprent box

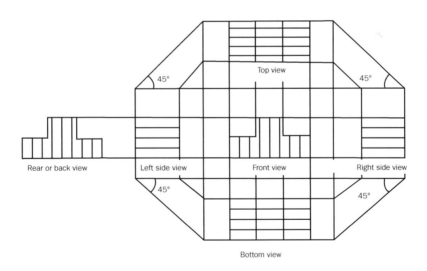

(b) Six view in third-angle projection method

Fig. 17.4 Third-angle projection method (six views)

17.4 Spacing of Views

Before commencing a drawing, it is of utmost importance to obtain an idea of the space required for different views on the drawing sheet. Generally, three views of the object are sufficient to describe it completely, so spacing for three views will be discussed both for first- and third-angle projection methods. In Fig. 17.5, the space A and *A* should be kept equal. Space B should be equal to or slightly greater than space A. Space C and *C* should be kept nearly equal. Space D should be equal to or slightly greater than space C. Outline for a three view drawing may be changed depending upon the requirement of the drawing available.

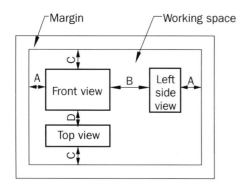

First-angle projection method

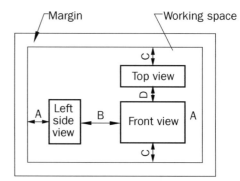

Third-angle projection method

Fig. 17.5 Spacing of views

17.5 Procedure for Preparing Orthographic Views

A systematic procedure is required for preparing the orthographic views, which will result in saving of time. The following basic steps to be noted while preparing the orthographic views of an object, as shown in Figs. 17.6 and 17.7.

- Determine overall dimensions of the required views. Select a suitable scale so that the required views can be accommodated conveniently on the drawing sheet.
- Draw rectangles for different views and keep sufficient space between them as per discussion in Section 17.4.
- Draw centre lines in all the views as per the requirement of the part or component of the object.
- Simultaneously, draw all the requisite details in the following order.
    (i) Circles and arc of circles.
    (ii) Straight lines for proper shape of the object.
    (iii) Straight lines, small curves, etc., for minor details.
- After the completion of the required views, erase all unnecessary lines completely and mark the outlines so as to provide nice appearances to the object.
- Provide the dimensions, the scale and the title along with the other required particulars such as notes, etc.
- Check the drawing carefully and see that it is complete in all respect.

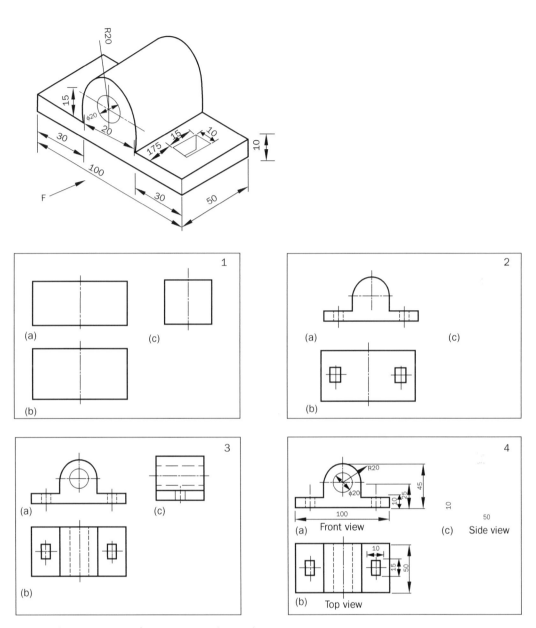

Fig. 17.6 Steps for preparing orthographic views

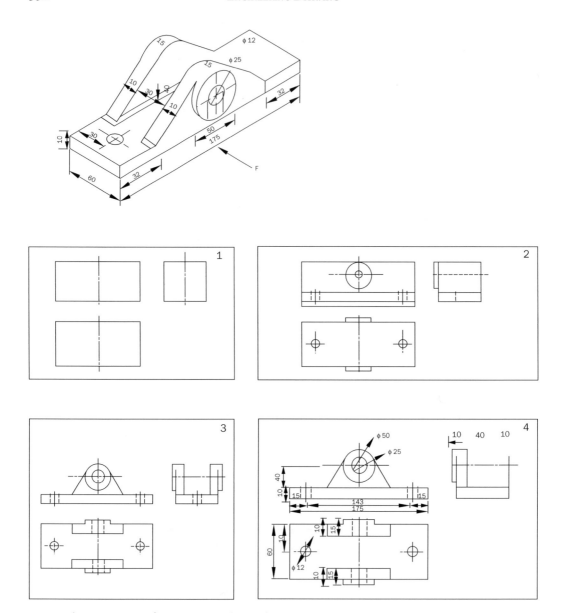

Fig. 17.7 Steps for preparing orthographic views

Problem 17.1 Draw the following views of the object shown pictorially in Fig. 17.8 (a). Use first-angle projection method (i) Front view; (ii) Top view; and (iii) Right side view.

Solution: For its solution, see Fig. 17.8 (*b*).

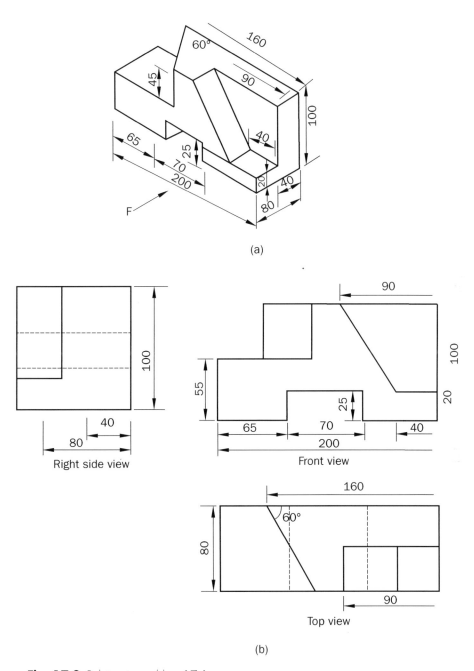

(a)

Right side view

Front view

Top view

(b)

Fig. 17.8 Solution to problem 17.1

Problem 17.2 Draw the following views of the object shown pictorially in Fig. 17.9 (a). (i) Front view; (ii) Top view; and (iii) Right side view.

Solution: See Fig. 17.9 (*b*).

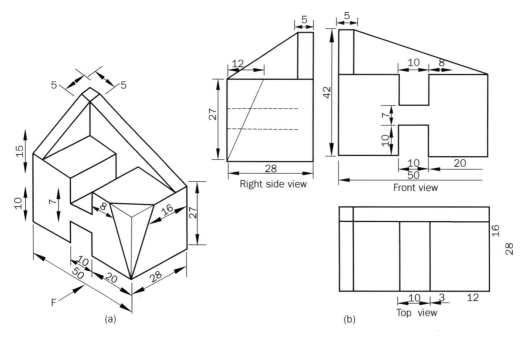

Fig. 17.9 Solution to problem 17.2

Problem 17.3 Fig. 17.10 (a) shows isometric projection of an object. Draw the following views: (i) Front view; (ii) Top view; (iii) Left side view; and (iv) Right side view.

Solution: Refer to Fig. 17.10 (*b*).

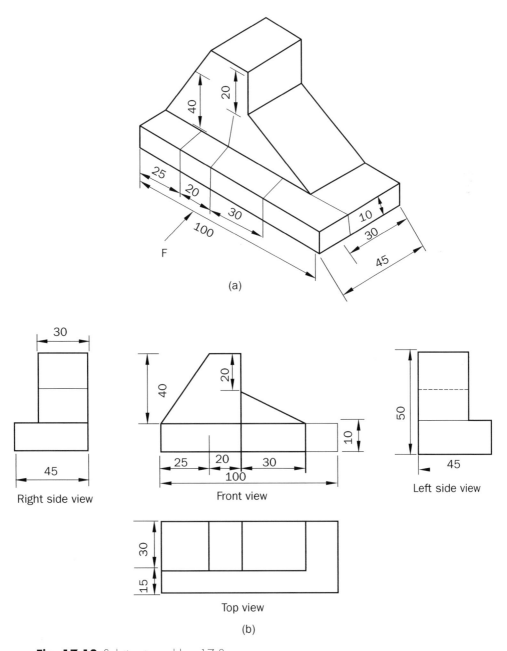

Fig. 17.10 Solution to problem 17.3

Problem 17.4 Draw the following views of the object shown pictorially in Fig. 17.11 (a).
(i) Front view, (ii) Top view, (iii) Right side view.

Solution: Refer to Fig. 17.11 (*b*).

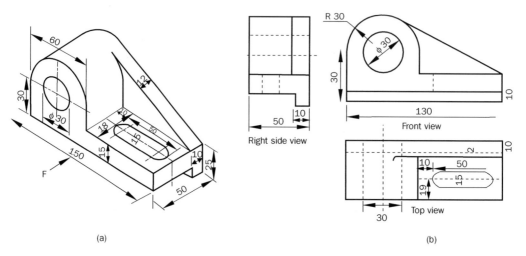

(a)   (b)

Fig. 17.11 Solution to problem 17.4

Problem 17.5 Draw the following views of the object shown pictorially in Fig. 17.12 (a).
(i) Front view (ii) Top view (iii) Left side view (iv) Right side view.

Solution: Refer to Fig. 17.12 (*b*).

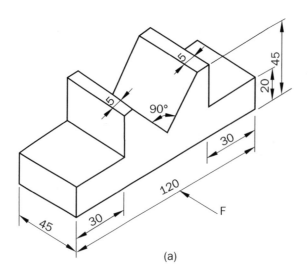

(a)

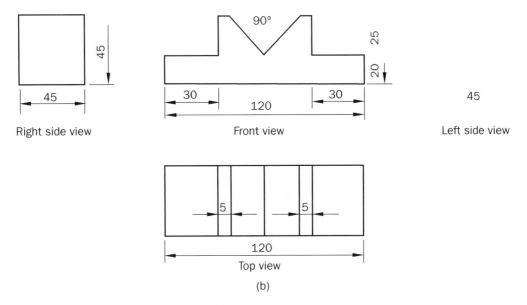

Fig. 17.12 Solution to problem 17.5

17.6 Identification of Surfaces

The art of making the corresponding surfaces of an object from pictorial views to orthographic views and vice-versa is known as the identification of surfaces. For this, surfaces of the object and their corresponding surfaces in the orthographic views are marked by letters or numerals such as *A*, *B*, *C*, etc., or 1, 2, 3, etc.

The identification of surfaces of an object can be had from:

- Pictorial view into orthographic projections
- Orthographic projections into pictorial view.

Problem 17.6 Figure 17.13 (a) shows the pictorial view of an object in which the various surfaces are marked by different letters. Identify and mark various surfaces from the pictorial view to the orthographic projections.

Solution: For its solution, see Fig. 17.13 (*b*).

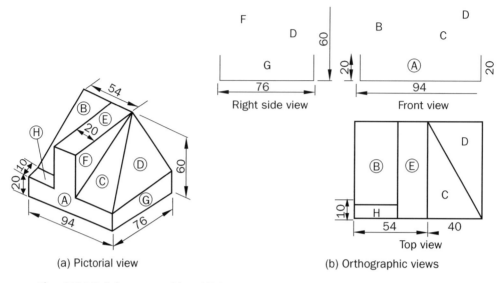

(a) Pictorial view   (b) Orthographic views

Fig. 17.13 Solution to problem 17.6

Problem 17.7 Figure 17.14 (a) shows the pictorial view of an object in which the various surfaces are marked by different alphabets. Identify the mark various surfaces from the pictorial view to the orthographic projections.

Solution: For its solution, see Fig, 17.14 (b).

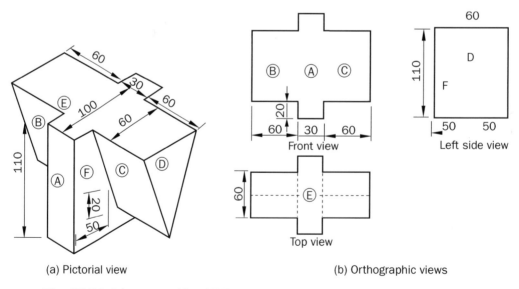

(a) Pictorial view   (b) Orthographic views

Fig. 17.14 Solution to problem 17.7

Problem 17.8 Figure 17.15 (a) shows the pictorial view of an object in which the various surfaces are marked by different letters. Identify the mark various surfaces from the pictorial view to the orthographic projections.

Solution: For its solution, see Fig. 17.15 (b).

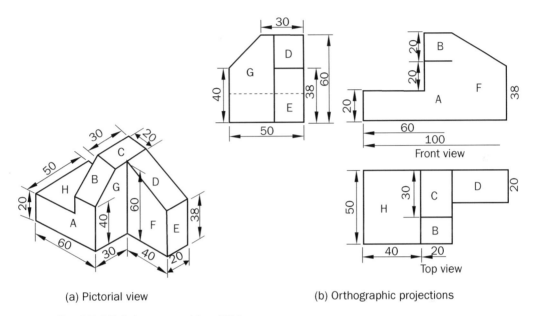

(a) Pictorial view

(b) Orthographic projections

Fig. 17.15 Solution to problem 17.8

17.7 Missing Lines and Missing Views

The lines which are to be added in the given orthographic projections of an object in order to complete the drawing of an object are called missing lines. Similarly, the view which is to be added in the given orthographic projections in order to complete the drawing of an object is called missing view. The principles involved are explained with the help of some solved problems.

Problem 17.9 Figure 17.16 (a) shows the incomplete orthographic projections of the object. Draw the missing lines and complete the orthographic projection.

Solution: Refer to Fig. 17.16 (b).

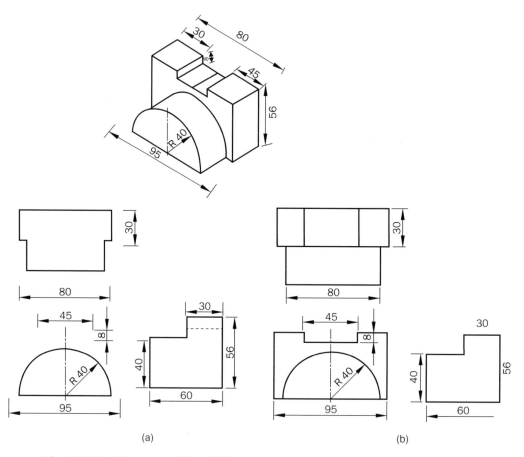

(a)   (b)

Fig. 17.16 Solution to problem 17.9

Problem 17.10 Figure 17.17 (a) shows the incomplete orthographic projections of the object. Draw the missing lines and complete the orthographic projections.

Solution: See Fig. 17.17 (*b*).

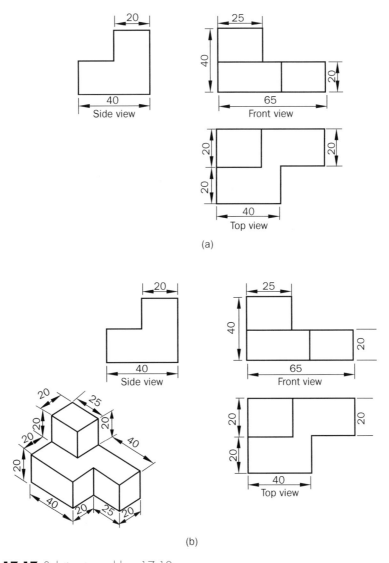

Fig. 17.17 Solution to problem 17.10

Problem 17.11 Figure 17.18 (a) shows the incomplete orthographic projections of the object. Draw the missing view and also draw its isometric projection.

Solution: Refer to Fig. 17.18 (*b*).

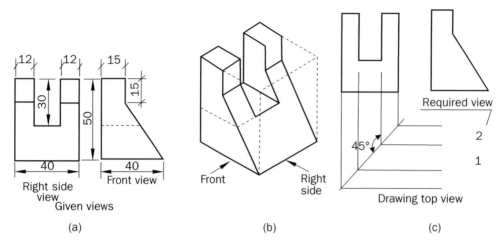

Fig. 17.18 Solution to problem 17.11

Problem 17.12 Figure 17.19 (a) shows the incomplete orthographic projections of the object. Draw the missing view and also draw its isometric projection.

Solution: Refer to Fig. 17.19 (*b*).

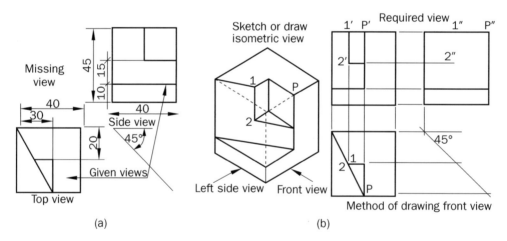

Fig. 17.19 Solution to problem 17.12

Problem 17.13 Figure 17.20 (a) shows the incomplete orthographic projections of the object. Draw the missing view and also draw its isometric projection.

Solution: Refer to Fig. 17.20 (*b*).

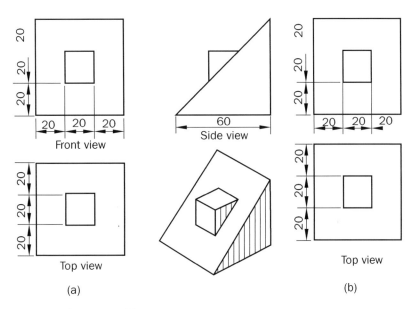

Fig. 17.20 Solution to problem 17.13

Additional Problems

Problem 17.14 Draw front view, top view and right side view of the object shown in Fig. 17.21 (a).

Solution: See Fig. 17.21 (b).

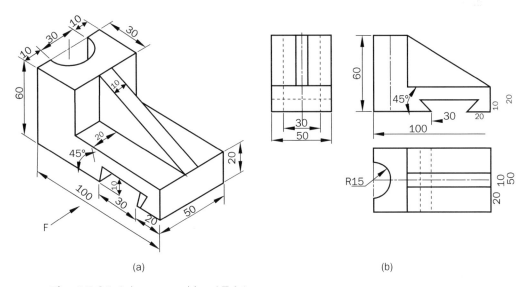

Fig. 17.21 Solution to problem 17.14

Problem 17.15 Draw front view, top view and left side view of the object shown in Fig. 17.22 (a).

Solution: See Fig. 17.22 (b).

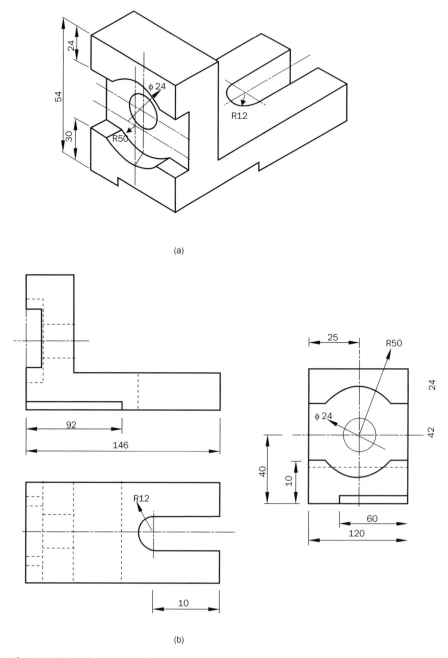

(a)

(b)

Fig. 17.22 Solution to problem 17.15

Problem 17.16 Draw the following views of the object shown pictorially in Fig. 17.23 (a). (i) Front view; (ii) Top view; and (iii) Right side view.

Solution: See Fig. 17.23 (b).

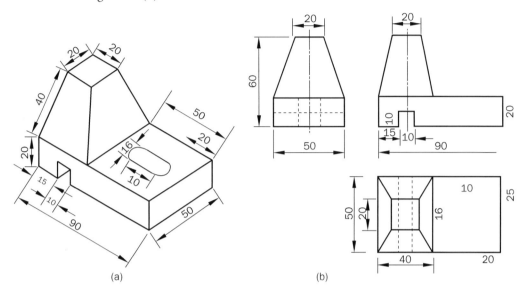

(a)   (b)

Fig. 17.23 Solution to problem 17.16

Problem 17.17 Draw front view, top view and left side view of the object shown in Fig. 17.24 (a).

Solution: Refer to Fig. 17.24 (b).

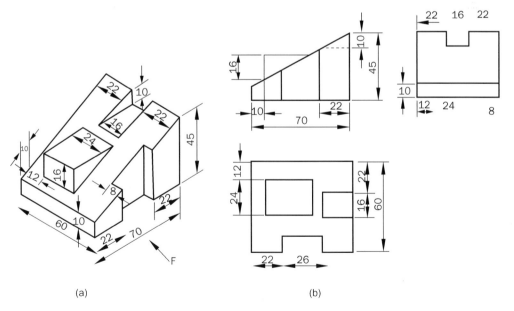

(a)   (b)

Fig. 17.24 Solution to problem 17.17

548 ENGINEERING DRAWING

Problem 17.18 Draw the following views of the object shown pictorially in Fig. 17.25 (a). (i) Front view; (ii) Top view; and (iii) Right side view.

Solution: Refer to Fig. 17.25 (*b*)

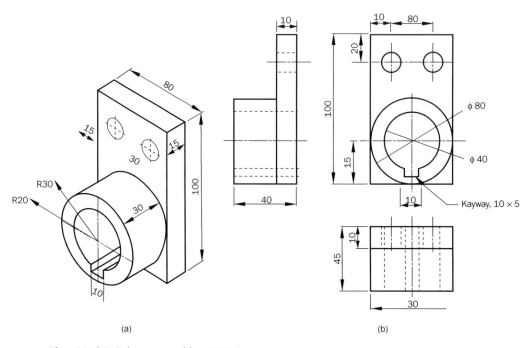

(a)   (b)

Fig. 17.25 Solution to problem 17.18

Problem 17.19 Draw the following views of the object shown in Fig. 17.26 (a). (i) Front view; (ii) Top view; and (iii) Right side view.

Solution: Refer Fig. 17.26 (b).

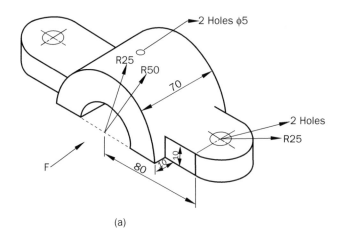

(a)

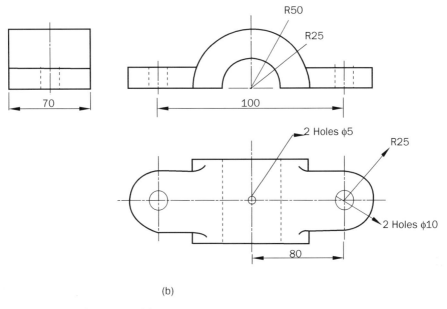

(b)

Fig. 17.26 Solution to problem 17.19

Problem 17.20 Draw the following views of the object shown pictorially is Fig. 17.27 (a). (i) Front view; (ii) Top view; and (iii) Left side view.

Solution: Refer to Fig. 17.27 (b).

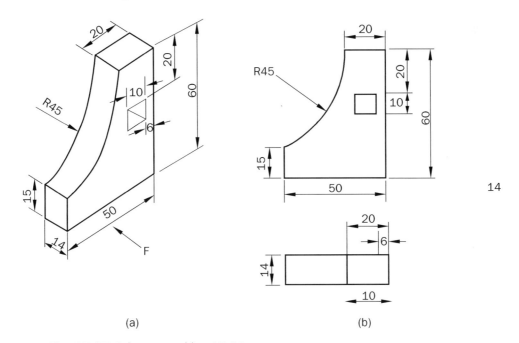

(a)   (b)

Fig. 17.27 Solution to problem 17.20

Problem 17.21 Draw the following views of the object shown pictorially in Fig. 17.28 (a). (i) Front view; (ii) Top view; and (iii) Left side view.

Solution: Refer to Fig. 17.28 (b).

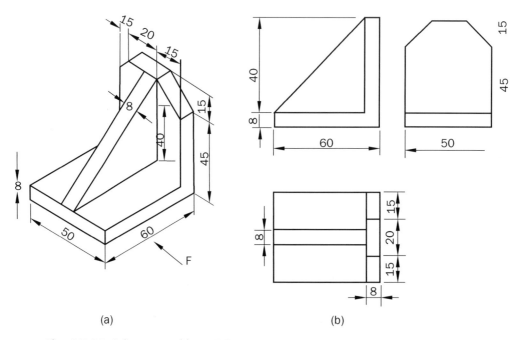

(a)   (b)

Fig. 17.28 Solution to problem 17.21

Problem 17.22 Draw the following views of the object shown pictorially in Fig. 17.29 (a). (i) Front view; (ii) Top view; and (iii) Right side view.

Solution: Refer to Fig. 17.29 (b).

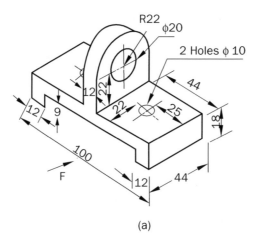

(a)

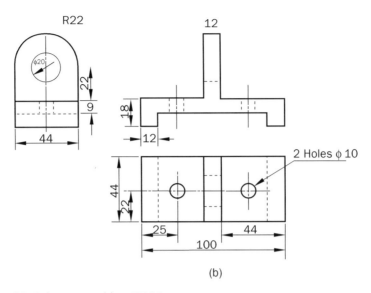

Fig. 17.29 Solution to problem 17.22

Problem 17.23 Draw the following views of the object shown pictorially in Fig. 17.29 (a). (i) Front view; (ii) Top view; and (iii) Left side view.

Solution: Refer to Fig. 17.30 (b).

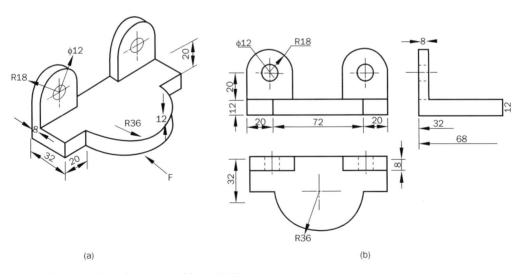

(a)   (b)

Fig. 17.30 Solution to problem 17.23

Exercises

17.1   Figure 17.31 to Fig. 17.43 show the isometric projections of the different objects. Draw the following views of each figure separately (*i*) Front view; (*ii*) Top view; (*iii*) Left side view; and (*iii*) Right side view.

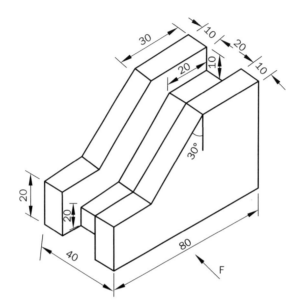

Fig. 17.31

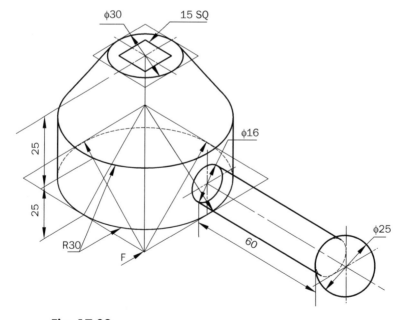

Fig. 17.32

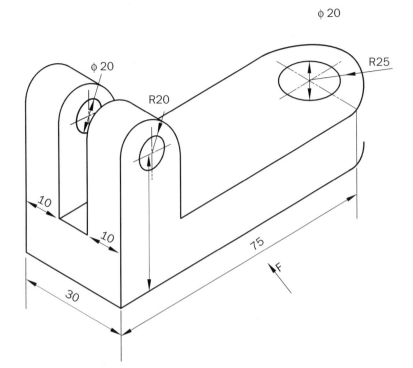

Fig. 17.33

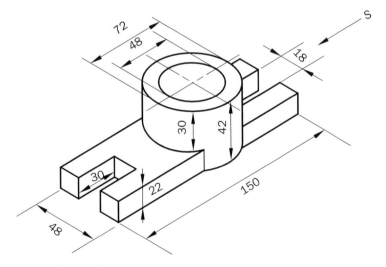

Fig. 17.34

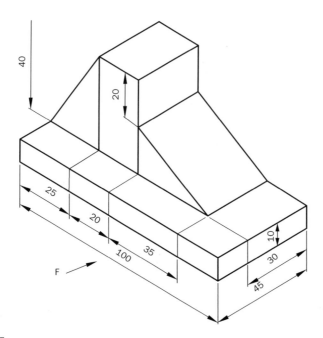

Fig. 17.35

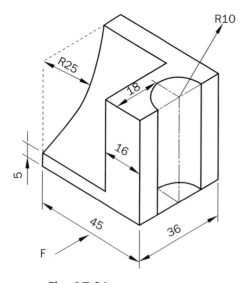

Fig. 17.36

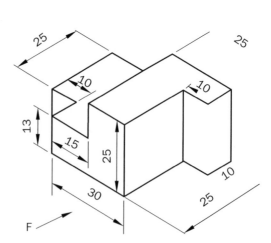

Fig. 17.37

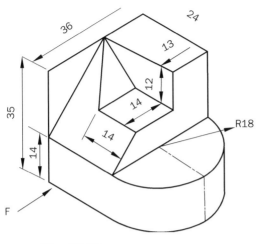

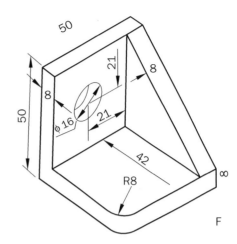

Fig. 17.38

Fig. 17.39

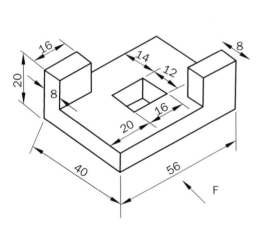

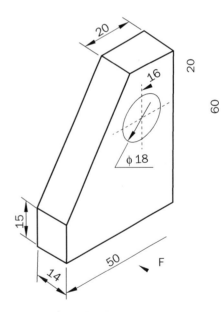

Fig. 17.40

Fig. 17.41

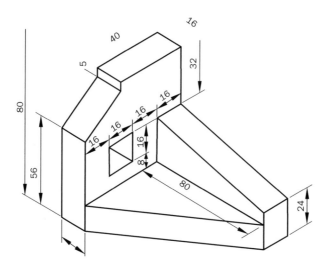

Fig. 17.42

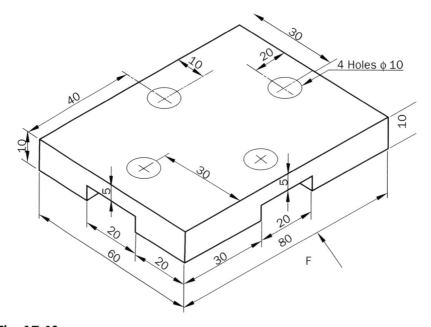

Fig. 17.43

17.2   Figure 17.44 to Fig. 17.47 show the pictorial view of the different objects in which the
       various surfaces are marked by different letters. Identify and mark the various surfaces
       from the pictorial views to the orthographic projections.

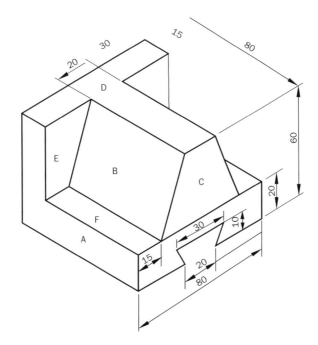

Fig. 17.44

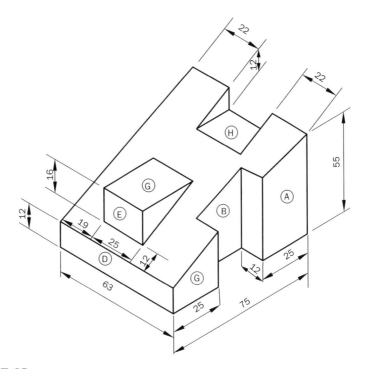

Fig. 17.45

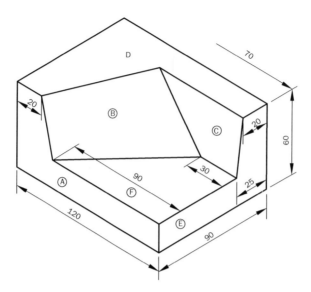

Fig. 17.46

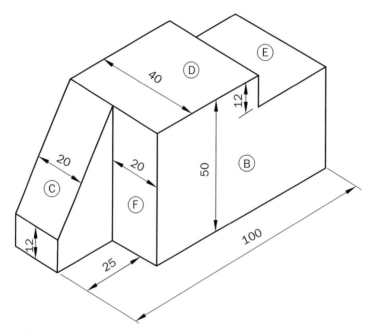

Fig. 17.47

17.3 Figure 17.48 to Fig. 17.52 show the incomplete orthographic projections of the object. Draw the missing views and complete the orthographic projections.

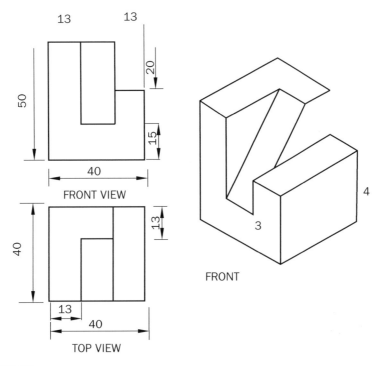

13   13

20

50

15

40

FRONT VIEW

40

13

13

40

TOP VIEW

4

3

FRONT

Fig. 17.48

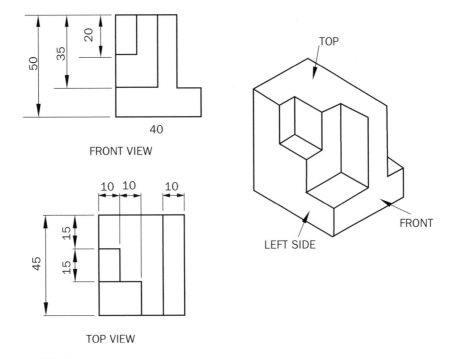

50

35

20

40

FRONT VIEW

10 10   10

45

15

15

TOP VIEW

TOP

LEFT SIDE

FRONT

Fig. 17.49

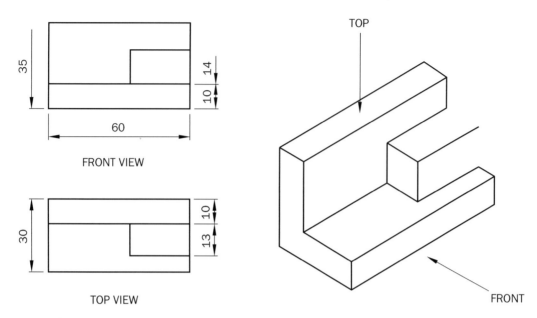

35

14

10

60

FRONT VIEW

30

10

13

TOP VIEW

Fig. 17.50

TOP

FRONT

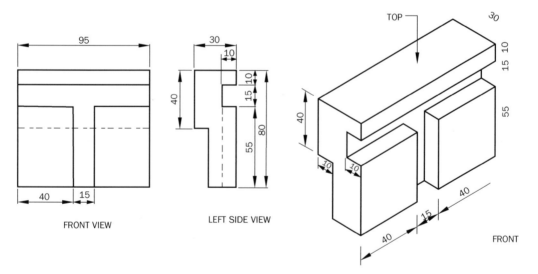

95

40   15

FRONT VIEW

30

10

40

10

15

80

55

LEFT SIDE VIEW

TOP

30

15 10

55

40

10   10

40   15

40

FRONT

Fig. 17.51

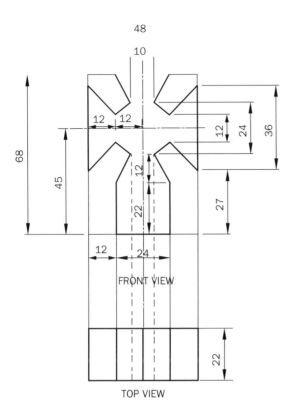

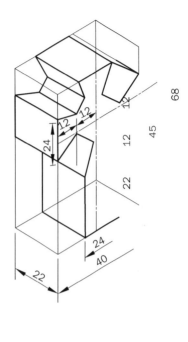

Fig. 17.52

Objective Questions

17.1 In orthographic projections, the lines of sight are .... to the plane of projection.

17.2 Every object is made of number of ....

17.3 What do you mean by identification of surfaces?

17.4 What do you mean by missing lines?

17.5 What is the need of drawing missing view?

17.6 What do you mean by spacing of views?

17.7 What are the different ways of projecting three view drawings?

17.8 What is the process of visualisation of an object?

Answers

17.1 Perpendicular   17.2 Parts

Chapter 18

Freehand Sketching

18.1 Introduction

Freehand sketching is the technique of making drawing without the use of drawing instruments. It is one of the efficient ways employed to communicate ideas and designs. A designer records his ideas critically in the form of sketches which are later converted into drawings.

It is usually assumed that skill in sketching may be acquired more easily than proficiency in instrumental drawings. However, a lot of effort and practice is required to sketch two parallel lines, a circle and an ellipse, etc. Proficiency in sketching can be achieved only with constant practice.

18.2 Sketching Materials

The following materials are required for making good sketches:

(i)   A fairly soft pencil, particularly HB or H
(ii) A soft eraser
(iii) A suitable paper

The pencil used for sketching should have a conical point. A good quality eraser should be used that will not spoil the paper. Depending upon the conditions of work and the purpose of a drawing, a variety of papers are used for sketching. Rectangular or square coordinate papers and isometric ruled papers are generally used by beginners. This helps the beginners in sketching straight lines and keeping a drawing to a good scale. As such, papers may not always be readily available; it is important to start sketching with plain paper too.

18.3 Uses of Sketches

These are as follows:

(i)   To help the designer in developing new ideas
(ii) To convey the ideas of the designer to the draughtsman and management
(iii) To provide a basis for discussion between engineers and workmen
(iv) To serve as a teaching aid in the classroom

18.4 Sketching Straight Lines

Generally, the shape of an object consists of straight and curved surfaces which are represented by straight and curved lines respectively. The lines may be either horizontal, vertical or inclined. Horizontal lines are sketched with the motion of the wrist and the forearm. These are sketched from left to right. Vertical lines are sketched downwards with the movement of the fingers. Inclined lines (which are nearly horizontal) are sketched from left to right, whereas inclined lines (which are nearly vertical) are sketched downward.

The first step in sketching any straight line is to mark the end points of a line. A light line with one or series of strokes may then be tried between the end points. Fig. 18.1 shows the directions in which these lines must be sketched to attain the straightness.

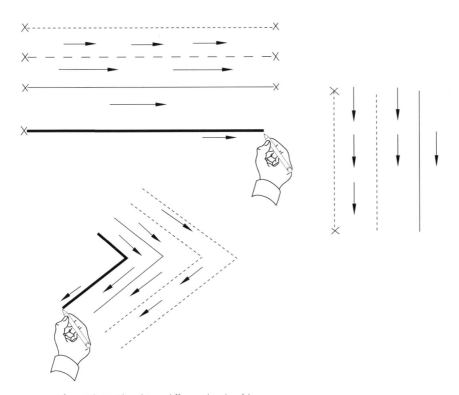

Fig. 18.1 Sketching different kinds of lines

18.5 Sketching Circles

In general, small circles are sketched very easily in one or two strokes but the large circles are sketched in a number of strokes. It is better to follow some systematic method for sketching circles. The best method for sketching a circle is to locate a number of points, through which the curve should pass. It is to be remembered that all points on the circumference of a circle are equidistant from its centre. The following methods are generally adopted for sketching circles:

(a) **Methods for sketching small circles**
Method I

(i)   Sketch horizontal and vertical centre lines and mark the centre of a circle.
(ii)   Sketch a square, with its side being equal to the diameter of a circle.
(iii) Sketch the diagonals of a square.
(iv)   Along these diagonals, plot the points, one on each side of the centre, at a distance of circle radius.
(v)   A smooth curve passing through these points results in the required circle. See Fig. 18.2.

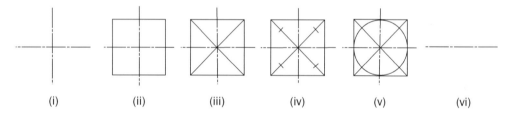

Fig. 18.2 Sketching small circles (Method I)

Method II

(i)   Sketch radial lines, preferably eight in number with the help of four straight lines.
(ii)   Along each radial line, the radius of the circle is estimated and marked off.
(iii) A smooth curve passing through these points result in the required circle. See Fig. 18.3

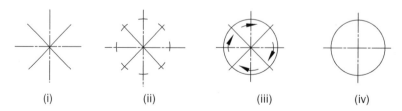

Fig. 18.3 Sketching small circles (Method II)

(b) **Methods for sketching large circles**

Method I

   (i)   First of all, draw centre lines of the circle.

   (ii)   Take piece of paper called trammel and mark the radius of the required circle on it.

   (iii) Fix one end of the trammel by a pin or hand at the centre.

   (iv) Take the pencil point on the marked point.

   (v)   Complete the circle as shown in Fig. 18.4.

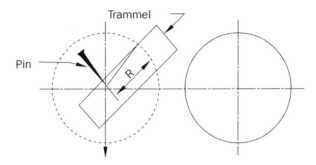

Fig. 18.4 Sketching large circles (Method I)

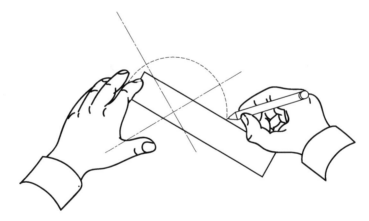

Fig. 18.5 Sketching large circles (Method II)

Method II

It is an efficient and quick method of sketching a circle in which hand is used as a compass.

   (i)   In this, the little finger is used as a point at the centre and the pencil is held stationary at the required radius from the centre.

   (ii)   Rotate the paper under the hand and pencil.

   (iii) Complete the circle as shown in Fig. 18.5.

18.6 Sketching an Ellipse

Following are the steps for sketching an ellipse in the orthographic views:

(i)   Sketch horizontal and vertical centre lines and mark by estimation, the semi-major OA (= OB) and semi-minor OC (= OD) axes.

(ii)   Complete the rectangle with major and minor axes as sides.

(iii)   Sketch the axes tangentially at the mid-points of the sides, resulting in required ellipse as shown in Fig. 18.6.

For a more accurate work, additional points on the curve may be obtained as shown in Fig. 18.7

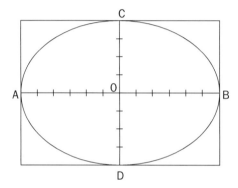

Fig. 18.6 Sketching an ellipse

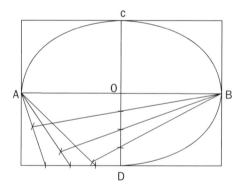

Fig. 18.7 Method of obtaining additional points on ellipse

18.7 Sketching Arcs and Curves

Arcs and curves are sketched in a manner similar to that adopted for sketching circles. Fig. 18.8 shows the different methods of sketching various types of arcs and circles.

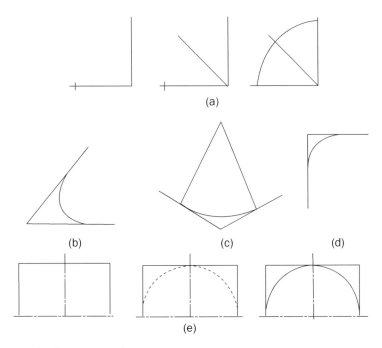

Fig. 18.8 Sketching arcs and curves

18.8 Sketching Angles

Figure 18.9 shows the different methods of sketching commonly used angles.

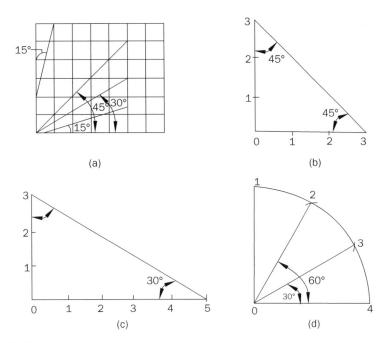

Fig. 18.9 Sketching angles

18.9 Types of Freehand Sketches

The freehand sketches may be classified as:

- Orthographic sketches
- Isometric sketches

18.10 Sketching Orthographic Views

Very often, when repairs are to be made or changes are proposed in an existing object or part, the production shop is asked to produce some parts. This is done by supplying shop drawings in the form of orthographic views. Orthographic sketching may require any combination of the six principal views of an object. The procedure followed in making a sketch is almost the same as that in drawing with instruments. Though the sketch is not made to scale, yet it should be fairly proportionate. The following are the steps in sketching the orthographic views and are illustrated in Fig. 18.10.

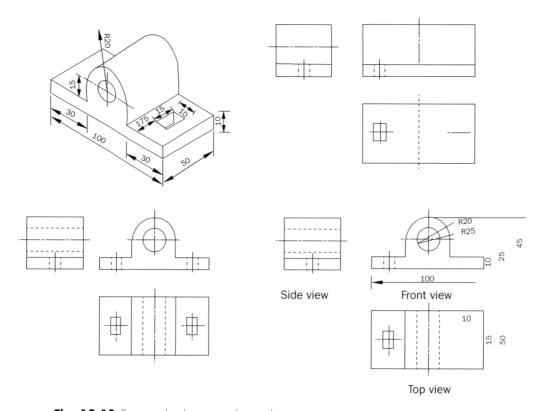

Fig. 18.10 Steps in sketching an orthographic view

(i)   Study the object critically untill its shape and function are understood
(ii)   Decide the necessary views
(iii) Estimate the proportions carefully
(iv)   Sketch lightly the overall size of each view as a rectangular or square block
(v)   Sketch in the required dotted lines for hidden features of the object
(vi)   Complete each view by darkening all the lines, forming the view
(vii) Dimension the views and add the necessary notes

18.11 Sketching Isometric Views

To make the drawing more understandable, several forms of one plane conventional or projection drawings are used to supplement the orthographic views. These one plane drawings which can be easily understood by everyone without any formal technical training are called isometric views. Though any of the method can be used in making isometric sketches, yet the box construction method is mostly used. The following procedure should be followed for making isometric sketches:

(i)   Study the orthographic views critically and decide the position in which it should be placed for making the isometric sketches
(ii)   Enclose the orthographic view in a rectangular or square box, to give length, breadth and height of the object
(iii) Sketch the isometric axes
(iv)   Estimate and make off the principal dimensions along the isometric axes
(v)   Lay off the distances in each face to locate all the features
(vi)   Sketch the lines through these points and parallel to the isometric axes. For non-isometric lines, first locate the end points to establish the line
(vii) Complete the view by darkening the required lines
(viii) Dimension the view and add the necessary notes as shown in Fig. 18.11

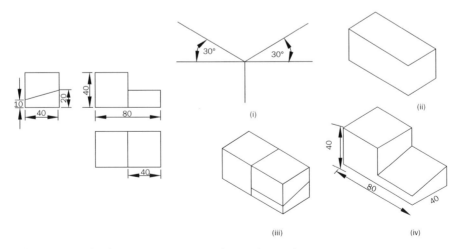

Fig. 18.11 Sketching an isometric view from orthographic views

Exercises

18.1 Sketch the following:
    (i)   Square block of 50 mm side
    (ii)   Rectangular block of 120 mm × 60 mm
    (iii) Circles of diameters 50, 100 and 120 mm.

18.2 Figure 18.12 shows the isometric view of an objects. Sketch the orthographic views of the given object.

18.3 Figure 18.13 shows the orthographic views of an object. Sketch the isometric view of the given object.

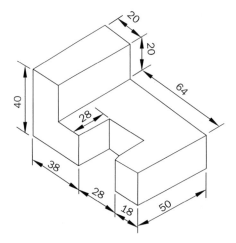

Fig. 18.12

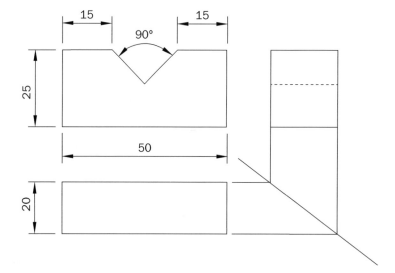

Fig. 18.13

18.4 Sketch any four types of workshop jobs prepared by you. Give the mean dimensions too.

18.5 Three cubes of 40 mm, 30 mm and 20 mm are placed centrally such that the biggest cube is at the bottom whereas the smallest is on the top. Sketch the isometric view of the solids.

18.6 What are the steps to be followed in sketching (*i*) a straight line; (*ii*) a rectangle; (*iii*) a circle; and (*iv*) an ellipse?

18.7 What are the steps to be followed in sketching orthographic views of an object. Explain it with a suitable diagram?

18.8 Using free hand sketching draw the projection of square prism assuming suitable dimensions both in first-angle projection method and third-angle projection method.

18.9 Draw a free hand sketch of a stool which you have seen in your drawing room.

18.10 What are the uses of free hand sketching in engineering field? Describe the steps to be followed in sketching a square?

Objective Questions

18.1 Horizontal lines are sketched from .... to ...., while vertical lines are sketched from .... to ....

18.2 The materials required for freehand sketching are ...., ... and ....

18.3 The drawing prepared by a pencil and without the use of drawing instruments is called ....

18.4 Freehand sketching is generally used for expressing and recording ....

18.5 A sketch is considered to be good when its features are shown in correct ....

18.6 What are the applications of freehand sketching?

Answers

18.1 Left right, top bottom

18.2 Paper, pencil, eraser or rubber

18.3 Sketching or Freehand sketching

18.4 New ideas

18.5 Proportions

Chapter 19

Computer Graphics

19.1 Introduction

The previous chapters of the book have described the fundamental concepts of manual drawing, using various drawing instruments and producing a hardcopy of the drawing. This traditional drawing, however, still remains the foundation for engineering communication, as the subject of computer aided drafting (CAD) has been developed on the fundamentals of traditional drawing. So the advent of CAD does not necessarily eliminate manual engineering graphics.

There are several reasons for implementing a computer aided design system. A few of them are:

- To increase the productivity of the designer
- To improve the quality of design
- To improve communication
- To create a database for manufacturing

19.2 Computer Graphics

The term 'computer graphics' is very generic in the sense that it has different meanings for different disciplines of study. In the operation of the graphics system by the user, a variety of activities take place, which can be divided into three categories: interaction with graphics terminal to create and alter images on the screen; construction of a model of something physical and of the image on the screen; and entering the model into computer memory. In working with the graphics system, the user performs these various activities in combination rather than sequentially.

19.3 Requirements for Computer Graphics

The basic requirements are a CAD workstation and requisite software. CAD workstation comprises a common PC with minimum Pentium configuration, a video display, a mouse or pen, and a plotter or printer. The graphics software can be divided into three modules: the graphic package, the application program and the application database. Various requirements of graphics software are:

- It should be easy to use.
- The package should operate in a consistent and predictable way, for the user.
- Graphics programs should be efficient.
- It should not be too expensive to make its use prohibitive.

This chapter explores the AutoCAD package for its basic drafting capabilities.

19.4 Getting Started with AutoCAD

Start AutoCAD by clicking on the Windows start button placed at the bottom left, then move the mouse to programs, click on AutoCAD as shown in Fig. 19.1 or it can directly be opened by double clicking on the desktop icon of AutoCAD.

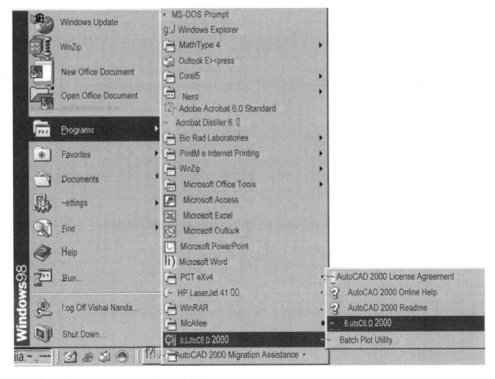

Fig. 19.1 Starting AutoCAD 2000

Open an Existing File Tab
Start from Scratch Tab
Use a Template Tab
Use a Wizard Tab

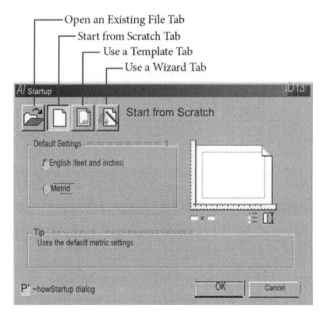

Fig. 19.2 Opening or creating an AutoCAD file

AutoCAD starts to load. In the process of start-up a dialog box, as shown in Fig. 19.2, giving various start-up options will be displayed. Select the option 'Start from Scratch'; click on 'Metric' radio button and 'OK'. Fig. 19.3 shows a typical initial AutoCAD window screen. The AutoCAD window screen has a number of important features:

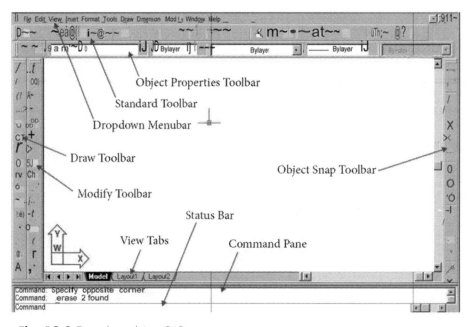

Fig. 19.3 Typical initial AutoCAD screen

- The top line displays the Windows pull-down menus for exiting a program and changing a program.
- The second line is the standard toolbar and contains a group of the most commonly used commands.
- The third line contains some command icons and an area that shows the current, clocked and object properties that are active.
- The line just above the drawing portion of the screen displays the name of the current drawing. Once a drawing name has been defined, it will appear at the top of the screen.
- The bottom left corner of the screen shows the coordinate display position.
- View tabs give access to different views of the current drawing. The model tab should be selected generally.
- Command pane is the place to type commands.
- The commands listed on the status line are displayed in light gray when they are off and in black when they are on.
- The large open area in the centre of the screen is called the drawing area.
- The two rectangular boxes of command icons, located along the left edge of the drawing screen, are the draw and modify toolbars.

19.5 Saving a Drawing

There are three ways to save a drawing:

- Go to the top left corner of the screen and click on 'File' option and then select 'Save', as shown in Fig. 19.4.
- The quickest way is to save a drawing by pressing Ctrl + S.
- Another way is to select the save icon from the standard toolbar.

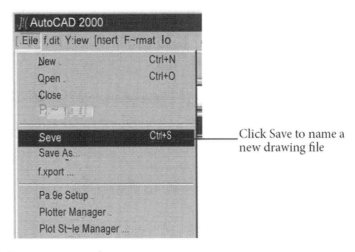

Fig. 19.4 How to save a drawing

19.6 Command Entry

There are four methods of giving a command. They are:

- First: on the command pane.
- Second: from the side screen.
- Third: from the toolbar.
- Fourth: from drop-down menus.

19.7 Drawing Limits

This is the first command that is to be given in AutoCAD. The lower left corner is normally kept at 0, 0 and the upper right corner depends upon the size of the figure to be drawn. By default, the size of the screen is 12 × 9. If the size of the figure is 200 × 200 units, then the size of the upper right corner should be more than that, in order to fit the figure on the screen.

To set the drawing limits
Command: _limits.

19.8 Units

This command will help in selecting appropriate units for the given figure i.e., mm, cm, m, inches, etc.

Command: _units; enter
See Figs. 19.5 and 19.6.

Select the appropriate units and also the precision, i.e., how many zeros after decimal point. Set the units and precision for angles too.

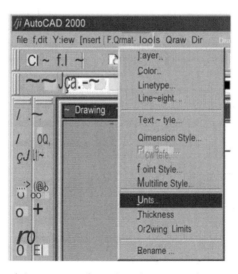

Fig. 19.5 Selecting of drawing units from drop-down menu bar

Select the Architectural units

Fig. 19.6 How to select the drawing units

19.9 Draw Commands

These commands are arranged in Draw drop-down menu bar or standard draw toolbar as shown in Fig. 19.7. Some of these commands will be discussed in subsequent paragraphs.

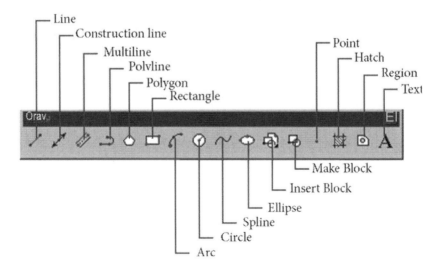

Fig. 19.7 Draw Toolbar

Line

There are different ways to define the length and locations of a line. These are: randomly selected points, entering the coordinate values for the start and end points or by specifying the first point, the length and the direction of the line.

(a) **Line using randomly selected points**

(i)   Select the line tool in the draw toolbar.
      The line command can also be accessed by typing the word 'line'. The following command sequence will appear in the command.
      Command: _line; enter

(ii) Place the cursor randomly on the drawing screen and press the left mouse button.
      Specify next point or [undo]:
      Move the cursor; a line extends from the designated point to the cursor.

(iii) Randomly pick another point on the screen.
      Specify next point or [undo]:
      Command pane will keep on asking for another point until the user presses either enter or the right mouse button.

(iv) If enter is pressed, the line command sequence will end. Pressing enter a second time will restart the line command.

(b) **Line using Coordinate values**

There are three ways to draw a line in AutoCAD.

Draw 50 mm × 50 mm rectangle starting at the 10, 10 coordinate point by using absolute, incremental and polar mode.

• **Absolute Mode**

(i)   Select the line tool from the draw toolbar
       Command: _line; enter
       Specify the first point:

(ii)   Type 10, 10; enter
       Specify next point or [undo]:

(iii) Type 60, 10; enter
       Specify next point or [clear/undo]:

(iv)   Type 60, 40; enter
       Specify next point or [close/undo]:

(v)   Type 10, 40; enter
       Specify next point or [close/undo]:

(vi)   Type C, enter. See Fig. 19.8

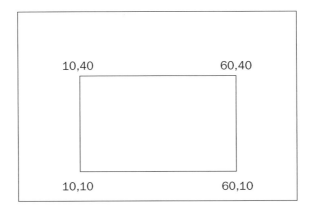

Fig. 19.8 Use of absolute mode

• **Incremental Mode**

(i)   Select the line tool from the draw toolbar
      Command: _line; enter
      Specify first point.
(ii) Type 10, 10; enter
      Specify next point or [undo]:
(iii) Type @ 50, 0; enter
      Specify next point or [close/undo]:
(iv) Type @ 0, 30; enter
      Specify next point or [close/undo]:
(v)   Type @ −50, 0; enter
      Specify next point or [close/undo]:
(vi) Type @ 0, −30, (or C); enter. See Fig. 19.9
      Absolute coordinate measures distance from the origin of AutoCAD system, whereas
      the relative coordinates assign the distances with respect to the current position.

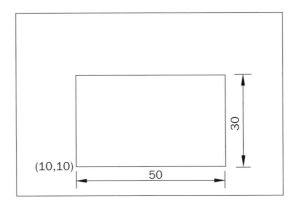

Fig. 19.9 Use of incremental mode

- **Polar Mode**

(i)   Select the line tool from the draw toolbar
      Command: _line; enter
      Specify first point.
(ii) Type 10, 10; enter
      Specify next point or [undo]:
(iii) Type @ 50 < 0; enter
      Specify next point or [close/undo]:
(iv) Type @ 30 < 90; enter
      Specify next point or [close/undo]:
(v)   Type @ 50 < 180; enter
      Specify next point or [close/undo]:
(vi) Type @ 30 < 270 (or @ 30 < –90 or C); enter. See Fig. 19.10

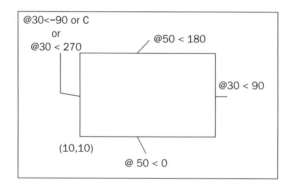

Fig. 19.10 Use of polar mode

Polyline

A polyline is a line made from a series of individual, connected, line segments that act as a single entity. Polylines are used to generate curves, splines, polygons, etc. They are also used in three-dimensional representations to produce solid objects.

To draw a polyline.

(i)   Select a polyline tool from the draw toolbar
      Command: _pline; enter
      Specify start point.
(ii) Select a start point.
      Specify next point or [arc/close/half with/length/undo/width]
(iii) Select a second point.
      Specify next point.
(iv) Select several more points.
      Specify next point.
(v)   Press the right mouse button; then enter. See Fig. 19.11

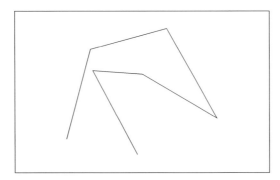

Fig. 19.11 Use of polyline

To draw a polyline-**arc**

- (i) Select a polyline tool from the draw toolbar
  Command: _pline
  Specify start point.
- (ii) Select a start point
  Specify next point or [arc/close/halfwidth/length/undo/width]:
- (iii) Type A; then enter
  Specify end point of arc or [angle/center/close/ direction/halfwidth/line/radius/ second point/ undo/width].
- (iv) Select the point.
  Specify endpoint of arc or [angle/center/close/ direction/halfwidth/line/radius/ second point/ undo/width].
- (v) Select another point.
- (vi) Press the right mouse button, then enter. See Fig. 19.12.

Similarly, other options of a polyline may be used depending upon the requirement.

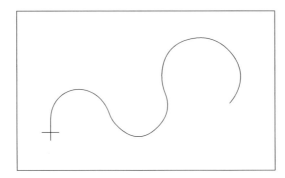

Fig. 19.12 Polyline drawn using the arc option

Circle

A circle can be drawn by any of the option as shown in Fig. 19.13. Depending on the requirement of the drawing any of these options may be selected. Some of them are as follows:

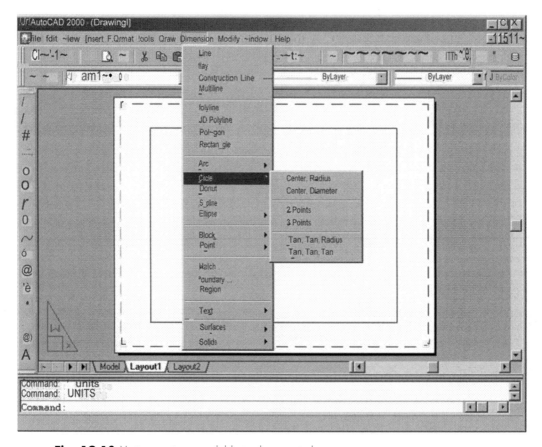

Fig. 19.13 Various options available to draw a circle

To draw a circle – **radius**

(i)   Select the circle tool from the draw toolbar
      Command: _circle; enter
      Specify a centre point for circle or [3P/2P/Ttr (tan tan radius)]

(ii) Select a centre point.
      Specify radius of circle or [diameter]:

(iii) Give value, then enter. See Fig. 19.14.

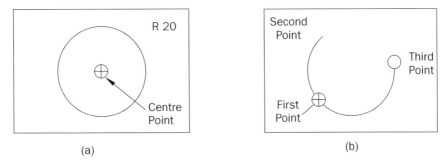

Fig. 19.14 Use of circle command specifying various option.

To draw a circle – **3 points**

  (i)   Select the circle tool from the draw toolbar
         Command: _circle; enter
         Specify the centre point for circle or [3P/2P/Ttr (tan tan radius)]:
 (ii)   Type 3 P; enter
         Specify first point of a circle:
 (iii)   Select a first point.
         Specify second point on circle:
 (iv)   Select a second point.
         Specify third point on circle.
  (v)   Select a third point.
         The student is advised to explore other methods for drawing a circle.

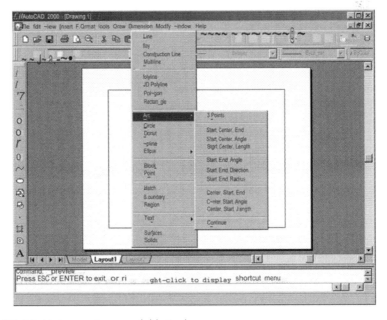

Fig. 19.15 Various options available to draw an arc

Arc

An arc can be drawn by any of the options as shown in Fig. 19.15. Depending on the requirement of the drawing any of these options may be used. Some of them are as follows:

To draw an arc – **3 points**

(i)   Select the arc tool from the draw toolbar.
      Command: _arc
      Specify start point of the arc or [enter]:
(ii) Select a start point.
      Specify second point of the arc or [center/end]:
(iii) Select a second point.
      Specify end point of the arc:
(iv) Select an end point. See Fig. 19.16.

The student is advised to explore other methods for drawing an arc.

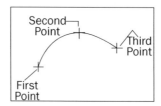

Fig. 19.16 An arc drawn through three points using arc command

Ellipse

An ellipse can be drawn by any of the options as shown in Fig. 19.17. Depending on the requirement of the drawing any of these options may be used. Some of them are as follows:

Fig. 19.17 Various options available to draw an ellipse

To draw an ellipse – **axis, end**

(i)   Select the ellipse tool from the draw toolbar.
      Command: _ellipse; enter
      Specify axis endpoint of ellipse or [arc/center]:
(ii) Select a start point for one of the axes
      Specify other end point of axis:
(iii) Select an endpoint that defines the length of the axis.
      Specify distance to other axis or (rotation):
(iv) Select a point that defines half the length of the other axis
      Points 1 and 2 define the major axis
      Points 3 and 4 define the minor axis

To draw an ellipse– **centre**

(i)   Select the ellipse tool from the draw toolbar.
      Command: _ellipse; enter
      Specify axis endpoint of ellipse or [arc/center]:
(ii) Type C; press enter
      Specify the centre point of the ellipse:
(iii) Select the centre point of the ellipse.
      Specify axis end point.
(iv) Select one of the endpoints of one of the axes.
      Specify distance to the other axis or (rotation):
(v)   Select a point that defines half the length of the other axis. See Fig. 19.18.

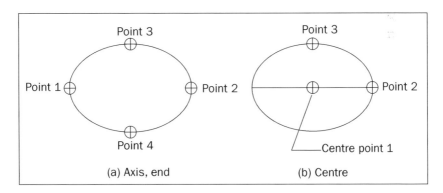

Fig. 19.18 Drawing an ellipse: specifying various options

Rectangle
This command is used to draw rectangles by specifying two diagonal points.
To draw a rectangle

(i)   Select the rectangle tool from the draw toolbar.
      Command: _rectangle; enter
      Specify first corner point or [chamfer/elevation/fillet/thickness/width]:

(ii)   Select a point.
        Specify other corner point:
(iii)   Select a point. See Fig. 19.19.

The student is advised to work with the different options mentioned above.

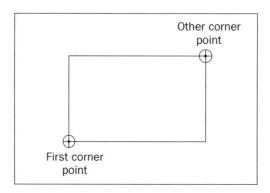

Fig. 19.19 Drawing a rectangle

Polygon

A polygon is a closed figure bounded by straight lines. Only regular polygons will be drawn here. A regular polygon with six equal sides is called a regular hexagon.

To draw a polygon– **centre point**

(i)   Select the polygon tool from the draw toolbar
        Command: _polygon
        Enter number of sides < 4 >:
(ii)   Type 6; press enter
        Edge 1 < centre of polygon >:
(iii)   Select centre point.
        Enter an option [inscribed in circle/circumscribed about a circle] < I >:
(iv)   Type C; press enter
        Specify radius of circle.
(v)   Type any value, press enter. See Fig. 19.20.

Similarly, other options for a polygon may be used depending on the requirement.

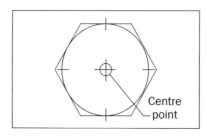

Fig. 19.20 Use of polygon command specifying centre point

Construction Line

This command is used to draw lines of infinite length. These lines are used to project various entities from one view to another. These lines may be drawn vertical, horizontal or inclined.

(i) Select the construction line tool from the draw toolbar.
     Command: _xline; enter
     Specify a point or [Hor./Ver/Ang/Bisect/Offset]:
(ii) Select a starting point.
     Specify through point:
     A line will pivot about the designated starting point and extend an infinite length in a direction through the cursor.
(iii) Select a through point.
     A line of infinite length is drawn through the two designated points.
(iv) Specify through point:
     Another infinite line will appear through the starting point through the cursor.
(v) Press enter. See Fig. 19.21

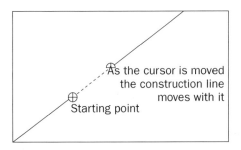

Fig. 19.21 Drawing a construction line

19.10 Modify Commands

There commands are used to carry out specific functions such as editing, transformation, etc., on an existing graphic. These commands are categorically arranged in Modify drop-down menu bar. A standard Modify toolbar is shown in Fig. 19.22.

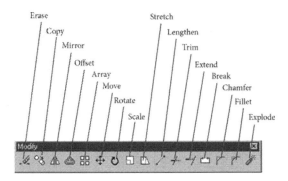

Fig. 19.22 Modify toolbar

Erase

This command is used to erase any number of entities that have been created in the current drawing. There are two ways to erase lines: select individual lines or window a group of lines.

To erase individual lines

(i)   Select the erase tool from the modify toolbar.
       Command: _erase; enter
       Select objects.

(ii)   Select the two open lines by placing the rectangular cursor on each line, one at a time and pressing the left mouse button.

(iii) Press the right mouse button or enter to complete the erase command.

To erase a group of lines simultaneously

(i)   Select the erase tool from the modify toolbar.
       Command: _erase; enter
       Select objects.

(ii)   Place the rectangular select cursor above and to the left of the lines to be erased and press the left mouse button.

(iii) Move the cursor and a window drag from the selected first point.

(iv)   When all the lines to be erased are completely within the window, press the left mouse button.

(v)   Press the right mouse button or enter to complete the erase command. See Fig. 19.23.

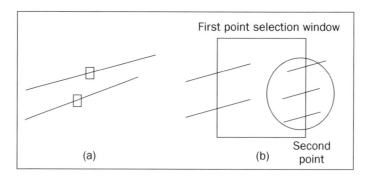

Fig. 19.23 Different methods to erase a group of lines

Copy

This command is used to make an exact copy of an existing line or object. This command can also be used to create more than one copy without reactivating the command.

To copy an object

(i)   Select the copy tool from the modify toolbar
       Command: _copy; enter
       Select objects.

(ii)   Window the entire object
       Select objects.

(iii) Press enter
       Specify base point or displacement or [multiple]:
(iv)   Select a base point
       Specify second point of displacement or <use first point of displacement:
(v)   Select a second displacement point.

The original object remains in its original location and a new object appears at the second displacement point as shown in Fig. 19.24.

Similarly, other options of a copy command may also be useful depending on the requirement.

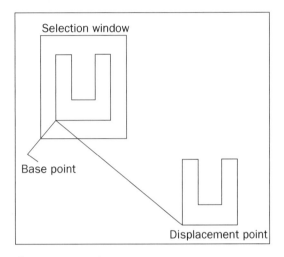

Fig. 19.24 Use of copy command.

Mirror

This command is very useful when drawing symmetrical objects. Only half of the object need to be drawn. The second half can be created using the mirror command.

To mirror an object

(i)   Select the mirror tool from the modify toolbar
       Command: _mirror; enter
       Select objects.
(ii)   Window the object
       Select objects:
(iii) Press enter.
       Specify first point of mirror line.
(iv)   Select a point on the mirror line.
       Specify second point of mirror line:
(v)   Select a second point on the mirror line.
       Delete source object ? [Yes/No] < N >:
(vi)   Press enter, see Fig. 19.25.

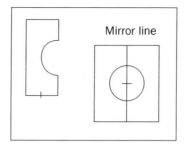

Fig. 19.25 Use of mirror command

Offset

This command is used to construct a new entity that is parallel to an existing entity.

To offset an object.

(i) Select the offset tool from the modify toolbar.
Command: _offset; enter
Specify offset distance or [Through]:

(ii) Set the value, enter.
Select object to offset or < exist >:

(iii) Select a line.
Specify point on side to offset:

(iv) Select a point to the right side of the line.
Specify object to offset or < exist >:

This process may be repeated by double clicking the right mouse button. See Fig. 19.26.

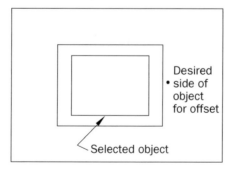

Fig. 19.26 Use of offset command

Array

This command produces multiple copies of selected objects arrayed in a rectangular or polar pattern.

To array an object: **rectangular option**

(i) Select the array tool from the modify toolbar
Command: _array; enter
Select objects.

(ii) Press enter.
Enter type of array [rectangular/polar] < R >:

(iii) Press enter.
Enter te number of rows (_) < 1 >:

(iv) Type any value, press enter.
Enter the number of columns (I I I) < 1 >:

(v) Type any value, press enter.
Enter the distance betwen rows or specify unit cell [_ _ _]:

(vi) Type any value, press enter.
Specify distance between columns (III):

(vii) Type any value, press enter. See Fig. 19.27.

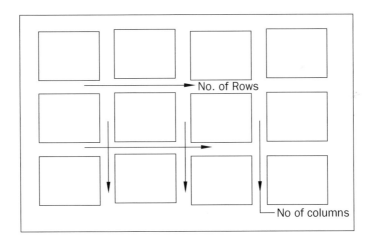

Fig. 19.27 Use of array command – rectangular option

To array an object: **polar option**

(i) Select the array tool from the modify toolbar
Command: _array; entry
Select objects.

(ii) Press enter.
Enter type of array [rectangular/polar] < R >:

(iii) Type P, press enter.
Specify the centre point of array:

(iv)   Select a centre point
       Enter the number of items in the array:
(v)   Type value, press enter
       Specify angle to fill (+ = ccw, – = cw) < 360 >:
(vi)   Press enter
       Rotate array objects ? [Yes/No] < Y >:
(vii) Press enter. See Fig. 19.28.

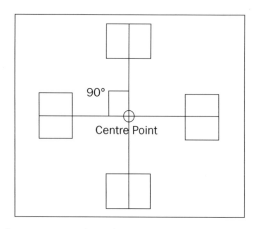

Fig. 19.28 Use of array command – polar option.

Move

This command is used to move a line or object to a new position on the drawing.
   To move an object

(i)   Select the move tool from the modify toolbar
       Command: _move; enter
       Select objects.
(ii)   Window the entire objects
       Select objects.
(iii) Press enter.
       Specify base point or displacement.
(iv)   Select a base point.
       Specify second point of displacement or < use first point of displacement >:
(v)   Select a second displacement point.
       Press enter. See Fig. 19.29.

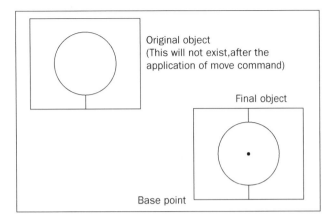

Fig. 19.29 Use of move command

Rotate

This command is used to rotate a group of entities about a base point through a given angle.

To rotate an object

(i) Select the rotate tool from the modify toolbar
Command: _rotate; enter
Select objects:

(ii) Window the objects
Select objects:

(iii) Press enter.
Specify base point:

(iv) Select a base point.
Specify rotation angle or (reference):

(v) Type given value, press enter. See Fig. 19.30.

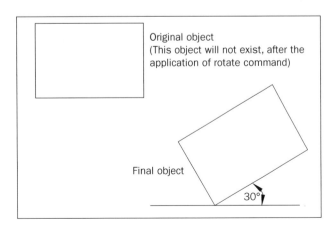

Fig. 19.30 Use of rotate command

Trim

This command is used to cut away excessively long lines.

To use the trim command.

(i) Select the trim tool from the modify toolbar.
Command: _trim; enter
Select cutting edges.
Specify objects.

(ii) Press enter
Select object to trim or [Project/Edge/Undo]:

(iii) Press enter. See Fig. 19.31.

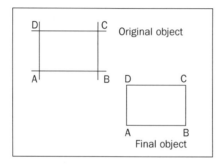

Fig. 19.31 Use of trim command

Extend

This command is used to extend the given lines. It produces the effect opposite to that by the trim command.

To use the extend command.

(i) Select the extend tool from the modify toolbar
Command: _extend; enter
Select boundary edges.
Select objects.

(ii) Select a line that can be used as a boundary edge.
Select object to extend or [Project/Edge/Undo]:

(iii) Select the lines to be extended, press enter.
See Fig. 19.32.

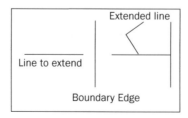

Fig. 19.32 Use of extend command

Break

This command is used to split an exisiting line, arc, circle into two parts.

To use the break command

(i) Select the break tool from the modify toolbar

Command: _break; enter

Select object

(ii) Select the line

Specify second break point or [First Point]:

(iii) Type *f*, press enter

Specify first break point:

(iv) Select the first point of the break.

Specify second break point:

(v) Select the second point of the break. See Fig. 19.33.

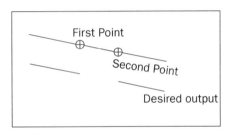

Fig. 19.33 Use of break command

Fillet

This command is used to connect two existing lines, circles or arcs by means of an arc of a given radius.

To use the fillet command.

(i) Select the fillet tool from the modify toolbar.

Command: _fillet; enter

Select first object or [Polyline/Radius/Trim]:

(ii) Type R, press enter

Specify fillet radius < 0.5000 >:

(iii) Type given value, press enter. See Fig. 19.34.

The student is advised to work with different options mentioned above.

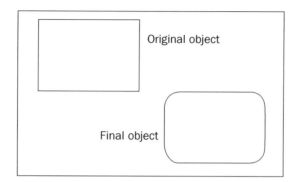

Fig. 19.34 Use of fillet command

Chamfer

This command is similar to the fillet except that, in chamfer, the corners are made with rectangular coordinates and give a straight edge. Here, the chamfer intersects the two edges at a specified distance from their point of intersection.

To use the chamfer command.

(i)   Select the chamfer tool from the modify toolbar.
      Command: _chamfer; enter
      Select first line or [Polyline/Distance/Angle/Trim] < Select first line >:

(ii) Type D, press enter.
      Specify first chamfer distance < 0.5000 >:

(iii) Type given value, press enter
      Specify second chamfer distance < 0.5000 >:

(iv) Type given value, press enter. See Fig. 19.35

The student is advised to work with different options mentioned above. Some important commands of draw and modify toolbar have already been described, which will help the students in making the drawings on computers.

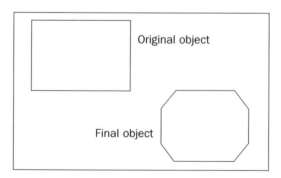

Fig. 19.35 Use of chamfer command

19.11 More Advanced Commands

Some more advanced commands like Osnap will be discussed in the subsequent paragraph.

Osnap

Object snap setting configures the movement of cursor from specific points like end point, point of intersection, mid-point, centre point of circle, etc. This software has a large selection of snap modes for this purpose. All snap modes are shown in Fig. 19.36. The object snap toolbar can be displayed by selecting the toolbar from the view drop-down menu and then clicking on object map, which is shown in Fig. 19.37

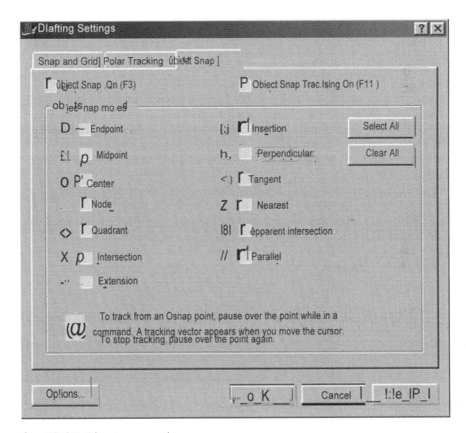

Fig. 19.36 Object snap modes

Osnap Options

End point:	It selects the nearest end point of a line or an arc.
Mid point:	It selects the mid point of a line or an arc.
Centre:	It selects the centre of an arc or a circle.
Intersection:	It selects the intersection of two lines in the drawing.
Perpendicular:	It selects a point on the line which is perpendicular to the selected object.

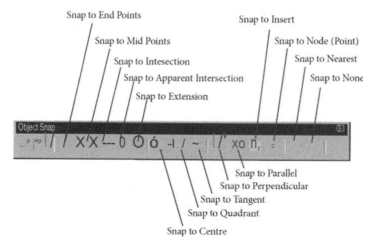

Fig. 19.37 Object snap toolbar

Tangent: It selects a point on an arc or a circle that forms a tangent to the picked arc or circle from the last point. See Fig. 19.38

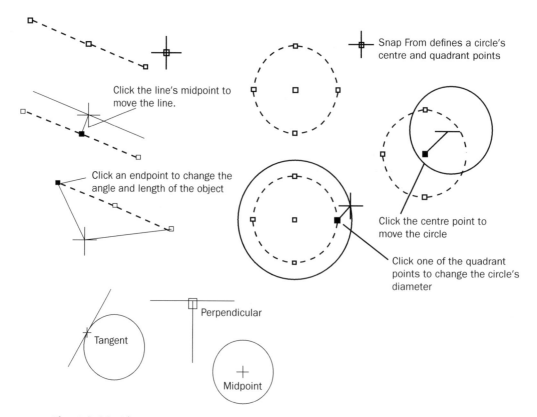

Fig. 19.38 Object snap options

Problem 19.1 Using graphic software, prepare the drawing as in Fig. 19.39. All dimensions are in mm.

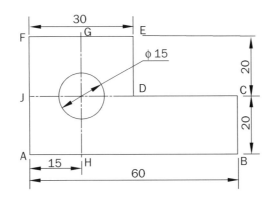

Fig. 19.39

Solution:

Command: _limits; press enter

Specify lower left corner or [ON/OFF] < 0.0000, 0.0000 >: 0, 0; enter

Specify upper right corner < 12, 9 >: 500, 300, enter

Command: _line, enter

Specify first point: 10, 10; enter

Specify next point or [undo]: @ 60 < 0, enter

Specify next point or [close/undo]: @ 20 < 90, enter

Specify next point or [close/undo]: @ 30 < 180, enter

Specify next point or [close/undo]: @ 20 < 90, enter

Specify next point or [close/undo]: @ 30 < 180, enter

Specify next point or [close/undo]: @ 40 < 270, or (@ 40 < – 90 or C); enter

Command: _offset, enter

Specify offset distance or (Through): 15, enter

Select object to offset or < exit >: Select line AF by clicking

Specify point on side to offset: click inside the figure; enter

Select object to offset or < exit >: enter

Command: _offset, enter

Specify offset distance or [Through]: 20, enter

Select object to offset or < exit >: Select line EF by clicking

Specify point on side to offset: click inside the figure; enter

Select object to offset or < exit >: enter

Command: _circle, enter

Specify centre point for circle or [3P/2P/Ttr (tan tan radius)]: Select centre by clicking at the intersection of two lines, enter

Specify radius of circle or [diameter]: 75, enter

Command: _change property, enter

Select object: click on JD, enter

Select object: click on GH, enter
Select object: 2 total, enter
Enter property to change [color/layer/Ltype/Lt scale/L weight/Thickness]: lt, enter
Enter new line type name < By layer >: centre, enter
Enter property to change [color/layer/L type/Lt scale/L weight/Thickness]: lt scale, enter
Enter new line type scale factor: 10, enter

Problem 19.2 Using graphic software, prepare the drawing as shown in Fig. 19.40. All dimensions are in mm.

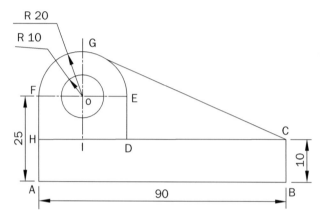

Fig. 19.40

Solution:
Command: _limits; press enter
Specify lower left corner or [ON/OFF] < 0.0000, 0.0000 >: 0, 0; enter
Specific upper right corner < 12, 9 >: 300, 200; enter
Command: _line, enter
Specify first point: 10, 10; enter
Specify next point or [under]: @ 90 < 0, enter
Specify next point or [close/undo]: @ 10 < 90, enter
Specify next point or [close/undo]: @ 90 < 100, enter
Specify next point or [close/undo]: @ 10 < 270° or (@ 10 < –90° or C); enter
Command: _line, enter
Click at pon 'H'
Specify next point or [undo]: @ 30 < 90, enter
Press enter twice to complete the command
Command: _offset, enter
Specific offset distance or (Through): 20, enter
Select object to offset or < exit >: Select line H by clicking
Specify point on side to offset: Click inside the figure; enter
Select object to offset or < exit >: Select line I by clicking
Specify point on side to offset: Click inside the figure; enter

Select object to offset or < exit >: enter
Command: _offset, enter
Specify offset distance or (Through): 15, enter
Select object to offset or < exit >: Select line HC by clicking
Specify point on side to offset: Click outside the figure; enter
Select object to offset or < exit >: enter
Command: _circle, enter
Specify centre point for circle or [3P/2P/Ttr (tan tan radius)]:
Select centre by clicking at the interaction of two lines (point O); enter
Specify radius of circle or [diameter]: 10, enter
Command: _circle, enter
Specify centre point for circle or [3P/2P/Ttr (tan tan radius)]:
Select centre by clicking at the interaction of two lines (point O), enter
Specify radius of circle or [diameter]: 20, enter
Command: _trim, enter
Select cutting edges: HJ, DK, FL, circle GI; enter
Select cutting edges: 4 total, enter
Select object to trim or [project/edge/undo]: FJ, EK, EL and arc FIE; enter
Command: _line, enter
Specify first point: Select point C by clicking, enter
Specify next point or [undo]: Select point G by clicking, enter
Command: _change property, enter
Select object: Click on IG, enter
Select object: Click on FE, enter
Select object: 2 total, enter
Enter property to change [color/layer/L type/Lt scale/L weight/thickness]: lt, enter
Enter new line type name < By layer >: centre, enter
Enter property to change [colour/layer/L type/Lt scale/L weight/thickness]: lt scale, enter
Enter new line type scale factor: 10, enter

Problem 19.3 Using graphics software, prepare the drawing as shown in Fig. 19.41. All dimensions are in mm.

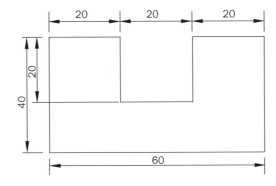

Fig. 19.41

Solution: Command: _limits; press enter
Specify lower left corner or [ON/OFF] < 0.0000, 0.0000 >: 0, 0; enter
Specify upper right corner <12.0000, 9.0000 >: 200, 200; enter

Absolute Mode
Command: _line; enter
Specify first point: 20, 20; enter
Specify next point or [undo]: 80, 20; enter
Specify next point or [close/undo]: 80, 60; enter
Specify next point or [close/undo]: 60, 60; enter
Specify next point or [close/undo]: 60, 40; enter
Specify next point or [close/undo]: 40, 40; enter
Specify next point or [close/undo]: 40, 60; enter
Specify next point or [close/undo]: 20, 60; enter
Specify next point or [close/undo]: 20, 20 or C; enter

Incremental Mode
Command: _line; enter
Specify first point: 20, 20; enter
Specify next point or [undo]: @ 60, 0; enter
Specify next point or [close/undo]: @ 0, 40; enter
Specify next point or [close/undo]: @ –20, 0; enter
Specify next point or [close/undo]: @ 0, –20; enter
Specify next point or [close/undo]: @ –20, 0; enter
Specify next point or [close/undo]: @ 0, 20; enter
Specify next point or [close/undo]: @ –20, 0; enter
Specify next point or [close/undo]: @ 0, –40 or C; enter

Polar Mode
Command: _line; enter
Specify first point: 20, 20; enter
Specify next point or [undo]: @ 60 < 0°; enter
Specify next point or [close/undo]: @ 40 < 90°; enter
Specify next point or [close/undo]: @ 20 < 180°; enter
Specify next point or [close/undo]: @ 20 < 270° (or 20 < –90°); enter
Specify next point or [close/undo]: @ 20 < 180°; enter
Specify next point or [close/undo]: @ 20 < 90°; enter
Specify next point or [close/undo]: @ 20 < 180°; enter
Specify next point or [close/undo]: @ 40 < 270° (or 40 < –90° or C); enter

Problem 19.4 Using graphics software, prepare the drawing as shown in Fig. 19.42. All dimensions are in mm.

Solution: Command: _limits; press enter
Specify lower left corner or [ON/OFF] < 0.000; 0.0000>; 0, 0; enter
Specify upper right corner <12.000; 9.000: 200, 200; enter
Command: _line; enter
Specify first point: 20, 20; enter
Specify next point or [undo]: 30, 20; enter
Specify next point or [close/undo]: 30, 30; enter
Specify next point or [close/undo]: 90, 30; enter
Specify next point or [close/undo]: 90, 20; enter
Specify next point or [close/undo]: 100, 20; enter
Specify next point or [close/undo]: 100, 100; enter
Specify next point or [close/undo]: 90, 100; enter
Specify next point or [close/undo]: 60, 70; enter
Specify next point or [close/undo]: 30, 100; enter
Specify next point or [close/undo]: 20, 100; enter
Specify next point or [close/undo]: 20, 20 or C; enter

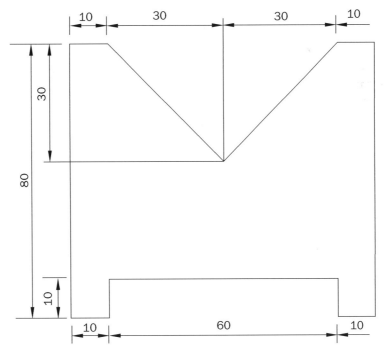

Fig. 19.42

Problem 19.5 Using graphics software prepare the drawing as shown in Fig. 19.43. All dimensions are in mm.

Solution: Command: _limits; press enter
Specify lower left corner or [ON/OFF] <0.0000, 0.0000>: 0, 0; enter
Specify upper right corner <12.0000; 9.0000>: 200, 200; enter
Command: _line; enter
Specify first point: 20, 20; enter
Specify next point or [undo]: 100, 20; enter
Specify next point or [close/undo]: 100, 60; enter
Specify next point or [close/undo]: 70, 60; enter
Specify next point or [close/undo]: 70, 40; enter
Specify next point or [close/undo]: 50, 40; enter
Specify next point or [close/undo]: 50, 90; enter
Specify next point or [close/undo]: 20, 90; enter
Specify next point or [close/undo]: 20, 20 or C; enter

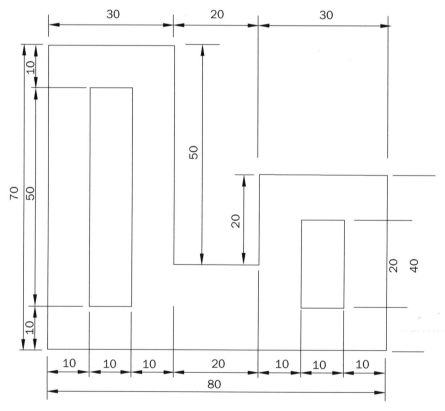

Fig. 19.43

Command: _line; enter
Specify next point or [close/undo]: 30, 30; enter
Specify next point or [undo]: 40, 30; enter
Specify next point or [close/undo]: 40, 80; enter
Specify next point or [close/undo]: 30, 80; enter
Specify next point or [close/undo]: 30, 30 or C; enter
Command: _line; enter
Specify first point: 80, 30; enter
Specify next point or [undo]: 90, 30; enter
Specify next point or [close/undo]: 90, 50; enter
Specify next point or [close/undo]: 80, 50; enter
Specify next point or [close/undo]: 80, 30 or C; enter

Problem 19.6 Using graphics software, prepare the drawing as shown in Fig. 19.44. All dimensions are in mm.

Solution: Command: _limits; press enter
Specify lower left corner or [ON/OFF] <0.0000, 0.0000>: 0, 0; enter
Specify upper right corner <12.0000; 9.0000>: 200, 200; enter
Command: _line; enter
Specify first point: 10, 10; enter
Specify next point or [undo]: @15 < 0; enter
Specify next point or [close/undo]: @15 < 120; enter
Specify next point or [close/undo]: @45 < 0; enter
Specify next point or [close/undo]: @15 < 210; enter
Specify next point or [close/undo]: @15 < 0; enter
Specify next point or [close/undo]: @45 < 90; enter
Specify next point or [close/undo]: @60 < 180; enter
Specify next point or [close/undo]: @45 < 270; or (@45 < –90 or C); enter

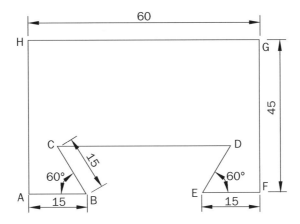

Fig. 19.44

Problem 19.7 Using graphics software, prepare the drawing as shown in Fig. 19.45. All dimensions are in mm.

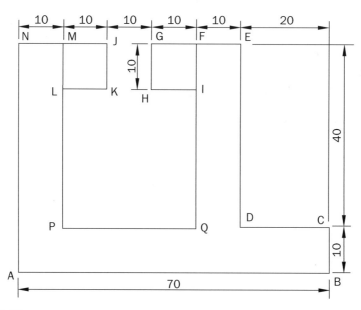

Fig. 19.45

Solution: Command: _limits; press enter
Specify lower left corner or [ON/OFF] <0.0000, 0.0000>: 0, 0; enter
Specify upper right corner <12.0000; 9.0000>: 200, 200; enter
Command: _line; enter
Specify first point: 10, 10; enter
Specify next point or [undo]: @70 < 0; enter
Specify next point or [close/undo]: @10 < 90; enter
Specify next point or [close/undo]: @20 < 180; enter
Specify next point or [close/undo]: @40 < 90; enter
Specify next point or [close/undo]: @10 < 180; enter
Specify next point or [close/undo]: @40 < 270; enter
Specify next point or [close/undo]: @30 < 180; enter
Specify next point or [close/undo]: @40 < 90; enter
Specify next point or [close/undo]: @10 < 180; enter
Specify next point or [close/undo]: @50 < 270 or (@50 < –90 or C); enter
Command: _line; enter
Specify first point: click on the point F; enter
Specify next point or [undo]: @10 < 180; enter
Specify next point or [close/undo]: @10 < 270; enter
Specify next point or [close/undo]: @10 < 0 or C; enter
Command: _line; enter

Specify first point: click on the point M; enter
Specify next point or [undo]: @10 < 0; enter
Specify next point or [close/undo]: @10 < 270; enter
Specify next point or [close/undo]: @10 < 180 or C; enter

Problem 19.8 Using graphics software, prepare the drawing as shown in Fig. 19.46. All dimensions are in mm.

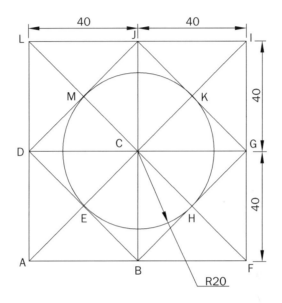

Fig. 19.46

Solution: Command: _limits; press enter
Specify lower left corner or [ON/OFF] <0.0000, 0.0000>: 0, 0; enter
Specify upper right corner <12.0000; 9.0000>: 200, 200; enter
Command: _line; enter
Specify first point: 10, 10; enter
Specify next point or [undo]: @40 < 0; enter
Specify next point or [close/undo]: @40 < 90; enter
Specify next point or [close/undo]: @40 < 180; enter
Specify next point or [close/undo]: @40 < 270 or (@40 < –90 or C); enter
Command: _line; enter
Specify first point: click on the point A; enter
Specify next point or [undo]: click on the point C; enter
Command: _line; enter
Specify first point: click on the point B; enter
Specify next point or [undo]: click on the point D; enter
Command: _mirror; enter
Select object: select all the objects; enter

Specify first point of mirror line: click on the point C; enter
Specify second point of mirror line: click on the point B; enter
Delete source object ? [Yes/No] <N>: N; enter
Command: _mirror; enter
Select object: select all the objects; enter
Specify first point of mirror line: click on the point D; enter
Specify second point of mirror line: click on the point G; enter
Delete source object ? [Yes/No] <N>: N; enter
Command: _circle; enter
Specify centre point for circle or [3P/2P/Ttr (tan tan radius)]:
Select centre by clicking at the point C; enter
Specify radius of circle or [diameter]: 20, enter

Problem 19.9 Using graphics software, prepare the drawing as shown in Fig. 19.47. All dimensions are in mm.

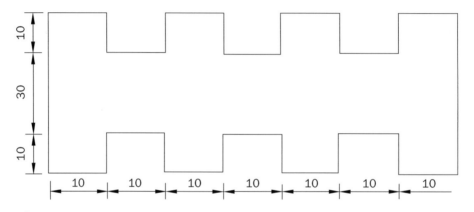

Fig. 19.47

Solution:
Command: _limits; press enter
Specify lower left corner or [ON/OFF] < 0.0000, 0.0000 >: 0, 0; enter Specify upper right corner < 12, 9 >: 200, 200, enter
Command: _line, enter
Specify first point: 10, 10; enter
Specify next point or [undo]: 20, 10, enter
Specify next point or [close/undo]: 20, 20, enter
Specify next point or [close/undo]: 30, 20, enter
Specify next point or [close/undo]: 30, 10, enter
Specify next point or [close/undo]: 40, 10, enter
Specify next point or [close/undo]: 40, 20, enter
Specify next point or [close/undo]: 50, 20, enter
Specify next point or [close/undo]: 50, 10, enter

Specify next point or [close/undo]: 60, 10, enter
Specify next point or [close/undo]: 60, 20, enter
Specify next point or [close/undo]: 70, 20, enter
Specify next point or [close/undo]: 70, 10, enter
Specify next point or [close/undo]: 80, 10, enter
Specify next point or [close/undo]: 80, 60, enter
Specify next point or [close/undo]: 70, 60, enter
Specify next point or [close/undo]: 70, 50, enter
Specify next point or [close/undo]: 60, 50, enter
Specify next point or [close/undo]: 60, 60, enter
Specify next point or [close/undo]: 50, 60, enter
Specify next point or [close/undo]: 50, 50, enter
Specify next point or [close/undo]: 40, 50, enter
Specify next point or [close/undo]: 40, 60, enter
Specify next point or [close/undo]: 30, 60, enter
Specify next point or [close/undo]: 30, 50, enter
Specify next point or [close/undo]: 20, 50, enter
Specify next point or [close/undo]: 20, 60, enter
Specify next point or [close/undo]: 10, 60, enter
Specify next point or [close/undo]: 10, 10 or C; enter

Problem 19.10 Using graphics software, prepare the drawing as shown in Fig. 19.48. All dimensions are in mm.

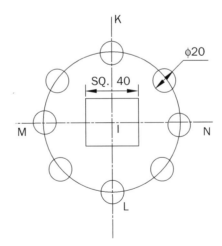

Fig. 19.48

Solution:
Command: _limits; press enter
Specify lower left corner or [ON/OFF} <0.0000, 0.0000>: 0, 0; enter
Specify upper right corner <12, 9>:: 300, 300; enter
Command: _circle; enter

Specify centre point for circle or [3P/2P/Ttr(tan tan radius)]:
Specify centre by clicking at the intersection of two lines I; enter
Specify radius of circle or [diameter]: 140, enter
Command: _circle; enter
Specify centre point for circle or [3P/2P/Ttr(tan tan radius)]:
Select centre by clicking at the intersection of point J; enter
Specify radius of circle or [diameter]: 20; enter
Command: _array; enter
Select objects; enter
Enter type of array [rectangular/polar] <R>: P; enter
Specify the centre point of array: Select centre by clicking at point J; enter
Enter the number of items in the array: 8; enter
Specify angle to fill (+ = ccw, – = cw) <360>: 360; enter
Command: _polygon; enter
Enter number of sides <4>: 4; enter
Edge 1<center of polygon>: 40; enter
Edge 2: 40; enter
Command: _change property; enter
Select object: click on KL; enter
Select object: click on MN; enter
Select object: 2 total; enter
Enter property to change [colour/layer/Ltype/Ltscale/Lweight/Thickness]: lt; enter
Enter new line type name<By layer>: centre; enter
Enter property to change[colour/layer/Ltype >Ltscale/Lweight/Thickness]: ltscale; enter
Enter new line type scale factor: 10; enter

Problem 19.11 Using graphics software, prepare the drawing as shown in Fig. 19.49. All
dimensions are in mm.

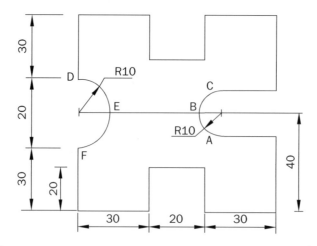

Fig. 19.49

Solution:

Command: _limits; press enter
Specify lower left corner or [ON/OFF] <0.0000, 0.0000>: 0, 0; enter
Specify upper right corner <12.0000; 9.0000>: 200, 200; enter
Command: _line; enter
Specify first point: 10, 10; enter
Specify next point or [undo]: 40, 10; enter
Specify next point or [close/undo]: 40, 30; enter
Specify next point or [close/undo]: 60, 30; enter
Specify next point or [close/undo]: 60, 10; enter
Specify next point or [close/undo]: 90, 10; enter
Specify next point or [close/undo]: 90, 40; enter
Specify next point or [close/undo]: 70, 40; enter
Command: _arc; enter
Specify start point of the arc or [center]: Select point A; enter
Specify second point of the arc or [center/end]; Select point B; enter
Specify end point of the arc: Select point C; enter
Command: _line, enter
Specify first point: click on point C; enter
Specify next point or [undo]: 90, 60; enter
Specify next point or [close/undo]: 90, 90; enter
Specify next point or [close/undo]: 60, 90; enter
Specify next point or [close/undo]: 60, 70; enter
Specify next point or [close/undo]: 40, 70; enter
Specify next point or [close/undo]: 40, 90; enter
Specify next point or [close/undo]: 10, 90; enter
Specify next point or [close/undo]: 10, 60; enter
Command: _arc; enter
Specify start point of the arc or [center]: Select point D; enter
Specify second point of the arc or [center/end]: Select point E; enter
Specify end point of the arc: Select point F; enter
Command: _line; enter
Specify first point: click on point F; enter
Specify next point or [undo]: 10, 10 or C; enter

Problem 19.12 Using graphics software, prepare the drawing as shown in Fig. 19.50. All dimensions are in mm.

Solution:

The solution to this problems is self-explanatory. Learners are requested to apply the commands learnt in previous problems.

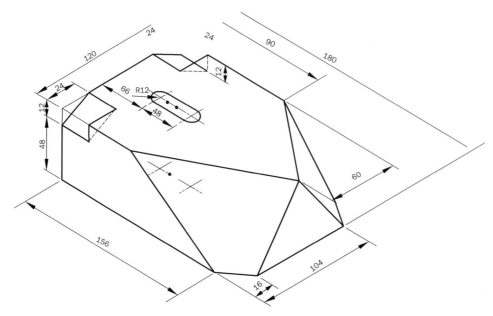

Fig. 19.50

Problem 19.13 Using graphics software, prepare the drawing as shown in Fig. 19.51. All dimensions are in mm.

Solution:

The solution to this problems is self-explanatory. Learners are requested to apply the commands learnt in previous problems.

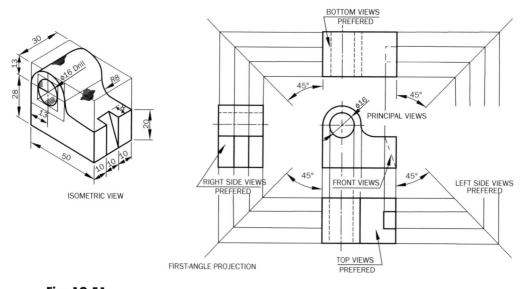

Fig. 19.51

Exercises

19.1 Discuss various types of methods by which a line can be drawn, with suitable diagrams.

19.2 List the various methods by which you can draw circles. Explain it with suitable diagrams.

19.3 Describe any five types of draw commands with suitable diagrams.

19.4 Describe any five types of modify commands with suitable diagrams.

19.5 Using graphics software, prepare the drawings as in Figs. 19.52, 19.53, 19.54, 19.55, 19.56, 19.57, 19.58, 19.59, 19.60, 19.61, 19.62, and 19.63. **Note:** All dimensions are in mm.

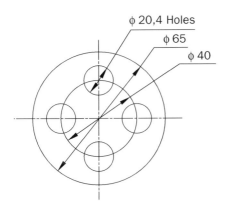

Fig. 19.52

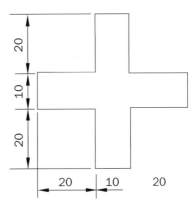

Fig. 19.53

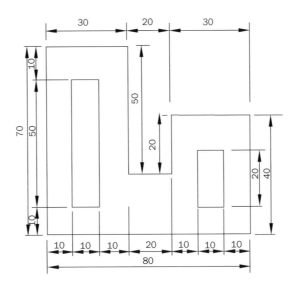

Fig. 19.54

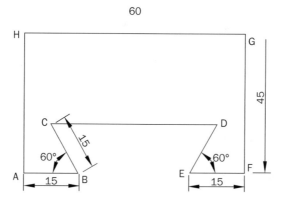

Fig. 19.55

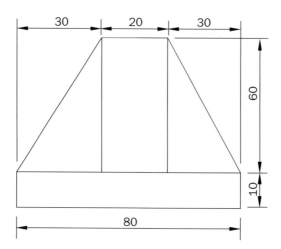

Fig. 19.56

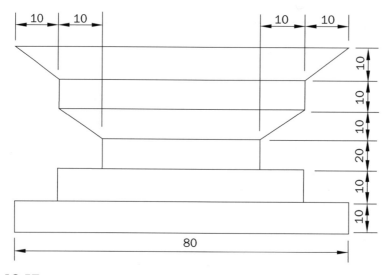

Fig. 19.57

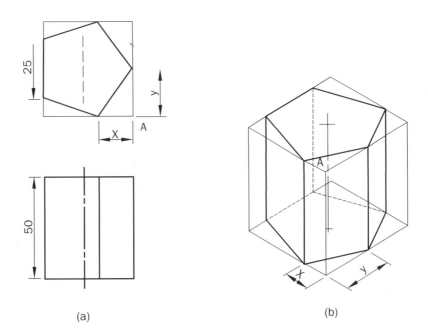

(a)   (b)

Right Regular Pentagonal Prism

Fig. 19.58

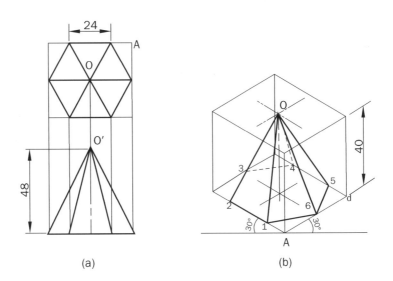

(a)   (b)

Right Regular Hexgonal Pyramid

Fig. 19.59

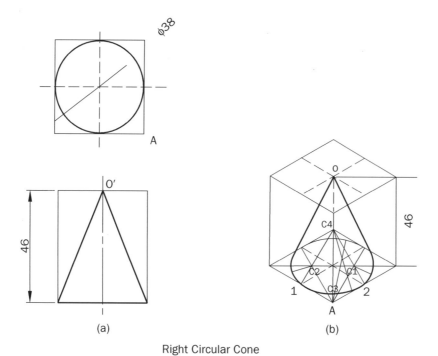

(a)   (b)

Right Circular Cone

Fig. 19.60

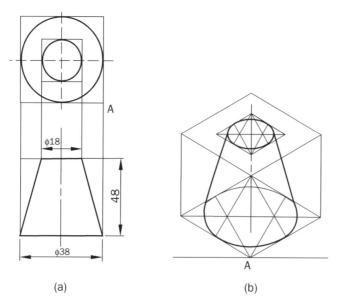

(a)   (b)

Frustum of a Cone

Fig. 19.61

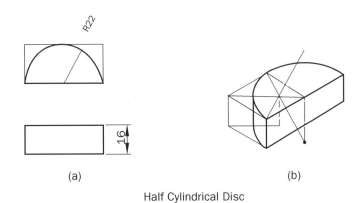

(a)   (b)

Half Cylindrical Disc

Fig. 19.62

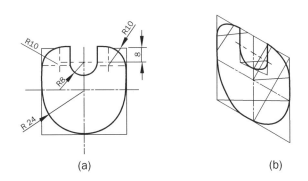

(a)   (b)

Sheet Metal Templet

Fig. 19.63

Objective Questions

19.1   Explain drawing limits.

19.2   List the various methods for command entry.

19.3   List the various methods for drawing a line.

19.4   How will you draw a rectangle?

19.5   Distinguish between manual drafting and computer aided drafting.

19.6   What is the use of an erase command?

19.7   What do you mean by the term 'CAD'?

19.8   List the various methods for drawing a circle.